Handbook of Supererogation

David Heyd
Editor

Handbook of Supererogation

Springer

Editor
David Heyd
Department of Philosophy
The Hebrew University of Jerusalem
Jerusalem, Israel

ISBN 978-981-99-3632-8 ISBN 978-981-99-3633-5 (eBook)
https://doi.org/10.1007/978-981-99-3633-5

This Springer imprint is published by the registered company Springer Nature Singapore Pte Ltd.
The registered company address is: 152 Beach Road, #21-01/04 Gateway East, Singapore 189721, Singapore

Contents

Introduction 1
David Heyd

Saints and Heroes 17
J. O. Urmson

Supererogation and Duty 29
F. M. Kamm

Supererogation, Conditional Obligation, and the All or Nothing Problem 51
Joe Horton

Latitude, Supererogation, and Imperfect Duties 63
Douglas W. Portmore

The Staircase Scene: Supererogation and Moral Attunement 87
Dale Dorsey

Must Virtue Be Heroic? Virtue Ethics and the Possibility of Supererogation 105
Rebecca Stangl

The Expected, the Contra-Expected, the Supererogatory, and the Suberogatory 119
Terry Horgan and Mark Timmons

Supererogation and Its Conceptual Neighborhood Through a DWE Lens 131
Paul McNamara

Supererogation and the Limits of Reasons 165
Nathaniel Baron-Schmitt and Daniel Muñoz

The Evaluative Condition for Supererogation 181
Claire Benn

Supererogation and Forgiveness 199
Christopher Cowley

Promise-Making and Supererogation 221
David Heyd

What's the Use of Non-moral Supererogation? 237
Alfred Archer

Going Above and Beyond: Non-moral Analogues of Moral Supererogation 255
Brian McElwee

Feminist Perspectives on Supererogation 271
Katharina Naumann, Marie-Luise Raters, and Karoline Reinhardt

Supererogation in Christianity 293
Dimitrios Dentsoras

Supererogation and Protestantism 315
Gregory Mellema

Does Judaism Recognize the Supererogatory? 329
Samuel Lebens

Islamic Narratives of Duty, Supererogation, and *Ithar* 349
Mustafa Abu Sway

Supererogation in Buddhism 361
Soraj Hongladarom

About the Editor

David Heyd received his Ph.D. from Oxford University in 1976. The dissertation's topic was supererogation, which is the subject of this handbook. He then became a tenured faculty member at the department of philosophy at the Hebrew University of Jerusalem from 1976 till his formal retirement in 2014 as a full professor and Chaim Perelman Chair in Philosophy. He is still teaching part-time at the university. He has been a visiting fellow and a visiting scholar in Harvard, Berkeley, Princeton, Columbia, and at the NIH (each time for a year). In 2017, he won the EMET Prize in philosophy, which is one of the two most prestigious academic awards in Israel. His main fields of expertise are moral philosophy, political philosophy, and bioethics.

Introduction

David Heyd

1 Urmson's "Saints and Heroes"

Only few philosophers are fortunate enough to start single-handedly a new topic in philosophy. The Oxford philosopher J. O. Urmson is one of them. But at the time he wrote his seminal article, 'Saints and Heroes' (1958), he was certainly not aware of it. Had he lived now and seen the vast literature written on supererogation he would have been both awe-inspired and puzzled. The breadth and depth, the sophistication, and the analytical acumen that have developed out of his theoretically modest but pioneering article would have surprised him. Strangely, we have direct evidence from Urmson's own writings that he would have been ambivalent with regard to his success record, since thirty years later he partly retracted his original positions and denied the existence of supererogatory acts as many of us understand them (Ursmon, 1988).[1] Still, his disavowal of supererogation, if it is at all known to scholars, has been forgotten and the number of citations of his formative article is steadily growing every year. Dostoevsky has allegedly claimed that all Russian writers 'came out from under Gogol's *Overcoat.*' In a way this is true of us, supererogationists, with regard to 'Saints and Heroes.' It seems therefore appropriate to reproduce this article as the

[1] Urmson's self-critique appears in an *apropos* manner in an article on R. M. Hare. He does not deny the existence of saintly and heroic acts but warns against the use of the concept of the supererogatory as a blanket term under which all actions (like small favors or considerateness) are included. In the original article Urmson explicitly includes acts of "kindness and generosity" as going beyond duty. As this collection attests, Urmson's warning has been mostly ignored and the discussion of the "blanket term of supererogation" is flourishing.

D. Heyd (✉)
Department of Philosophy, The Hebrew University of Jerusalem, Jerusalem, Israel
e-mail: david.heyd@mail.huji.ac.il

D. Heyd (ed.), *Handbook of Supererogation*,
https://doi.org/10.1007/978-981-99-3633-5_1

opening of this collection of articles which are all new and original and provide an updated review of the subject of supererogation 65 years later.[2]

The term 'supererogatory' is mentioned only once in Urmson's article. It is also noteworthy that although he mentions St. Francis in one of his examples, he completely ignores the long Catholic tradition and the theological theorizing of supererogation in the Middle Ages. From a historical perspective, we can say that Urmson did not *invent* the idea of supererogation but only *revived* it after centuries in which the idea was dormant due to the critical blow it received from the Reformation (see Mellema[3]). Urmson's revival gave the subject a secular twist and made it for the first time in the history of ethics a theoretical term worth investigating as a separate deontic category. Thus, Urmson's main thesis is that the traditional threefold classification of actions into the obligatory, the permitted, and the forbidden is not exhaustive and should be complemented by the supererogatory, which phenomenologically we find it hard to deny, especially in extreme cases of saintly and heroic actions. We must acknowledge the existence of acts which go beyond the call of duty by being particularly good and praiseworthy but which are nevertheless not obligatory.

The reclassification of the tripartite deontic map still draws critical examination today. For example, should a fifth category, a logical mirror image of the fourth category of supererogation, be recognized? This fifth category is referred to by contemporary writing as *suberogation.* Some philosophers believe it aptly describes small offenses or rudeness, acts that are bad to do but not morally prohibited. Others deny the existence of suberogation and hold that acting beyond the call of duty and acting below the call of duty are not symmetrical since the last category should always be classified as wrong and forbidden. Horgan and Timmons offer a new pair of categories which lies between the obligatory and the fully supererogatory—namely the morally expected and the morally contra-expected. Supererogatory acts are fully optional, but suberogatory acts like rudeness or unkindness are not optional; even though they are not prohibited, they are morally contra-expected. Hence they deny the symmetry between supererogation and suberogation.

But Urmson's mapping of the field also includes the distinction between actions that are obligatory but their omission *excused* due to their difficulty and saintly and heroic actions which are by no way and under no circumstances obligatory (I developed this distinction in contrasting qualified and unqualified conceptions of supererogation, respectively, in my book which was the first comprehensive monograph on supererogation; Heyd, 1982). Furthermore, Urmson inspired many philosophers who have been writing on the way supererogation can be accommodated by traditional ethical theories, particularly utilitarianism and Kantianism. For example, much interpretive work has been done on the relation between imperfect duties and

---

[2] The editor wishes to express his gratitude to the University of Washington Press (Seattle) for granting the permission to reproduce Urmson's article originally published in *Essays on Moral Philosophy* (ed. A. I. Melden), 1958, pp. 198–216.

[3] When no date of publication is given, the reference in this introduction is to an article included in this volume. When dates are noted, the reference is to the bibliography of this introduction.

supererogation (see, for example, Kamm). Although being a relatively marginal category of actions, Urmson had the original insight that the question of supererogation is a fruitful test for ethical theories in general (in his case arguing for the superiority of utilitarianism). Later philosophers added virtue ethics as a way to address the problem of supererogation, a theory which Urmson does not consider in his article (see Stangl).

Although in the beginning of his article Urmson declares that his concern is primarily 'to draw attention to the facts' (that relate to the way we respond to heroic and saintly acts), in the latter part he suggests some normative justifications for breaking the exclusivity of the threefold classification of moral actions. Here too Urmson's work created a master plan for further research. He distinguishes between the minimum conditions for a tolerable social life and the 'higher flights or morality'; he turns our attention to the question which is referred to today as the 'over-demandingness' of morality that should be avoided by distinguishing between duty and beyond duty; he distinguishes between what we can claim or rightly expect and what we can 'merely hope for and receive with gratitude'; and perhaps most importantly, he highlights the moral significance of acting out of freedom and discretion which is contrasted to the constraint of acting out of duty and social pressure (see Baron-Schmitt and Muñoz who call this freedom the 'prerogative' or the agent's rights).

So we see that Urmson drew the road map for the study of supererogation, but had he witnessed the contemporary discussions he would have hardly recognized the terrain. Paul McNamara expands the four-fold and five-fold classification to a twenty-one (or even eighty-four) classification of possible positions regarding supererogation and the relevant terms in its neighboring semantic fields. He uses the much enriched conceptual field to define more subtly the concept of supererogation to solve some of the problems involved in the counter-intuitive implications of its application (see McNamara). Basically, McNamara suggests that we distinguish between 'beyond the call' and 'supererogation' since the former applies also to doing just beyond the minimum required while we want supererogation to characterize doing more than that. This augmented conceptual map enables McNamara to introduce an agent-relative element (beyond the axiological element of goodness and the deontic element) refining the theory of supererogation. Since Urmson was well aware that his discussion only skims over these deep questions, he would have been gratified to see how detailed and sophisticated the analysis of each one of them has become. In the rest of this introduction, I will try to review some of the extensions of Urmson's road map in the six decades following 'Saints and Heroes.' The introduction will not summarize the main arguments of the contributors to this volume since each article is preceded by a helpful Abstract for the reader's convenience in the online version of the book. But it will show how each author contributes to the drawing of the evolving chart of the study of supererogation.

2 Deontic Logic and Its Limits

The first post-Urmson development in the study of supererogation was the attempt to formalize Urmson's deontic classification of moral actions. Already in the early 1960s Chisholm (1963) developed an elegant reductionist schema for such formalization. The tripartite classification of the obligatory, the permitted, and the prohibited was described as actions that are good to do and bad not to do, actions that are neither good to do nor bad not to do, and actions that are bad to do and good not to do (respectively). This allowed Chisholm to account for supererogatory action as that which is good to do but not bad not to do, and even to add a fifth deontic category which was not mentioned by Urmson—*offence*, which is the mirror image of the supererogatory, namely that which is bad to do but not good not to do. But as we have already mentioned, this new category, which is today referred to as *suberogation*, gave rise to some controversy, particularly whether it is indeed symmetrical to that of supererogation.

Chisholm's formal account widened the scope of the supererogatory. Urmson (like the medieval Catholic theoreticians of supererogation) was thinking of saintly and heroic acts, but the description of 'good to do but not bad not to do' applies also to common acts of giving, doing favors, kindness, and (maybe) forgiveness. This extension created a threat of the trivialization of supererogation that would include every act of politeness or considerateness. But it tried (although I believe with no success) to make sense of the symmetry of supererogation and suberogation, since a small favor is good to do but not bad not to do, while refusing to do such a favor is bad to do but not good not to do. This symmetry is even harder to accept when it comes to saintly and heroic acts. For what could be the negative counterpart of a particularly virtuous and beneficent act? If it is a particularly vicious and harmful act, can it be 'not bad not to do', or in deontic language—not prohibited?

The formal schema, despite its elegant simplicity, proved to be lacking in the explanation of the phenomenon of actions beyond the call of duty. The reduction to a pair of contrary concepts (good but not bad not to do, or praiseworthy but not blameworthy not to do) could not do justice to the complexity of supererogation, for example, cases in which a supererogatory act is bad not to do even if lies within the bounds of one's rights or 'privilege' (see Baron-Schmitt and Muñoz). Or, praiseworthiness cannot define supererogatory action since the fulfillment of a particularly burdensome duty is more praiseworthy than doing a small supererogatory favor. It seems that adding another normative concept like right, or morally expected (see Horgan and Timmons) is necessary for the definition of the supererogatory, which is constituted by the combination of both the deontic and the axiological.

Furthermore, the 'good-but-optional' schema does not take into account the *comparative* nature of supererogatory action: it must be *better* than alternative permissible actions, at least if these alternative actions are not more costly. Claire Benn offers a way to combine the condition of goodness with the condition of the cost or risk (which was ignored by deontic logicians) in defining whether an act is supererogatory: supererogatory actions are better than all the less costly permissible alternatives (see Benn, and see also Horton who similarly combines the variables of the degree of sacrifice required of the agent and the measure of goodness achieved by the act).

3 Duties to Act Supererogatorily?

A different reductionist account of supererogation tried to address the 'paradox of supererogation' in terms of duty rather than goodness. How can a particularly good action not be obligatory? There have been philosophers who denied the very possibility of supererogatory action. In a strictly duty-based moral theory, it is claimed, there is no room for actions that lie beyond duty. It is thus only natural that attempts to explain, or explain away, supererogatory action have focused on Kant. One Kantian way to deny the possibility of supererogatory action is to argue that it violates duties *to oneself*. But the very idea of moral duties to oneself is controversial, especially in liberal autonomy-based theory, according to which a person can relieve oneself from such alleged duties and act supererogatorily. And it is also unintuitive to claim that giving a particularly generous birthday present goes against the donor's duties to herself (such as keeping enough for her own living). Another way is to try to identify supererogatory action as the fulfillment of *imperfect* duties. This is not the place to go into this complex interpretive debate about Kant's view, since he has different ways of drawing the distinction between perfect and imperfect obligations and consequently of the way supererogation can still find space in Kantian ethics. But one interesting way to understand supererogation in terms of imperfect duties is to follow Kant's idea that imperfect duties apply to the *ends* moral agents ought to adopt (rather than to *acts* which they have to perform as a matter of their perfect duties), and that supererogatory action is a matter of *attitude* rather than of action (see Portmore). Thus, according to Douglas Portmore we can surpass our imperfect duties by adopting attitudes that are morally better than those which are minimally required of us to fulfill those imperfect duties. For example, we can develop our power of empathy that is constitutive of being helpful to others, which is a moral end that we ought to adopt.

Much of the literature on supererogation assumes that duty has priority over supererogation: before going beyond duty, one is supposed to fulfill her duty. If you owe me \$100 you have to pay me first, before giving the \$100 to charity, or even before you give \$200 to charity. But the relations between duty and supererogation turn out to be more complex. There are cases in which the supererogatory act, for example, saving the life of someone in danger, has priority over the fulfillment of a promise (to help you prepare for an exam), at least in the sense that saving life *justifies* the breaking of a promise (see Kamm). Furthermore, once a person decides to engage in a supererogatory action, this might impose on him or her a duty to do another supererogatory act. Parfit (1982) argued that after deciding to enter a building which is on fire to save a stranded child (which involves a risk of death), and finding that there is another stranded child there, you have a duty to save the other child as well, *if* that does not involve an extra cost or risk to you. This is an example of the way a supererogatory act can become obligatory, or at least entail an obligatory act (see Kamm). Another case in which a supererogatory act conceptually involves an obligation is promise-making: making a promise is in most cases a supererogatory act but keeping it is almost always obligatory (see Heyd). Or take the problematical practice of forgiveness. According to some accounts, showing forgiveness is a conditional obligation: if the offender undergoes a process of repentance (and satisfies other conditions) then forgiving is a duty of the offended party; but if these conditions are not satisfied, and forgiveness is granted nevertheless, many philosophers see the act as supererogatory. Still, Christopher Cowley warns us that it is not easy to distinguish between a morally noble act of unconditional forgiveness and a morally wrong act of forgiving the unforgivable. The famous case of the dying Nazi officer asking a Jewish boy for forgiveness told in Simon Wiesenthal's *The Sunflower* illustrates this difficulty (see Cowley).

The complexity of the relations between supererogation and duty is well illuminated by the so-called All-or-Nothing Problem (see Horton), which has a similar structure to that of promise-making. To take Parfit's case mentioned above, you are permitted (and it is not morally wrong) not to save the two children due to the high risk in entering a burning house. It is wrong however to save only one child since saving the second does not impose and extra risk on you. Hence it seems that if you are not going to save the two children, you ought to save neither and just stay out of the house. This sounds counter-intuitive. The solution to this problem introduces 'conditional obligations': if you are prepared to save one child, you ought to save the other too; but if you are not prepared to save the two, you still ought to save one, i.e., do the second best (Horton). Horton adds that although once you are willing to do the risky act it is indeed your duty to do it, your willingness to do it is supererogatory (and makes the whole act supererogatory). This helps also to make sense of the typical report of the hero after completing his heroic deed, "I just did what I felt to be my duty" (see Horton, and see also Kamm and her critique). It is not false modesty or a moral error in categorizing the deontic status of the act. The solution to the All-or-Nothing problem appeals to the element of beneficent intention in supererogation: the way to justify the omission of supererogatory action is the risk or loss involved in the altruistic act; and since there is no difference between the loss or risk involved in the saving of one child and the saving of two children, all that remains is the good achieved by the act which is of course doubly higher in saving the two. Still, saving only one child does more good than saving neither.

One possible implication of the more complex cases of supererogatory action, like the saving of the two children, is that once you decide to give money to charity, you ought to choose among the accessible charities the one which would produce better results than all the rest. This conclusion is directly relevant to the widespread contemporary discussion of effective altruism. However, unlike the case of the two stranded children, in choosing the charity to which I am going to donate a large sum of money, am I under a duty to donate it to the charity that will bring maximal utility (welfare, happiness)? Doesn't the discretion of the supererogatory donation itself also include the choice of the recipient? There are purposes or causes which are dearer to us than others regardless of the sub-optimal capacity to promote overall utility or welfare. Unlike the absolute priority of saving life over keeping a promise, the priority of donating to the most effective cause over donating to a cause one identifies with seems to appeal only to strict utilitarians. The same applies to the gifts that we give to particular people rather than to complete strangers.

4 Reason-Based Accounts

One other reductionist account of supererogation is *reason-based*. By appealing already in the 1970s to the distinction between first-order and second-order reasons for action, Joseph Raz suggested that duty can be described as a second-order reason to ignore or exclude first-order reasons which would otherwise guide our action. A moral duty to return a book to its owner from whom I borrowed it is a second-order reason which excludes any first-order reason to keep the book (such as its utility for the borrower). Raz used the same conceptual tool to account for supererogatory action. *Exclusionary permissions* are second-order licenses to disregard first-order reasons to do the most beneficial or the morally best action (Raz, 1975). This account, however, leaves open the question of the relation between the first- and second-order reasons, and although Raz claims that the second-order supererogatory permissions are derived from the value of the autonomy of the individual, he does not explain why (and when) reasons of autonomy should be given the exclusionary priority over reasons of the welfare of others rather than be directly balanced against each other on the same level.

The reason-based approach to the study of supererogation took another turn by suggesting the distinction between moral and rational reasons (Portmore, 2003, 2008). Following Raz, this approach attempts to ground supererogation within a theory of practical reason in general rather than in the narrower moral theory. Roughly speaking, supererogatory action may be favored by moral reasons but may be in conflict with non-moral reasons of rationality. Camus' Dr. Rieux is morally commended for volunteering in a plague-stricken city to treat dying patients, but he acts contrary to the overall balance of reasons which includes the value of his life, the great risk of his own death, and maybe the limited good that he can do on the whole. In many cases we can even say that acting 'heroically' is irrational. Thus, non-moral reasons can block the requiring force of moral reasons which all-things-considered are not always over-riding (Dorsey, 2013). However, it is not the case that all supererogatory acts are irrational, since most of them are not saintly or heroic and do not involve irrational risk or sacrifice. It would be strange to hold a moral theory which recommends irrational acts (Postow, 2005). To solve this puzzle, we can argue that extreme altruistic sacrifices violate one's duty to oneself and hence ought to be avoided. But that would still lead to the denial of the paradigm cases of supererogation.

But what is the normative force of reason in practical reasoning? If a reason for action means that one ought or is under an obligation to act according to that reason, we are entangled in the 'paradox of supererogation,' which makes the (overall) good action obligatory, thus leaving no room for supererogation. And the paradox won't be resolved if we replace 'a reason' with 'the conclusive reason,' since that would equally make supererogatory action either really obligatory or irrational.

One way to overcome the threat of paradox is to distinguish between reasons which give rise to duties and obligations, on the one hand, and recommending reasons, justifying reasons, or merit conferring reasons, on the other. Thus, the reason to risk one's life by entering a burning house to save a child is just a recommendation to the agent which also justifies the act, or makes the act particularly praiseworthy. However, these distinctions cannot in themselves account for the supererogatory since they call for a non-reason-based element which would explain *why* one does not have to act from the recommending reason or the reason which justifies the act rather than from competing reason which *requires* the agent to do something. Or in other words, if we have a justifying reason not to do the supererogatory act (e.g., its high cost or risk) and a favoring reason to do the act (its goodness), how do we weigh those reasons and choose to act on them? After all, reasons by definition *favor* a certain action and that should include justifying reasons; but in the analysis of Baron-Schmitt and Muñoz justifying reasons do not 'favor' an action but rather justify the *failure* to do it. This leads Baron-Schmitt and Muñoz to the conclusion that we need a *non*-reason-based principle to justify the omission of acts beyond duty. They refer to it as prerogative (others would call it autonomy) of the agent (see Baron-Schmitt and Muñoz). One of the earliest Catholic formulators of the concept of supererogation, Tertullian, called it *licentia.*

There are however philosophers who do not ground supererogation in a principle (like prerogative, rights, autonomy, or freedom) but in more pragmatic and psychological terms. It is said that the distinction between supererogation and obligation is more effective in making people act morally. Expecting everybody to act supererogatorily as a matter of duty is not only unrealistic but would reduce people's motivation to act even according to their duties (see Archer). This is usually called the argument from over-demandingness. The norms of morality must be subjected to the ought-implies-can principle. According to this view, saintly and heroic acts are expected only of the very few who have strong character and will as well as a deep altruistic commitment. Archer believes that this argument applies also in non-moral contexts (see below). Drawing the line between what is required and what would be best helps us succeed in meeting at least the minimal demands in any normative domain.

5 New Domains of Supererogation

Another aspect in which the discussion of supererogation has made significant advances is the *scope* of contexts in which it is discussed. The list of examples in which the concept is nowadays used goes much beyond Urmson's conception (and indeed equally beyond the original domains which were the focus of the Christian concept, namely, the three Evangelical Counsels of poverty, chastity, and obedience).

Urmson concentrated on saintly and heroic acts which he thought were a direct challenge to the 'tripartite classification' of the deontic categories of action—the obligatory, the permitted, and the prohibited. Indeed he explicitly mentions any action of 'going the second mile' as equally beyond the call of duty, and hence includes under this new category acts of kindness and generosity, forgiveness, or assistance; but he does not analyze any of them. But Urmson himself was ambivalent about the extension of the concept of the supererogatory to these minor cases of non-obligatory well-doings in his later self-critique, as we mentioned in the beginning of this introduction. The fact that he warned against the use of the supererogatory as a 'blanket term' is indeed a challenge to the discussion of the limits of the concept, whose borderlines are often fuzzy. But the analysis of *particular* cases of what looks like supererogatory behavior has proven to be fruitful for the theory of supererogation itself. So let's mention some of them.

Much of the vast literature on forgiveness in the last few decades deploys the concept of supererogation. The question is whether one ought (or even must) forgive, and under what circumstances. Does repentance on part of the offender make forgiveness obligatory or is it still optional? And when no remorse is shown, is it even permissible to forgive, that is to say, are some offenses *unforgivable*? In other words, is forgiveness conditional or unconditional (see Cowley)? Isn't forgiveness of serious offenses particularly touching exactly *because* it lies beyond the duty of the offended party? Giving, like forgiving, seems to be in many cases supererogatory. But that raises the question of gratitude. If I am the recipient of a gift does that impose on me the duty to reciprocate by a similar giving? But if that is the case, uncalled gifts—or indeed any supererogatory beneficence—may often turn out to be a burden to the recipient who has involuntarily incurred a duty (that was the reason Nietzsche rejected the goodness of gifts). By their nature, supererogatory acts of giving or forgiving can never be truly reciprocated, as Derrida has argued, since in the cycle of giving and giving back, the original giver will always have the moral superiority of the one who initiated the cycle by acting supererogatorily (Derrida, 1992). So although Hobbes and many others hold that gratitude is a strict duty (and ingratitude—a sin), others, like Horgan and Timmons, believe that it is not required but only morally expected (and its omission blameworthy).

Toleration is a further example of supererogatory forbearance (Benbaji & Heyd, 2003). Even if tolerating the practices of a minority within society is a political duty of the state in a liberal society, tolerating the noise made by my neighbor past midnight seems to be supererogatory. I could have called the police, but didn't. Not acting on my rights when that could cause frustration or harm to another person is a case of supererogation. Volunteering is an interesting case of supererogation for by definition it is both free (optional) and praiseworthy (good). Some hard or dangerous job must be done by a group of people but no particular member of the group can be said to be under the duty to do it. A lottery is a morally fair selection procedure, but volunteering is morally superior (cf. Dorsey and his analysis of an individual's supererogatory behavior in the context of collective duties, and Heyd, 1982, 150–152). Another minor though very widespread supererogatory practice which has been completely overlooked is promising, which we have already mentioned above.

Although there is a large corpus of philosophical writing on promise-keeping as an obligation, there is nothing about the supererogatory deontic status of promise-*making*. There are only a few contexts in which promising is prohibited or obligatory, and it is also rarely a morally indifferent act (since it creates value for the promisee). So it seems that it should be included as a case of supererogation.

A more radical broadening of the scope of the supererogatory has been suggested in the last decade or two—exporting the concept of supererogation to *non-moral* contexts (see Archer and McElwee). This extension is analogous to the generalization of the specifically moral reasons for action to any normative reasoning, which is typically associated with the work of Tim Scanlon. Thus, one can speak of supererogatory standards in epistemology when someone who claims to know something comes forward with more evidence than is standardly required in her field of research (e.g., a double amount of data than is expected for a statistically valid methodology, or for a responsible practical decision). Other non-moral domains in which the normative dimension plays a role and therefore opens the door for supererogation are prudence, sports, and aesthetics. Supererogation is not an easy case for prudential reasoning since it is not clear how one can do more than is required in taking care of one's own interests. If the doctor tells me that for my health I have to walk *at least* half an hour each day and I decide to walk one full hour, that may be considered doing more than I owe myself. But the borderline between prudential supererogation and irrationality is not easy to draw. If I save 90% of my salary for old age and lead a miserable life in the present, I surely have crossed that line. The limits of prudence are not only defined by the moral duties that we have toward others but also by our own overall interests which ought to be balanced between health and other interests we have or between money that we will need in the future and money that we need now.

Sport is an interesting case. Alfred Archer brings the example of the soccer player who the referee believes has been fouled and who consequently awards a penalty kick to the player's team. The soccer player, knowing that he was not fouled, appeals to the referee to correct his mistake and cancel the penalty, thus losing an important advantage to his team. Since such an advantage is fully legal according to the rules of the game, the player's behavior can be treated as praiseworthy, not-to-be-expected and hence supererogatory. But such cases raise the question about the nature of fairness in sport. Is fair play a requirement *within* the game or also about the way the game is conducted in terms of values external to the game? Should refraining from appealing to the referee be seen as wrong and blameworthy, as suberogatory, or as fully permissible, or maybe even as obligatory?

Maybe the most controversial example of non-moral supererogation is the aesthetic case (see Archer). There are surely aesthetic norms regarding dress codes and guests can fall below these standards but also surpass them (buying a new suit rather than wearing an old one for an important event). One can also behave supererogatorily by tending a beautiful public garden in one's neighborhood, making it more pleasant than the municipality's gardener does. These are debatable cases of supererogation since they seem to be moral rather than aesthetic: they aim at making the hosts of the party or the residents of the neighborhood happier. Had they been performed by a Robinson Crusoe on a desert island they would have been regarded as self-serving rather than supererogatory because it is doubtful whether a Robinson Crusoe has any aesthetic duties to begin with.

Brian McElwee expands the list of non-moral supererogation even further and includes beyond the above-mentioned list etiquette, collegiality, and professionalism (see McElwee). In all these spheres the supererogatory is that act or behavior the omission of which is not blameworthy or open to critical reaction. We only have to identify in each sphere the standards which are obligatory or required falling below which merits a critical reaction.

So we see that the extension of the list of examples of putative supererogatory actions contributes to the better understanding of the concept of supererogation as does the enrichment of the conceptual analysis of the deontic categories to which we have referred above.

6 Virtue Ethics, Feminism and Collectives

We should also note recent discussions of supererogation from new perspectives which initially looked irrelevant to the concept. Urmson and his early followers treated supererogatory action as combining axiological and deontic dimensions and dealt with the question of how to adapt utilitarianism and Kantian ethics to good actions that lie beyond the call of duty. It was often argued that the notion of supererogation is alien to virtue ethics. For one cannot be more virtuous than the virtuous person or more balanced in one's psychological dispositions than the well-balanced character. Roger Crisp (Crisp, 2013) argues that we should read Aristotle's theory of virtue as involving an explicit *duty* to become virtuous and act virtuously. Crisp is thus led to a clearly anti-supererogationist view. But other views hold that supererogation is compatible with virtue ethics and can be accounted by it. Kawall (2009) suggests that a supererogatory action is that which the perfectly virtuous person would judge as such. Rebecca Stangl uses the same Chisholm schema for defining supererogation, namely that which is virtuous to do but not vicious not to do (Stangl, 2016). Since virtue is a 'range-based' concept, one can be virtuous in having a generous character but without necessarily donating to charity in each and every case or opportunity. This opens the door to supererogation. Stangl offers an opposite reading of Aristotle to Crisp's: actions are right because they are virtuous, whereas for Crisp actions are virtuous only if they are right. If right is taken as what is obligatory, we understand that only on Stangl's reading there is a place for the supererogatory in neo-Aristotelian virtue ethics. Thus, according to Stangl, a doctor treating *his* patients in a plague-ridden city is acting virtuously and in the right way, and had he not done so, he would have been blamed for acting viciously or unjustly. But the doctor volunteering to go to a faraway plague-ridden city to treat unknown patients is acting virtuously and supererogatorily, since had she not volunteered no one could have blamed her for acting viciously or unjustly (see Stangl).

Another recent angle from which the idea of supererogation has been critically challenged is feminist theory. Beyond gender biases which abound in substantive moral views, it is argued that moral theory and concepts often reflect an androcentric approach. Accordingly, the debate on supererogation—its methodology and paradigm examples—must be subjected to a careful scrutiny from a gender perspective. Katharina Naumann, Marie-Luise Raters, and Karoline Reinhardt argue that the association of supererogation with saintly and heroic acts reflects a gender prejudice by implying that the extremely praiseworthy actions are the domain of men. The sacrifice of women, especially within the family, and the limitless giving of mothers to children are considered to be their natural duty and are expected as a matter of natural affection, in contrast to the male moral ideal of military sacrifice based on principle (such as patriotism) and which goes beyond duty. This contrast serves to stereotype genders and marginalizes women by excluding their action from the realm of the supererogatory. A feminist approach to supererogation should emphasize empathy rather than abstract altruism, social rather than individual good, love rather than rational respect. Furthermore, add the three authors, women should not be expected to make sacrifices as their duty—such as taking care of an ailing parent or carrying an embryo to term—when parallel sacrifices of men are considered supererogatory. Moral praise should be distributed equally to men and women (see Naumann, Raters, and Reinhardt).

Finally, a more minor expansion of the discussion of supererogation has been its possible application in the public and collective sphere. Can a group or a collective act supererogatorily? It is doubtful that states can act altruistically and we understand their assistance to poorer states in terms of their duties rather than their discretion, good intention, or empathy. It is the state's duty to help countries in need because their individual citizens express their belief through the democratic process that this is what their country ought to do. Furthermore, a state cannot act supererogatorily toward its own citizens. It has duties toward them, but these cannot be exceeded. The main reason is that supererogatory action toward one individual or one sector in society would mean the discrimination of other individuals or sectors. Similarly, public officials (representing the state) are prohibited from acting supererogatorily toward individual citizens, e.g., in granting special privileges to some without doing the same to others. Judges are expected to act strictly by law and must not go beyond it in a supererogatory way. Legal pardon is an interesting borderline case since if granted with no rules of desert it may be seen as discriminatory. Forgiveness in interpersonal relations is not subjected to the principle of universalizability, but pardon by the state must be.

Although groups and collectives cannot act beyond their duty, the duties that they have must be carried out by individuals. As mentioned above there are various procedures by which these individuals can be selected, but asking for volunteers has a particular moral attraction (Dorsey). Military operations are a rich reservoir of examples of this kind but so are daily situations in which *people* are expected to lend help to a woman needing help in a train station and expecting *someone* to help her. The group is under an obligation to lend this help but the volunteer who actually responds to this duty is acting supererogatorily. Unlike all others in the crowd she shows particular attention which is beyond what we usually expect of people (see Dorsey). In the Muslim tradition we can also find the idea of the collective duty of a community which is fulfilled by individuals acting beyond the call of duty (see Abu Sway).

One non-moral sphere of collective supererogation which we haven't mentioned is business ethics. We may claim that by donating money to charities firms act beyond their duty, i.e., beyond being responsible, law-abiding, and fair in their dealings. This is a problematic example because usually the aim of such donations is not the good of others but part of a self-serving aim of promoting the public image of the firm. Furthermore, donations may violate the primary duty of the firm to its stock holders who expect the maximization of the firm's profit. Of course the charitable donations of the firm can be decided by the individual stock holders but then it is not the firm as such which acts supererogatorily.

7 Religions and the Challenge of Supererogation

The concept of supererogation was the invention of the Catholic Church, first articulated by the Church Fathers. It was never used in non-theological contexts till Urmson. But the modern revival of the concept in secular ethical theory gave rise to an interest of religious thinkers, particularly of non-Catholic denominations, and the way that their respective religious doctrines contend with this category of action. The final part of this Handbook is devoted to attempts to assess the role of supererogation in religious thought in the light of the modern analysis of the concept. It is only natural to start with the Catholic religion and the development of its concept of supererogation which is mostly less known to contemporary theorists and which proves to contain many elements of the post-Urmsonian definitions of the concept, as is shown in Dimitrios Dentsoras' comprehensive survey (see Dentsoras). He shows how the concept of supererogation began in the New Testament, reached its peak in Thomas Aquinas' work, and gradually declined till it suffered a deadly attack by the Reformation. Since according to the Reformers the Christian believer cannot even hope to have discharged his *duties* to God, there obviously remains no space for actions *beyond* duty. Gregory Mellema shows what made Luther, Calvin, and Melanchthon so passionate in their objection to supererogation which probably was the cause for the very little interest in the concept for four hundred years. Beyond the theological case against the idea of supererogation, it was the corruption of the Catholic practice of Indulgences (directly based on the extra merit of supererogatory action) which motivated Luther in his revolution. But surprisingly, some twentieth-century Protestant theologians have re-examined the traditional rejection of supererogation and acknowledged the possibility of actions beyond duty. For example, it is held that born-again believers perform good works in gratitude to God rather than in obedience to God's commandments. The possibility of supererogatory action is also implied in Reinhold Niebuhr's statement that "sacrificial love is, therefore, a form of love which transcends the limits of law" (see Mellema). In a way, by making the distinction between love and law, contemporary Protestants return to the old Catholic distinction between counsels and precepts, or between the 'new [Christian] law' and the 'old [Mosaic] law.'

Both Judaism and Islam are rigorously deontological. They are based on the law which consists of a list of commandments by which the believer must abide. Like Kantian ethics, their deontological character makes it hard for them to recognize actions beyond the call of duty, for duty is usually held as exhausting the realm of religious value. However, as Samuel Lebens shows, the Jewish tradition has a concept of 'going beyond the letter of the law' which breaks this rigid deontological character of the law and opens the way to some form of supererogation. Room is left to the ethical appraisal of the legal commandments and is often highly commended and recommended. Moreover, looking for a way to incorporate supererogation in Jewish thought may trigger a fresh assessment of the role of law in the *Halakha* in general (see Lebens).

Islam seems at first sight to be even less forthcoming with regard to supererogation. When we think of human beings' duties to God, it is indeed incomprehensible how we can do more than is required. But regarding duties of humans to humans, one can find some recognition of supererogatory action. Mustafa Abu Sway provides rich evidence from the Quar'an and other Muslim sources for the special value ascribed to cases of extreme charity (giving one's meagre resources to the needy, leaving almost nothing to the donor's family), forgiveness (requiring controlling one's justified anger), going on a pilgrimage to Mecca (*hajj*) in the place of someone who finds it difficult and beyond his capacity, or risking one's life in rescuing people in danger of death. Islam has a special term, *mandub*, which refers to actions that are recommended by God although not strictly required. This term can serve as a basis for reconstructing the idea of supererogation in Muslim ethics (see Abu Sway).

Finally, there are religious cultures which are typically non-deontological and consequently less receptive of the idea of supererogation. Buddhism presents its followers with ideals of the right way of life rather than with laws, duties, and commandments. Nevertheless, by careful scrutiny of Buddhist texts Soraj Hongladarom reveals certain elements which are clearly supererogatory. Buddhism offers some minimal norms which can be exceeded. For example, choosing the life of a monk (*bodhisattva*) is completely voluntary and highly praised because one of the tasks of a monk is descending to hell in order to save human beings there. There is a striking similarity between the saintly ideal of a Christian monk and that of a Buddhist monk: both demonstrate a supererogatory choice of a life completely devoted to spiritual purification and in adopting that end do more than what is minimally required according to the norms of the respective cultures. But Hongladarom emphasizes that Buddhist culture recognizes also common acts of special assistance and kindness as supererogatory (see Hongladarom).

So we see that religion which historically gave rise to the concept of supererogation is now critically examined and challenged by it.

Naturally this Handbook could not cover all the aspects of the discussion of supererogation. But it is hoped that the multifaceted collection of articles in this volume will provide the reader with an up-to-date picture of this notion and its undeniable contribution to the enrichment of ethical theory.

As we have seen, gratitude for supererogatory acts is owed as a duty, or at least is morally expected. My wife Milly surely went beyond her duty in encouraging me to embark on that project and in being actively and wisely involved in all its stages.

References

Benbaji, H., & Heyd, D. (2003). The charitable perspective: Forgiveness and toleration as supererogatory. *Canadian Journal of Philosophy, 31*, 567–586.

Chisholm, R. (1963). Supererogation and offence: A conceptual scheme for ethics. *Ratio, 5*, 1–14.

Crisp, R. (2013). Supererogation and virtue. In *Oxford Studies in Normative Ethics* (vol. 3), M. Timmons (ed.), Oxford: Oxford University Press.

Derrida, J. (1992). *Given Time: I. Counterfeit Money* (trans. Peggy Kamuf), University of Chicago Press.

Dorsey, D. (2013). The supererogatory, and how to accommodate it. *Utilitas, 25*, 355–382.

Heyd, D. (1982). *Supererogation: Its status in ethical theory.* Cambridge University Press.

Kawall, J. (2009). Virtue theory, ideal observers, and the supererogatory. *Philosophical Studies, 146*, 179–196.

Parfit, D. (1982). Future generations: Further problems. *Philosophy and Public Affairs, 11*, 131.

Portmore, D. (2003). Position-relative consequentialism, agent-centered options, and supererogation. *Ethics, 113*, 303–332.

Postow, B. C. (2005). Supererogation again. *Journal of Value Inquiry, 39*, 245–253.

Raz, J. (1975). Permissions and supererogation. *American Philosophical Quarterly, 12*, 161–168.

Stangl, R. (2016). Neo-aristotelian supererogation. *Ethics, 126*, 339–365.

Urmson, J. O. (1988). Hare on intuitive moral thinking. In *Hare and Critics*, D. Seanor, & N. Fotion (eds.), Oxford: Clarendon Press.

Saints and Heroes

J. O. Urmson

Moral philosophers tend to discriminate, explicitly or implicitly, three types of action from the point of view of moral worth. First, they recognize actions that are a duty, or obligatory, or that we ought to perform, treating these terms as approximately synonymous; second, they recognize actions that are right in so far as they are permissible from a moral standpoint and not ruled out by moral considerations, but that are not morally required of us, like the lead of this or that card at bridge; third, they recognize actions that are wrong, that we ought not to do. Some moral philosophers, indeed, could hardly discriminate even these three types of action consistently with the rest of their philosophy; Moore, for example, could hardly recognize a class of morally indifferent actions, permissible but not enjoined, since it is to be presumed that good or ill of some sort will result from the most trivial of our actions. But most moral philosophers recognize these three types of action and attempt to provide a moral theory that will make intelligible such a threefold classification.

To my mind this threefold classification, or any classification that is merely a variation on or elaboration of it, is totally inadequate to the facts of morality; any moral theory that leaves room only for such a classification will in consequence also be inadequate. My main task in this paper will be to show the inadequacy of such a classification by drawing attention to two of the types of action that most conspicuously lie outside such a classification; I shall go on to hazard some views on what sort of theory will most easily cope with the facts to which I draw attention, but the facts are here the primary interest.

We sometimes call a person a saint, or an action saintly, using the word 'saintly' in a purely moral sense with no religious implications; also we sometimes call a person

This article was first published in J. O. Urmson, "Saints and Heroes", in A. I. Melden (ed.), *Essays in Moral Philosophy* (Seattle: University of Washington Press, 1958), pp. 198–216. The editor wishes to express his gratitude to the Press for the permission to reproduce the article in this volume.

J. O. Urmson
(1915–2012), Oxford, England

D. Heyd (ed.), *Handbook of Supererogation*,
https://doi.org/10.1007/978-981-99-3633-5_2

a hero or an action heroic. It is too clear to need argument that the words 'saint' and 'hero' are at least normally used in such a way as to be favorably evaluative; it would be impossible to claim that this evaluation is always moral, for clearly we sometimes call a person a saint when evaluating him religiously rather than morally and may call a person the hero of a game or athletic contest in which no moral qualities were displayed, but I shall take it that no formal argument is necessary to show that at least sometimes we use both words for moral evaluation.

If 'hero' and 'saint' can be words of moral evaluation, we may proceed to the attempt to make explicit the criteria that we implicitly employ for their use in moral contexts. It appears that we so use them in more than one type of situation, and that there is a close parallel between the ways in which the two terms 'hero' and 'saint' are used; we shall here notice three types of situation in which they are used which seem to be sufficiently different to merit distinction. As the first two types of situation to be noticed are ones that can be readily subsumed under the threefold classification mentioned above, it will be sufficient here to note them and pass on to the third type of situation, which, since it cannot be subsumed under that classification, is for the purposes of this paper the most interesting.

A person may be called a saint (1) if he does his duty regularly in contexts in which inclination, desire, or self-interest would lead most people not to do it, and does so as a result of exercising abnormal self-control; parallel to this a person may be called a hero (1) if he does his duty in contexts in which terror, fear, or a drive to self-preservation would lead most men not to do it, and does so by exercising abnormal self-control. Similarly for actions: an action may be called saintly (1) if it is a case of duty done by virtue of self-control in a context in which most men would be led astray by inclination or self-interest, and an action may be called heroic (1) if it is a case of duty done by virtue of self-control in a context in which most men would be led astray by fear or a drive for self-preservation. The only difference between the saintly and the heroic in this sort of situation is that the one involves resistance to desire and self-interest; the other, resistance to fear and self-preservation. This is quite a clear difference, though there may be marginal cases, or cases in which motives were mixed, in which it would be equally appropriate to call an action indifferently saintly or heroic. It is easy to give examples of both the heroic and the saintly as distinguished above: the unmarried daughter does the saintly deed of staying at home to tend her ailing and widowed father; the terrified doctor heroically stays by his patients in a plague-ridden city.

A person may be called a saint (2) if he does his duty in contexts in which inclination or self-interest would lead most men not to do it, not, as in the previous paragraph, by abnormal self-control, but without effort; parallel to this a person may be called a hero (2) if he does his duty in contexts in which fear would lead most men not to do it, and does so without effort. The corresponding accounts of a saintly (2) or heroic (2) action can easily be derived. Here we have the conspicuously virtuous deed, in the Aristotelian sense, as opposed to the conspicuously self-controlled, encratic deed of the previous paragraph. People thus purged of temptation or disciplined against fear may be rare, but Aristotle thought there could be such; there is a tendency today to think of such people as merely lucky or unimaginative, but Aristotle thought more highly of them than of people who need to exercise self-control.

It is clear that, in the two types of situation so far considered, we are dealing with actions that fall under the concept of duty. Roughly, we are calling a person saintly or heroic because he does his duty in such difficult contexts that most men would fail in them. Since for the purposes of this paper I am merely conceding that we do use the term 'saintly' and 'heroic' in these ways, it is unnecessary here to spend time arguing that we do so use them or in illustrating such uses. So used, the threefold classification of actions whose adequacy I wish to deny can clearly embrace them. I shall therefore pass immediately to a third use of the terms 'heroic' and 'saintly' which I am not merely willing to concede but obliged to establish.

I contend, then, that we may also call a person a saint (3) if he does actions that are far beyond the limits of his duty, whether by control of contrary inclination and interest or without effort; parallel to this we may call a person a hero (3) if he does actions that are far beyond the bounds of his duty, whether by control of natural fear or without effort. Such actions are saintly (3) or heroic (3). Here, as it seems to me, we have the hero or saint, heroic or saintly deed, par excellence; until now we have been considering but minor saints and heroes. We have considered the, certainly, heroic action of the doctor who does his duty by sticking to his patients in a plague-stricken city; we have now to consider the case of the doctor who, no differently situated from countless other doctors in other places, volunteers to join the depleted medical forces in that city. Previously we were considering the soldier who heroically does his duty in the face of such dangers as would cause most to shirk—the sort of man who is rightly awarded the Military Medal in the British Army; we have now to consider the case of the soldier who does more than his superior officers would ever ask him to do—the man to whom, often posthumously, the Victoria Cross is awarded. Similarly, we have to turn from saintly self-discipline in the way of duty to the dedicated, self-effacing life in the service of others which is not even contemplated by the majority of upright, kind, and honest men, let alone expected of them.

Let us be clear that we are not now considering cases of natural affection, such as the sacrifice made by a mother for her child; such cases may be said with some justice not to fall under the concept of morality but to be admirable in some different way. Such cases as are here under consideration may be taken to be as little bound up with such emotions as affection as any moral action may be. We may consider an example of what is meant by 'heroism' (3) in more detail to bring this out.

We may imagine a squad of soldiers to be practicing the throwing of live hand grenades; a grenade slips from the hand of one of them and rolls on the ground near the squad; one of them sacrifices his life by throwing himself on the grenade and protecting his comrades with his own body. It is quite unreasonable to suppose that such a man must be impelled by the sort of emotion that he might be impelled by if his best friend were in the squad; he might only just have joined the squad; it is clearly an action having moral status. But if the soldier had not thrown himself on the grenade would he have failed in his duty? Though clearly he is superior in some way to his comrades, can we possibly say that they failed in their duty by not trying to be the one who sacrificed himself? If he had not done so, could anyone have said to him, “You ought to have thrown yourself on that grenade”? Could a superior have decently ordered him to do it? The answer to all these questions is plainly negative. We clearly have here a case of a moral action, a heroic action, which cannot be subsumed under the classification whose inadequacy we are exposing.

But someone may not be happy with this conclusion, and for more respectable reasons than a desire to save the traditional doctrine. He may reason as follows: in so far as that soldier had time to feel or think at all, he presumably felt that he ought to do that deed; he considered it the proper thing to do; he, if no one else, might have reproached himself for failing to do his duty if he had shirked the deed. So, it may be argued, if an act presents itself to us in the way this act may be supposed to have presented itself to this soldier, then it is our duty to do it; we have no option. This objection to my thesis clearly has some substance, but it involves a misconception of what is at issue. I have no desire to present the act of heroism as one that is naturally regarded as optional by the hero, as something he might or might not do; I concede that he might regard himself as being obliged to act as he does. But if he were to survive the action only a modesty so excessive as to appear false could make him say, “I only did my duty,” for we know, and he knows, that he has done more than duty requires. Further, though he might say to himself that so to act was a duty, he could not say so even beforehand to anyone else, and no one else could ever say it. Subjectively, we may say, at the time of action, the deed presented itself as a duty, but it was not a duty.

Another illustration, this time of saintliness, may help. It is recorded by Bonaventura that after Francis of Assisi had finished preaching to the birds on a celebrated occasion his companions gathered around him to praise and admire. But Francis himself was not a bit pleased; he was full of self-reproach that he had hitherto failed in what he now considered to be his duty to preach to the feathered world. There is indeed no degree of saintliness that a suitable person may not come to consider it to be his duty to achieve. Yet there is a world of difference between this failure to have preached hitherto to the birds and a case of straightforward breach of duty, however venial. First, Francis could without absurdity reproach himself for his failure to do his duty, but it would be quite ridiculous for anyone else to do so, as one could have done if he had failed to keep his vows, for example. Second, it is not recorded that Francis ever reproached anyone else for failure to preach to the birds as a breach of duty. He could claim this action for himself as a duty and could perhaps have exhorted others to preach to the birds, but there could be no question of reproaches for not so acting.

To sum up on this point, then, it seems clear that there is no action, however quixotic, heroic, or saintly, which the agent may not regard himself as obliged to perform, as much as he may feel himself obliged to tell the truth and to keep his promises. Such actions do not present themselves as optional to the agent when he is deliberating; but since he alone can call such an action of his a duty, and then only from the deliberative viewpoint, only for himself and not for others, and not even for himself as a piece of objective reporting, and since nobody else can call on him to perform such an act as they can call on him to tell the truth and to keep his promises, there is here a most important difference from the rock-bottom duties which are duties for all and from every point of view, and to which anyone may draw attention. Thus we need not deny the points made by our imaginary objector in order to substantiate the point that some acts of heroism and saintliness cannot bc adequately subsumed under the concept of duty.

Let us then take it as established that we have to deal in ethics not with a simple trichotomy of duties, permissible actions, and wrong actions, or any substantially similar conceptual scheme, but with something more complicated. We have to add at least the complication of actions that are certainly of moral worth but that fall outside the notion of a duty and seem to go beyond it, actions worthy of being called heroic or saintly. It should indeed be noted that heroic and saintly actions are not the sole, but merely conspicuous, cases of actions that exceed the basic demands of duty; there can be cases of disinterested kindness and generosity, for example, that are clearly more than basic duty requires and yet hardly ask for the high titles, 'saintly' and 'heroic.' Indeed, every case of 'going the second mile' is a case in point, for it cannot be one's duty to go the second mile in the same basic sense as it is to go the first—otherwise it could be argued first that it is one's duty to go two miles and therefore that the spirit of the rule of the second mile requires that one go altogether four miles, and by repetition one could establish the need to go every time on an infinite journey. It is possible to go just beyond one's duty by being a little more generous, forbearing, helpful, or forgiving than fair dealing demands, or to go a very long way beyond the basic code of duties with the saint or the hero. When I here draw attention to the heroic and saintly deed, I do so merely in order to have conspicuous cases of a whole realm of actions that lie outside the trichotomy I have criticized and therefore, as I believe, outside the purview of most ethical theories.

Before considering the implications for ethics of the facts we have up to now been concerned to note, it might be of value to draw attention to a less exalted parallel to these facts. If we belong to a club there will be rules of the club, written or unwritten, calling upon us to fulfill certain basic requirements that are a condition of membership, and that may be said to be the duties of membership. It may perhaps be such a basic requirement that we pay a subscription. It will probably be indifferent whether we pay this subscription by check or in cash—both procedures will be 'right'—and almost certainly it will be quite indifferent what sort of hat we wear at the meetings. Here, then, we have conformity to rule which is the analogue of doing one's duty, breach of rule which is the analogue of wrongdoing, and a host of indifferent actions, in accordance with the traditional trichotomy. But among the rule-abiding members of such a club what differences there can be! It is very likely that there will be one, or perhaps two or three, to whose devotion and loyal service the success of the club is due far more than to the activities of all the other members together; these are the saints and the heroes of the clubs, who do more for them by far than any member could possibly be asked to do, whose many services could not possibly be demanded in the rules. Behind them come a motley selection, varying from the keen to the lukewarm, whose contributions vary in value and descend sometimes to almost nothing beyond what the rules demand. The moral contribution of people to society can vary in value in the same way.

So much, then, for the simple facts to which I have wished to draw attention. They are simple facts and, unless I have misrepresented them, they are facts of which we are all, in a way, perfectly well aware. It would be absurd to suggest that moral philosophers have hitherto been unaware of the existence of saints and heroes and have never even alluded to them in their works. But it does seem that these facts have been neglected in their general, systematic accounts of morality. It is indeed easy to see that on some of the best-known theories there is no room for such facts. If for Moore, and for most utilitarians, any action is a duty that will produce the greatest possible good in the circumstances, for them the most heroic self-sacrifice or saintly self-forgetfulness will be duties on all fours with truth-telling and promise-keeping. For Kant, beyond the counsels of prudence and the rules of skill, there is only the categorical imperative of duty, and every duty is equally and utterly binding on all men; it is true that he recognizes the limiting case of the holy will, but the holy will is not a will that goes beyond duty but a will that is beyond morality through being incapable of acting except in accordance with the imperative. The nearest to an equivalent to a holy will in the cases we have been noting is the saintly will in the second sense we distinguished—the will that effortlessly does its duty when most would fail—but this is not a true parallel and in any case does not fall within the class of moral actions that go beyond duty to which our attention is primarily given. It is also true that Kant recognized virtues and talents as having conditional value, but not moral value, whereas the acts of heroism and saintliness we have considered have full moral worth, and their value is as unconditional as anyone could wish. Without committing ourselves to a scholarly examination of Kant's ethical works, it is surely evident that Kant could not consistently do justice to the facts before us. Intuitionism seems to me so obscurantist that I should not wish to prophesy what an intuitionist might feel himself entitled to say, but those intuitionists with whose works I am acquainted found their theories on an intuition of the fitting, the prima facie duty or the claim; the act that has this character to the highest degree at any time is a duty. While they recognize greater and lesser, stronger and weaker, claims, this is only in order to be able to deal with the problem of the conflict of duties; they assign no place to the act that, while not a duty, is of high moral importance.

Simple utilitarianism, Kantianism, and intuitionism, then, have no obvious theoretical niche for the saint and the hero. It is possible, no doubt, to revise these theories to accommodate the facts, but until so modified successfully they must surely be treated as unacceptable, and the modifications required might well detract from their plausibility. The intuitionists, for example, might lay claim to the intuition of a nonnatural characteristic of saintliness, of heroism, of decency, of sportingness, and so on, but this would give to their theory still more the appearance of utilizing the advantages of theft over honest toil.

Thus as moral theorists we need to discover some theory that will allow for both absolute duties, which, in Mill's phrase, can be exacted from a man like a debt, to omit which is to do wrong and to deserve censure, and which may be embodied in formal rules or principles, and also for a range of actions which are of moral value and which an agent may feel called upon to perform, but which cannot be demanded and whose omission cannot be called wrongdoing. Traditional moral theories, I have suggested, fail to do this. It would be well beyond the scope of this paper, and probably beyond my capacity, to produce here and now a full moral theory designed to accommodate all these facts, including the facts of saintliness and heroism. But I do think that of all traditional theories utilitarianism can be most easily modified to accommodate the facts, and would like before ending this paper to bring forward some considerations tending to support this point of view.

Moore went to great pains to determine exactly the nature of the intrinsically good, and Mill to discover the *summum bonum,* Moore's aim being to explain thereby directly the rightness and wrongness of particular actions and Mill's to justify a set of moral principles in the light of which the rightness or wrongness of particular actions can be decided. But, though there can be very tricky problems of duty, they do not naturally present themselves as problems whose solution depends upon an exact determination of an ultimate end; while the moral principles that come most readily to mind—truth-telling; promise-keeping; abstinence from murder, theft, and violence; and the like—make a nice discrimination of the supreme good seem irrelevant. We do not need to debate whether it is Moore's string of intrinsic goods or Mill's happiness that is achieved by conformity to such principles; it is enough to see that without them social life would be impossible and any life would indeed be solitary, poor, nasty, brutish, and short. Even self-interest (which some have seen as the sole foundation of morality) is sufficient ground to render it wise to preach, if not to practice, such principles. Such considerations as these, which are not novel, have led some utilitarians to treat avoidance of the *summum malum* rather than the achievement of the *summum bonum* as the foundation of morality. Yet to others this has seemed, with some justification, to assign to morality too ignoble a place.

But the facts we have been considering earlier in this paper are surely relevant at this point. It is absurd to ask just what ideal is being served by abstinence from murder, but on the other hand nobody could see in acts of heroism such as we have been considering a mere avoidance of antisocial behavior. Here we have something more gracious, actions that need to be inspired by a positive ideal. If duty can, as Mill said, be exacted from persons as a debt, it is because duty is a minimum requirement for living together; the positive contribution of actions that go beyond duty could not be so exacted. It may, however, be objected that this is a glorification of the higher flights of morality at the expense of duty, toward which an unduly cynical attitude is being taken. In so far as the suggestion is that we are forgetting how hard the way of duty may be and that doing one's duty can at times deserve to be called heroic and saintly, the answer is that we have mentioned this and acknowledge it; it is not forgotten but irrelevant to the point at issue, which is the place of duty in a moral classification of actions, not the problem of the worth of moral agents. But I may be taken to be acquiescing in a low and circumscribed view of duty which I may be advised to enlarge. We should, it may be said, hitch our wagons to the stars and not be content to say: you must do this and that as duties, and it would be very nice if you were to do these other things but we do not expect them of you. Is it perhaps only an imperfect conception of duty which finds it not to comprise the whole of morality? I want to examine this difficulty quite frankly, and to explain why I think that we properly recognize morality that goes beyond duty; for it seems to me incontestable that properly or improperly we do so.

No intelligent person will claim infallibility for his moral views. But allowing for this one must claim that one's moral code is ideal so far as one can see; for to say, "I recognize moral code A but see clearly that moral code B is superior to it," is but a way of saying that one recognizes moral code B but is only prepared to live up to moral code A. In some sense, then, everybody must be prepared to justify his moral code as ideal, but some philosophers have misunderstood this sense. Many philosophers have thought it necessary, if they were to defend their moral code as ideal, to try to show that it had a superhuman, a priori validity. Kant, for example, tried to show that the moral principles he accepted were such as any rational being, whether man or angel, must inevitably accept; the reputedly empiricist Locke thought that it must be possible to work out a deductive justification of moral laws. In making such claims such philosophers have unintentionally done morality a disservice; for their failure to show that the moral code was ideal in the sense of being a rationally justifiable system independent of time, place, circumstance, and human nature has led many to conclude that there can be no justification of a moral code, that moral codes are a matter of taste or convention.

But morality, I take it, is something that should serve human needs, not something that incidentally sweeps man up with itself, and to show that a morality was ideal would be to show that it best served man—man as he is and as he can be expected to become, not man as he would be if he were perfectly rational or an incorporeal angel. Just as it would be fatuous to build our machines so that they would give the best results according to an abstract conception of mechanical principles, and is much more desirable to design them to withstand to some extent our ham-fistedness, ignorance, and carelessness, so our morality must be one that will work. In the only sense of 'ideal' that is of importance in action, it is part of the ideal that a moral code should actually help to contribute to human well-being, and a moral code that would work only for angels (for whom it would in any case be unnecessary) would be a far from ideal moral code for human beings. There is, indeed, a place for ideals that are practically unworkable in human affairs, as there is a place for the blueprint of a machine that will never go into production; but it is not the place of such ideals to serve as a basic code of duties.

If, then, we are aiming at a moral code that will best serve human needs, a code that is ideal in the sense that a world in which such a code is acknowledged will be a better place than a world in which some other sort of moral code is acknowledged, it seems that there are ample grounds why our code should distinguish between basic rules, summarily set forth in simple rules and binding on all, and the higher flights of morality of which saintliness and heroism are outstanding examples. These grounds I shall enumerate at once.

1. It is important to give a special status of urgency, and to exert exceptional pressure, in those matters in which compliance with the demands of morality by all is indispensable. An army without men of heroic valor would be impoverished, but without general attention to the duties laid down in military law it would become

a mere rabble. Similarly, while life in a world without its saints and heroes would be impoverished, it would only be poor and not necessarily brutish or short as when basic duties are neglected.

2. If we are to exact basic duties like debts, and censure failure, such duties must be, in ordinary circumstances, within the capacity of the ordinary man. It would be silly for us to say to ourselves, our children, and our fellow men, "This and that you and everyone else must do," if the acts in question are such that manifestly but few could bring themselves to do them, though we may ourselves resolve to try to be of that few. To take a parallel from positive law, the prohibition laws asked too much of the American people and were consequently broken systematically; and as people got used to breaking the law a general lowering of respect for the law naturally followed; it no longer seemed that a law was something that everybody could be expected to obey. Similarly in Britain the gambling laws, some of which are utterly unpractical, have fallen into contempt as a body. So, if we were to represent the heroic act of sacrificing one's life for one's comrades as a basic duty, the effect would be to lower the degree of urgency and stringency that the notion of duty does in fact possess. The basic moral code must not be in part too far beyond the capacity of the ordinary men on ordinary occasions, or a general breakdown of compliance with the moral code would be an inevitable consequence; duty would seem to be something high and unattainable, and not for 'the likes of us.' Admirers of the Sermon on the Mount do not in practice, and could not, treat failure to turn the other cheek and to give one's cloak also as being on all fours with breaches of the Ten Commandments, however earnestly they themselves try to live a Christian life.
3. A moral code, if it is to be a code, must be formulable, and if it is to be a code to be observed it must be formulable in rules of manageable complexity. The ordinary man has to apply and interpret this code without recourse to a Supreme Court or House of Lords. But one can have such rules only in cases in which a type of action that is reasonably easy to recognize is almost invariably desirable or undesirable, as killing is almost invariably undesirable and promise-keeping almost invariably desirable. Where no definite rule of manageable complexity can be justified, we cannot work on that moral plane on which types of action can be enjoined or condemned as duty or crime. It has no doubt often been the case that a person who has gone off to distant parts to nurse lepers has thereby done a deed of great moral worth. But such an action is not merely too far beyond average human capacity to be regarded as a duty, as was insisted in (2) above; it would be quite ridiculous for everyone, however circumstanced, to be expected to go off and nurse lepers. But it would be absurd to try to formulate complicated rules to determine in just what circumstances such an action is a duty. This same point can readily be applied to such less spectacular matters as excusing legitimate debts or nursing sick neighbors.
4. It is part of the notion of a duty that we have a right to demand compliance from others even when we are interested parties. I may demand that you keep your promises to me, tell me the truth, and do me no violence, and I may reproach you if you transgress. But however admirable the tending of strangers in sickness

may be it is not a basic duty, and we are not entitled to reproach those to whom we are strangers if they do not tend us in sickness; nor can I tell you, if you fail to give me a cigarette when I have run out, that you have failed in your duty to me, however much you may subsequently reproach yourself for your meanness if you do so fail. A line must be drawn between what we can expect and demand from others and what we can merely hope for and receive with gratitude when we get it; duty falls on one side of this line, and other acts with moral value on the other, and rightly so.

5. In the case of basic moral duties we act to some extent under constraint. We have no choice but to apply pressure on each other to conform in these fundamental matters; here moral principles are like public laws rather than like private ideals. But free choice of the better course of action is always preferable to action under pressure, even when the pressure is but moral. When possible, therefore, it is better that pressure should not be applied and that there should be encouragement and commendation for performance rather than outright demands and censure in the event of nonperformance. There are no doubt degrees in this matter. Some pressure may reasonably be brought to persuade a person to go some way beyond basic duty in the direction of kindliness and forbearance, to be not merely a just man but also not too hard a man. But, while there is nothing whatever objectionable in the idea of someone's being pressed to carry out such a basic duty as promise-keeping, there is something horrifying in the thought of pressure being brought on him to perform an act of heroism. Though the man might feel himself morally called upon to do the deed, it would be a moral outrage to apply pressure on him to do such a deed as sacrificing his life for others.

These five points make it clear why I do not think that the distinction of basic duty from other acts of moral worth, which I claim to detect in ordinary moral thought, is a sign of the inferiority of our everyday moral thinking to that of the general run of moral theorists. It in no way involves anyone in acquiescing in a second best. No doubt from the agent's point of view it is imperative that he should endeavor to live up to the highest ideals of behavior that he can think of, and if an action falls within the ideal it is for him irrelevant whether or not it is a duty or some more supererogatory act. But it simply does not follow that the distinction is in every way unimportant, for it is important that we should not demand ideal conduct from others in the way in which we must demand basic morality from them, or blame them equally for failures in all fields. It is not cynicism to make the minimum positive demands upon one's fellow men, but to characterize an act as a duty is so to demand it.

Thus we may regard the imperatives of duty as prohibiting behavior that is intolerable if men are to live together in society and demanding the minimum of cooperation toward the same end; that is why we have to treat compliance as compulsory and dereliction as liable to public censure. We do not need to ask with Bentham whether pushpin is as good as poetry, with Mill whether it is better to be Socrates dissatisfied or a fool satisfied, or with Moore whether a beautiful world with no one to see it would have intrinsic worth; what is and what is not tolerable in society depends on no such nice discrimination. Utilitarians, when attempting to justify the main rules of duty in terms of a *summum bonum,* have surely invoked many different types of utilitarian justification, ranging from the avoidance of the intolerable to the fulfillment of the last detail of a most rarefied ideal.

Thus I wish to suggest that utilitarianism can best accommodate the facts to which I have drawn attention, but I have not wished to support any particular view about the supreme good or the importance of pleasure. By utilitarianism I mean only a theory that moral justification of actions must be in terms of results. We can be content to say that duty is mainly concerned with the avoidance of intolerable results, while other forms of moral behavior have more positive aims.

To summarize, I have suggested that the trichotomy of duties, indifferent actions, and wrongdoing is inadequate. There are many kinds of action that involve going beyond duty proper, saintly, and heroic actions being conspicuous examples of such kinds of action. It has been my main concern to note this point and to ask moral philosophers to theorize in a way that does not tacitly deny it, as most traditional theories have. But I have also been so rash as to suggest that we may look upon our duties as basic requirements to be universally demanded as providing the only tolerable basis of social life. The higher flights of morality can then be regarded as more positive contributions that go beyond what is universally to be exacted; but while not exacted publicly they are clearly equally pressing *in foro interno* on those who are not content merely to avoid the intolerable. Whether this should be called a version of utilitarianism, as I suggest, is a matter of small moment.

Supererogation and Duty

F. M. Kamm

Abstract This chapter considers the relation between supererogation and duties (also here referred to as obligations) from a nonconsequentialist point of view. It first considers whether supererogation may sometimes take precedence over positive and negative duties and how this relates to personal costs (including efforts) required to perform one's duty. It then considers how acquiescence to having large costs imposed on one (even permissibly) can be supererogatory. Finally, it considers how what are usually duties can become supererogatory and how what is usually supererogatory can become a duty. The relation between these topics and the trolley problem, the so-called 'all or nothing problem,' and the issue of abortion are examined.

Keywords Supererogation · Duty · Personal costs · Justification · Personal prerogative · Supererogatory acquiescence · Secondary permissibility · All-or-nothing problem · Trolley problem · Nonconsequentialism

Supererogation involves going beyond the call of duty to achieve something worthwhile (and in that sense not doing so is permissible). While promoting self-perfection or aesthetic achievement might also involve supererogation, the focus of this discussion will be on achieving what is morally worthwhile when not doing so would be morally permissible. Supererogatory acts (including omissions) that achieve a morally worthwhile end need not themselves be morally worthwhile. For example, suppose it is supererogatory to give up one's life to save one hundred strangers. Suppose further that one has a duty to save lives at small cost to oneself even without any prior agreement to do so. If instead of supererogatorily giving up one's life to save the hundred one knows one could dutifully save the same one hundred by easily snapping one's fingers, then the supererogatory act of saving the one hundred is not morally worthwhile since one could achieve the same outcome at less personal cost. While it may be morally permissible to waste one's efforts there is no moral worth in doing so. An act that goes beyond duty to produce a morally worthwhile end might

F. M. Kamm (✉)
Rutgers University, New Jersey, NJ, USA
e-mail: frances.kamm@philosophy.rutgers.edu

D. Heyd (ed.), *Handbook of Supererogation*,
https://doi.org/10.1007/978-981-99-3633-5_3

also not be morally worthwhile if the end does not justify the supererogation. For example, it could be morally worthwhile to help someone else study for an exam but (barring extraordinary circumstances) this end would not justify giving up one's life. The motive or ultimate intention with which one does the supererogatory act can also make it not praiseworthy even though (unlike the cases just mentioned) objective factors make the act both morally worthwhile and justified. For example, suppose one acts on the intention to save a hundred people at great cost to oneself but only because this is the only act that would frighten one's parent; if it didn't frighten one's parent, one wouldn't do it. This case satisfies the requirement that supererogation involves an intention to do something worthwhile but because the motivation for doing this is to frighten one's parent, the act is not praiseworthy.[1]

It is not only unjustified supererogatory acts that are not required. For example, an act could be justified by what it would achieve and still be supererogatory and not morally required because one does not always have a duty to produce a greater good.[2] In this case, the possibility of moral supererogation depends on the rejection of the act consequentialist view that one always has a moral duty to do what produces the best outcome. This rejection can depend, in some cases, on the nonconsequentialist view that one has a morally sanctioned personal prerogative not to make large sacrifices in order to bring about the best outcome. It can also depend, in some cases, on the nonconsequentialist view that making large personal sacrifices to bring about an equal or lesser good (e.g., giving up one's leg to either save another's leg or only his foot) is a morally sanctioned personal prerogative and can sometimes also be justified. By contrast, from an impartial act consequentialist perspective there is no moral gain in producing only an equal good and moral reason not to produce less good. More generally, in cases where supererogation would not involve major sacrifices (e.g., acts of kindness or minimal generosity) the permission not to supererogate depends simply on a right to retain personal autonomy.

This chapter will consider the relation between supererogation and duties (also here referred to as obligations). Among its topics are whether supererogation may take precedence over positive and negative duties and how this relates to personal costs (including efforts) required to perform one's duty. It will also consider the relation between having large costs imposed on one and making supererogatory efforts, how what are usually duties can become supererogatory, and whether what is usually supererogatory can become dutiful.

---

[1] Some (e.g., Mark Timmons and Terry Horgan) conceive of supererogatory acts as only those done from a morally worthwhile motive or intention and deserving praise. (See their "Untying A Knot From the Inside Out: Reflections on the "Paradox of Supererogation," *Social Philosophy & Policy* 2010, 29–63.) In my example, there is a morally worthwhile intention to save someone else and I think the act is supererogatory but the motive (or ultimate intention) for which it is done (only to frighten one's parent) makes the act not worthy of praise.

[2] Joshua Gert uses the example of moral supererogation to discuss reasons in general that justify but do not require. See, for example, his "Practical Rationality, Morality, and Purely Justificatory Reasons," *American Philosophical* Quarterly 37, 227–243.

1 Supererogation and Positive Duties

1. Must one perform one's duty before doing what is supererogatory in the sense that one should not forego doing one's duty in order to do what is beyond the call of duty?[3] Consider a case in which you have promised to meet someone for lunch and so have a duty to meet them. On your way you pass someone else who is drowning in shark-infested waters. You are willing to incur the risk of losing your limb to save the person's life and doing so would be justified though supererogatory. That doing so would interfere with performing your duty is not a sufficient moral reason not to do the supererogatory act; it is morally permissible to do the supererogatory act instead of one's duty. Indeed, in this particular case it could be said that one would be a 'moral idiot' to give priority to doing one's duty.

It might be counterargued that in this Lunch Case the person to whom the duty is owed would himself have a duty to waive it, obviating any conflict between duty and supererogation. However, suppose the promise is to a business client who will incur a large financial loss if you miss your appointment in order to supererogatorily save the person in shark-infested waters. There may be some size loss at which the business client would not be required to waive his right to your promised act and would not agree to waive it so that you may do the supererogatory rescue without conflict. Nevertheless, you may still be morally permitted to do the rescue rather than perform your duty. (Call this the Business Client Case.) This means that it is supererogatory for the business client to forego the performance of your duty to him and he does not agree to supererogate. Yet, it is being argued, you may do what results in this cost to him if the good achieved justifies the cost to him even if it doesn't require it. Indeed, you may do so even if the cost he would bear is greater than the cost you would undertake or be willing to undertake for the supererogatory act. That is, if the cost to you were as great as the cost to the client (e.g., $10,000), you would permissibly not bear it to save a life though your bearing it would be justified.

Now suppose that you could save someone's life at small cost to yourself and so this would be a duty (without prior commitment) or instead you could save one hundred other people at large though not fatal cost to yourself. The single person whose life you could easily save would not have a duty to give up your aid and thus his life in order to save the hundred. Even if you are unwilling to give up your own life to save the hundred, you may supererogatorily save them while leaving the one to die. Your decision to withdraw aid from one in order to save others involves a loss to the un-aided person of a type that you would not be willing to suffer yourself to achieve the same end.

It is not unreasonable to believe that there is no conflict between supererogation and duty in this last case. This is because what might be in play here is the permissibility of choosing whether to (i) perform an imperfect duty to which there is no correlative right to be aided had by the one person or (ii) supererogatorily save

[3] I first discussed the issues raised in Sect. 1 in my "Supererogation and Obligation," *Journal of Philosophy* 82(3), 1985: 118–138. A revised version appears in my *Morality, Mortality*, vol. 2 (New York: Oxford University Press, 1996).

one hundred people which would be a better way to carry out the imperfect duty to aid were the cost to oneself small. Hence, there is reason to limit discussion of supererogation taking precedence over duty to cases in which the duties are perfect (i.e., those that give a correlative right to the person to be provided with something and that usually demand a specific performance, such as those due to promises or agreements) rather than imperfect (which give one more options as to performance).

In the Business Client Case the fact that you acted permissibly in doing the supererogatory act rather than your duty need not mean that you would not owe compensation to the business client if this were possible. This is because his right to your performance has been infringed (even if permissibly and there is no wrongful violation of his right.)[4] However, the inability to provide compensation need not mean that omitting to perform the duty is impermissible. Indeed, if paying compensation should be added to what it costs one to do a supererogatory act, someone may decide to supererogate only when compensation will be impossible. When you fail to perform an imperfect duty to save one person in order to save more people supererogatorily when not all can be saved, no compensation to the single person is called for. The fact that one would infringe someone's right in failing to perform a perfect positive duty (i.e., a duty to provide some good to which another has a right) may itself contribute to making an alternative act supererogatory. That is, for someone who takes morality seriously, having to give up on performing one's perfect duty is itself a personal sacrifice (independent of having to provide compensation) that could contribute to making another act supererogatory (though it might not be sufficient to make an otherwise costless act supererogatory).

Another way to avoid the conclusion that the supererogatory act may take precedence over doing one's duty is if the *willingness* to do whatever makes an act supererogatory makes what would ordinarily be supererogatory become a duty.[5] Then one would face a conflict of duties in which one duty takes precedence over another rather than a conflict where supererogation may take precedence over a duty. For example, suppose that saving a life at small cost to oneself is a duty and it is only a large cost that would make saving supererogatory. Some claim that if one doesn't mind the large cost (and it is justified), the original duty to save is still present but at much larger cost. Consider a person who is willing to make a large justified sacrifice to save someone's life when avoiding the sacrifice would have been a sufficient justification for not saving. Then the person discovers that he would be saving someone who is black and for morally unacceptable racist reasons refuses to save her. The argument we are considering implies that he fails in a duty to save her; he does not act permissibly merely because objectively speaking there is a factor present (large

---

[4] Judith Thomson distinguished between violation and permissible infringement. See her *The Realm of Rights* (Cambridge, MA: Harvard University Press, 1990).

[5] Joe Horton claims this in his "The All or Nothing Problem" *Journal of Philosophy* 114(2), 2017: 94–104.

sacrifice) that could make the act supererogatory if he does not object to the sacrifice per se and there is no other morally acceptable justification for not saving.[6]

This argument for what is ordinarily supererogatory becoming a duty is problematic. First, can one be commended for saving a life from a morally worthwhile motive if willingness to make the large sacrifice implies one will have merely done one's duty? Joe Horton argues that one should only be commended for being willing to make the large sacrifice; the willingness is supererogatory rather than the act to which it leads.[7] Even on this account, there could be a conflict between supererogation and one's duty if the conflict was between acts it is *supererogatory to be willing* to do and acts one has a *duty to be willing* to do, and doing the former may sometimes take precedence over doing the latter.

However, the view that willingness to make large sacrifices to save someone can render doing it a duty (in the absence of an adequate justification for not saving) could still be objectionable. This is because it implies that one's prerogative to do something can be forfeited due to inadequate justification. But should I lose a prerogative to chew gum and have a duty not to do so if I chew only to unjustifiably annoy someone? Not having an adequate justification for exercising a prerogative need not result in forfeiting the prerogative. In addition, suppose that I do not object to the efforts that could make saving someone supererogatory but I wish to pursue a personal project instead of saving him. It is possible that pursuing this project (a) could not permissibly take precedence over dutifully saving someone at minor personal cost but (b) could permissibly take precedence over saving someone at large personal cost even though it is not the large cost per se to which I object. If my willingness to pay the large cost makes saving the person my duty, shouldn't it be equally impermissible to pursue the personal project in both cases? The difference would be explicable if one's willingness to pay the large cost did not erase the moral significance of the objective fact of the large cost. It would still give one the latitude, lacking in the case of saving at low cost, to make choices that result in a lesser good (e.g., pursuing a personal project).[8]

A final objection to the view that willingness to supererogate eliminates the conflict between justified supererogation and duty comes from cases in which it is not a large cost that turns duty into supererogation. For example, it is supererogatory, not a duty even at low cost, to provide great pleasure to many people.[9] Yet it might sometimes be permissible to supererogatorily provide this pleasure instead of

[6] See Horton 2017 and Tina Rulli "Conditional Obligations," *Social Theory and Practice* 46 (2), 2020, 365–90. Also see Theron Pummer's discussion in his *The Rules of Rescue: Cost, Distance, and Effective Altruism*, (New York: Oxford University Press, 2022).

[7] Horton's own argument for a duty to make a large sacrifice in virtue of one's willingness to make it (absent any other justifying reason for not making it) is part of his solution to what he calls "the all or nothing problem" considered in Sect. 4 (2).

[8] In Sect. 4 we will consider another way in which what is ordinarily supererogatory could become a duty.

[9] Horgan and Timmons call cases where supererogation does not involve a "prima facie" duty to aid that is overridden by the cost of aiding "unqualified" supererogation since there is no justification such as great costs needed for not seeking the greater good.

fulfilling one's duty to a business client who wouldn't be obligated to release one from the duty for this purpose.

2. The other important aspect of the life-saving cases in (1) is that the cost one would (i) have a duty and (ii) be willing to bear to perform the perfect positive duty could be much greater than the cost one would (i) have a duty and (ii) be willing to bear to save a life. Hence, effort/cost that is supererogatory for saving a life need not be supererogatory for performing the duty. Yet one could permissibly forego a duty that morally requires efforts that are the same as or greater than those one would make in instead doing the supererogatory act. This suggests that there are two measures for the stringency of a duty such as keeping a promise: (i) the cost/effort measure and (ii) the priority measure. By the cost/effort measure, doing one's business duty could be more stringent than saving a life; by the priority measure, doing one's business duty need not be more stringent than supererogatorily saving a life. This implies that the two measures of stringency can come apart and helps explain why a supererogatory act could have permissible priority over a duty.[10] (Strict duties could differ in these ways as well. For example, the efforts morally required to save a life could be less than the efforts morally required to fulfill the duty to one's client, and yet making the small effort to save a life—which is a duty—could take precedence over making the large effort—also a duty—to meet the obligation to one's client. Further, this is not because one should minimize the cost to oneself in selecting among duties to fulfill.)

What explains this? Is it that the loss to one in not saving another's life would make the personal cost of fulfilling one's business commitment go beyond what is required by the cost/efforts test? For example, suppose not saving the life would destroy one's mental health. However, in some cases there might be no limit on the personal cost that one would be willing and morally required (by agreement) to expend to fulfill a perfect positive duty were there no conflict with supererogation and yet one could permissibly perform the supererogatory act instead.[11] Furthermore, one might suffer a greater personal loss in not fulfilling a business commitment than in not saving the life and yet still permissibly save the life because it is objectively more important than the purpose of the commitment.

---

[10] Note that "permissible priority" ("may take precedence over") is not the same as simple priority ("must take precedence over").

[11] By contrast, Daniel Muñoz in his "Three Paradoxes of Supererogation" (Noûs 55 (3), 2021: 699–716) claims, "I have a prerogative to *Do Nothing* rather than *Save 1*, because saving the life is more harmful to me; but I don't have any such prerogative to *Do Nothing* rather than *Keep Promise*, since the harm to me is the same either way." However, I believe (and used cases in Kamm 1985 to show) that I lack the prerogative not to keep the promise even when the harm to me is greater to keep the promise than to do nothing. Munoz ultimately comes to accept such a view. He also claims (p. 208) that "if two options are equally harmful to me, I have no harm-based prerogative to pick one over the other." This suggests that the choice must be driven by the goodness of the outcome. But suppose I would lose a leg in either saving a life or doing my duty that does less good. I think I could still permissibly perform the duty rather than save the life in part *because* I would suffer a large harm and I could refuse to suffer it for what is not my duty. A harm-based reason could be about what the harm is for. For a case involving such a choice between types of supererogation see the Monet Case discussed in Sect. 3 (2).

The permissible priority of the supererogatory in these cases reveals an important relation between personal efforts (cost), personal duties, and the impartially considered greater good. Namely, avoiding large personal efforts may justifiably take precedence over achieving a greater good, achieving that greater good may take precedence over fulfilling a personal duty, and yet avoiding large personal efforts may not take precedence over fulfilling the personal duty. That is, where '>' is 'may take precedence over,' P is 'avoiding personal costs,' GG is 'greater good,' and D is 'personal duty,' P>GG, GG>D, -(P>D). That the greater good may take precedence over the duty implies that achieving the greater good can justify foregoing the duty though it may not morally require it.[12] It is not required, for example, when a large personal cost to achieve the greater good is not morally required.

The relation between P, GG, and D has been described above in three pair-wise comparisons (i.e., P>GG, GG>D, -(P>D)). However, even if P>GG on its own, when someone actually chooses to forego D in order to do GG, GG>P. That is, someone should be either doing his duty or performing the supererogatory act; he may not simply refuse to do either for the sake of his own interests. Furthermore, on a nonconsequentialist view, the greater good sought by supererogation must be more than minimally greater than the good foregone when a perfect positive duty is omitted. By contrast when the duty is imperfect only a slight difference in the good achievable by supererogation might be sufficient.

The relation between P, GG, and D described in the three pair-wise comparisons has the appearance of an intransitivity but it is explicable because there is no single factor between the items that account for the precedence relation in each step. In the first step, P may take precedence over GG not because it is a yet greater good but because producing the greater good at the large personal expense is beyond what morality requires. By contrast, fulfilling personal duties at large personal expense can be a requirement of morality (as represented in the final step) because it is not as a function of how much good is produced by doing one's duty that the requirement to make personal sacrifices to do it is determined. Sometimes achieving the greater good may take precedence over performing duties one owes to others if the duties do not generate as much good; it is the goodness of the outcome of the duty and the supererogatory act that get compared in deciding which may or must be done even if the goodness of the outcome does not determine how much personal effort/cost is morally demandable for each. Finally, since its being a yet greater good is not the reason why personal good may take precedence over achieving the greater good, there is no reason to think that transitivity demands that it (like the greater good) may take precedence over personal duty.

Notice that the good that is greater than what would be achieved in fulfilling one's duty need not be greater than the good lost in making personal efforts in order to

[12] This point made in Kamm 1985 seems structurally (though not substantively) similar to the point Joshua Gert makes in his "Requiring and Justifying: Two Dimensions of Normative Strength," *Erkenntnis*, July 2003: achieving a goal that can justify but not require a personal sacrifice may permissibly take precedence over a requirement of practical rationality not to make a certain sacrifice contrary to self-interest. See also his "Practical Rationality, Morality, and Purely Justificatory Reasons".

achieve the greater good. So one might justifiably supererogatorily give up one's life to save someone from paralysis (and this could be a morally worthwhile act) while the commitment one fails to keep is overridden because it would achieve something less significant than saving someone from paralysis. This implies that if we include the personal price paid in doing the supererogatory act, there may be less overall good produced than if the duty were performed. Hence, from a nonconsequentialist point of view, the good produced by supererogation that is compared with the good produced by the duty need not include the cost of supererogation itself. Furthermore, a greater good could justify not performing one's duty and yet not justify the personal cost to supererogatorily achieve it. An example involves saving an animal which could justify missing one's business obligation but not justify risking one's life. Nevertheless, it might be morally permissible to save the animal because (i) taking even an unjustified risk can be protected as an exercise of personal autonomy and (ii) the animal being saved is a good that justifies not doing one's duty. Earlier, we noted that not all supererogatory acts might be justified and we gave as an example giving up one's life to help someone study. But our current discussion shows that justification of the cost to the supererogator and justification of the cost of a foregone duty to those to whom the duty is owed are separate matters. The latter justification is more important if those who would suffer the cost of the foregone duty have not exercised a personal prerogative to suffer an unjustified cost to themselves.

If the supererogatory act is for the sake of an equal or lesser good than could be achieved in fulfilling the duty it may be wrong to forego the duty (without prior consent of the right bearer). If it were not wrong, one would need a different explanation than has been given for the permissible priority of supererogation over duty for in that explanation it is permissible to forego duty because a sufficiently greater good is achieved by supererogation. And it must be remembered that the cost to the supererogator may still be less than the loss to the person owed the duty when it is not fulfilled; the supererogator may permissibly pursue the greater good at a greater cost to others than he would (even be willing to) bear himself consistent with the good achieved being sufficiently greater than the good foregone in nonperformance of the duty.

2 Supererogatory Acquiescence, Negative Duties, and the Trolley Problem

1. The 'overdemandingness objection' claims that consequentialism morally requires too-great personal efforts (costs) to achieve greater goods and so has no room for the supererogatory. Some 'semi-consequentialists' make use of results described in Sect. 1 to avoid the overdemandingness objection. They accept a personal prerogative not to make great personal sacrifices to achieve the greater good impersonally considered, accepting one part of nonconsequentialism and making way for supererogation. (Possibly they could also support the personal prerogative to suffer a greater

personal loss to achieve a lesser good.) However, they argue against another aspect of nonconsequentialism, namely restrictions (also called side-constraints) on harming people to achieve the greater good. Section 1 discussed supererogation permissibly taking precedence over fulfilling a positive duty (i.e., a duty to provide some service) to another. Semi-consequentialists could accept that (but depart from a nonconsequentialist view in holding that even a slightly greater good than the good achieved by the duty may take priority over that duty). However, they would also claim that supererogation justified by a greater good may take precedence over fulfilling a negative duty not to harm another. For example, Derek Parfit argued for all these views, holding that it is morally permissible to kill someone (whether as a side effect or as a mere means) in order to save the lives of other people even though the person sacrificed has no duty to give up his life and does not consent to the loss being imposed on him for that end. He also seemed to hold that all this is consistent with having a duty to make large personal efforts not to harm others when only one's own good is at stake. Hence, his views imply that as with (perfect) positive duties, by the efforts test the negative duty not to harm others is more stringent than a (uncontracted, nonpromissory) duty to produce a greater good, yet by the priority test achieving a greater good even supererogatorily could permissibly take priority over not harming others.[13]

If this were true, it would imply that though a mere bystander needn't give up his life to save five people from a runaway trolley, it would be permissible to forego the ordinarily stringent duty not to kill someone else when this is necessary (e.g., by toppling someone into the trolley to stop it) in order to save the five people. This could be true even though the supererogatory price that the bystander would be willing to pay to save the five is less than death and therefore less than the harm that would come to another person as a result of the bystander's stopping the trolley from hitting the five.

On this semi-consequentialist view the relation between greater good, personal costs, and performancc of duty described earlier for positive duties could then also hold for negative duties (D(n)). That is, P>GG (hence aiding could be supererogatory), GG>D(n), but –(P>D(n)) (so personal sacrifices could be morally required not to harm others when only one's own interests are at stake).

2. Judith Thomson (ultimately) denied that it was permissible for a bystander to even turn a trolley (let alone topple someone into it) in order to stop it from killing five people when it would then kill one other person.[14] By contrast, many nonconsequentialists think that redirecting the trolley is permissible even though they deny Parfit's semi-consequentialist view that other ways of harming the one other person to save the five (e.g., toppling him in front of the trolley to stop it) are permissible. Thomson argued that it would be supererogatory for the bystander to save

---

13 See his *On What Matters*, Volume 3 (Oxford: Oxford University Press, 2017) and my discussion of it in "Parfit on the Irrelevance of Deontological Distinctions" in Mark Timmons (ed.), *Oxford Studies in Normative Ethics* Volume 10 (Oxford: Oxford University Press, 2020).

14 In Judith Jarvis Thomson, "Turning the Trolley," *Philosophy & Public Affairs* 36(4), 2008: 359–374.

five strangers by turning the trolley on himself and thereby cause his own death.[15] However, in Thomson's view doing this would not be morally worthy, commendable supererogation because it shows failure to sufficiently value one's own life. (By contrast, Parfit holds there is always sufficient reason to do what produces the greater good by making a supererogatory sacrifice.[16]). Thomson goes on to argue that if it is supererogatory to give up one's life to save five strangers, it is impermissible to impose such a loss on someone else (by turning the trolley on him) without his consent because he too is not required to be 'an altruist' (a form of supererogation). On Thomson's view, a bystander's saving the five by turning the trolley cannot take precedence over his negative duty not to harm another when it would be supererogatory for the other to volunteer to suffer the harm and he does not consent to do so. Hence, abiding by the duty not to harm someone requires not only personal sacrifice but sacrifice of the opportunity to produce the greater good (by saving the five). In essence, she denies the self-other asymmetry involved in claiming that though P>GG, it is possible for GG>D(n) thus causing another to suffer what one would not suffer for GG.[17]

Consider the following problems with Thomson's argument: (i) Since it would be supererogatory to give up one's leg to save five strangers from the trolley, her argument implies that one may not turn the trolley from killing five people when it will then hit another person removing only his leg. I think this is an implausible conclusion. (ii) Thomson's argument does not distinguish between someone donating something for the greater good (which he may have no duty to do) and someone having a loss imposed upon him as an effect of what others do for the sake of a greater good. As noted, she speaks of making the person to whom the trolley is turned be an 'altruist.' But altruism requires a decision to donate on the part of the altruist (as all forms of supererogation require a decision of an agent to go beyond duty); supererogation does not occur merely because one suffers a loss that one would not have a duty to suffer for a greater good. The consistency of P>GG and GG>D(n) depends on distinguishing the morality of donation from the morality of nonconsensual imposition (which can account for a self-other asymmetry).

However, a connection between supererogation and nonconsensual imposition of a loss on someone can be found if it is permissible for someone to resist the imposition on him. Suppose the one person who would be killed by the redirected trolley could physically and morally send it back to whoever sent it to him or even to the five who would thus lose only the benefit they received (of being unthreatened by the trolley) as a result of the trolley being redirected. (Call this the Sending Back

---

[15] Note that it could also be supererogatory for him to turn the trolley away when the effort to do so itself would seriously harm him.

[16] Though in personal remarks he referred to people who would give up their life for five strangers as "crazies" which seems inconsistent with his official position.

[17] Thomson does not discuss cases like Business Client, so I am not sure if she would judge it impermissible to fail in the positive duty to one's client in order to save the five when failing results in a loss to the client that would be supererogatory for him to donate to save five strangers and to which he does not consent.

Case.)[18] Then the person is refusing to 'supererogatorily acquiesce' in the imposition when he can resist it. Suppose one rejects Thomson's view that it would be wrong to redirect the trolley (i.e., one accepts that at least in the case of redirection of a threat GG>D(n)). One can still accept that (i) imposing the loss infringes the one person's negative right even if permissibly so and (ii) in virtue of that it would be permissible for him to resist the great imposition by sending the trolley back. This also means the one person need not die in tolerating the imposition in order to avoid harming the bystander who imposed it on him. Hence permissibly sending the trolley back implies that sometimes nonconsequentialism allows that P>Duty(n). It also means that sometimes P>GG even when this involves harming the five by undoing a benefit they received from the trolley being turned by turning the trolley back. (This is consistent with it being impermissible for the one person to send a threat that was originally headed to him to other people.)

The Sending Back Case also helps explain why a bystander's turning the trolley and harming the one person could be supererogatory. If the one person can both physically and morally send back the trolley toward the person who turned it on him, the bystander is at risk. But suppose the one person physically cannot send back the trolley. The fact that the bystander would be doing something permissible against which the one person would still be morally permitted to self-defend were he physically able to makes clear the morally fraught nature of the bystander's turning the trolley. This in itself could help make it supererogatory to do so. (This is an extension to the case of negative rights infringements of the earlier point that infringing a perfect duty may itself be a great cost to someone concerned with morality).[19]

Nevertheless, the imposition to which it is supererogatory to acquiesce might differ from the effort/cost it is supererogatory to donate. For example, it could be supererogatory of me to pay $10,000 for a device to save the five strangers from the trolley. However, suppose someone redirects the trolley away from the five and toward my $10,000 located on the other track (which seems permissible). It would not be permissible for me to redirect the trolley away from the money back toward either the bystander or the five thereby killing them.

Thomson also famously discusses supererogation (in a form she calls 'splendid Samaritanism') in her Violinist Case in which in order to save a violinist his friends attach him (without his knowledge) to the body of another person who never harmed the violinist and who does not consent to the attachment.[20] This is not a permissible infringement but a violation of a negative right. Thomson thinks it is permissible for the person impermissibly imposed on as well as those who can help him to detach

---

[18] I discussed Sending Back Cases in *Intricate Ethics* (New York: Oxford University Press, 2007), *The Trolley Problem Mysteries* (New York: Oxford University Press, 2015), and *Rights and Their Limits: In Theory, Cases, and Pandemics* (New York: Oxford University Press, 2022).

[19] Put in Hohlfeldian terms, the permissibility of turning the trolley is somewhat like having a liberty right to impose on someone without his having a correlative duty not to interfere with the exercise of that liberty, showing that the bystander and the five do not have a claim right to impose on the one person that would give him a duty not to resist.

[20] See her "A Defense of Abortion," *Philosophy & Public Affairs,* 1(1), 1971: 47–66 and "Rights and Deaths," Philosophy & Public Affairs, 2(2), 1973: 146–159.

or directly kill the violinist (who would be considered a 'morally innocent threat' because he imposes but is not responsible for this) to stop a prolonged imposition. She uses this case to argue that even if the fetus were a person, a woman pregnant by rape would be a splendid Samaritan in agreeing to let the fetus use her body for nine months and it would be permissible for her to kill it or have it killed in order to stop its use of her body. These cases involve the use of someone's body rather than their death, the person who would be killed is himself imposing on someone else's body, and removing him may require directly attacking him as a means to end the imposition, though he thereby loses only continued retention of his life that he would not have without the *impermissible* imposition.[21] By contrast, in the Sending Back Case only the bystander, not the five strangers, is a threat to the one other person, but if the trolley is sent back toward the five or the bystander they are not directly attacked as a means to stopping the *permissible* imposition on the one. (This is because preventing the imposition on the one by sending back the trolley would cause the deaths as a side effect.) Here is a somewhat similar revised violinist case still involving an *impermissible* imposition: A heavy link imposed on someone else's body without their consent delivers life support to the violinist in his home. The link will kill the person on whom it is imposed but detaching it will cause a lethal electric shock to the violinist. In this case, the violinist is not a threat (since he is not what imposes on someone else's body) and he is not killed as a means to prevent harm to the other person. It may be permissible to cause this violinist's death given that he loses only the life he gains from a lethal imposition (of the link) on someone else's body.

The trolley, violinist, and abortion cases suggest that many different factors (such as size of loss to person imposed on, whether the beneficiary of the imposition is also a threat and what he would lose if imposition ends, whether the imposition is permissible or not, and how the imposition would be stopped) could affect whether acquiescence is supererogatory or required.[22]

---

[21] My *Creation and Abortion* (New York: Oxford University Press, 1992) emphasizes the elimination of only the benefit of impermissible imposition which Thomson had not specifically employed as an explanation of her cases. It also considers how abortion might differ morally from the Violinist Case (and variations on it) because pregnancy involves the *creation* of a person.

[22] The difference between not donating aid and acquiescing in imposition may also come to the fore when the benefit of imposition on one person is someone else (i) retaining what he would ordinarily rightfully have as opposed to (ii) getting something he would not ordinarily rightfully have. This distinction between (i) and (ii) could imply that directly taking away a device that is necessary to save someone's life (the device not being rightfully his) could be morally different from directly killing someone who is being saved by the device (thinking of his life as something he would ordinarily rightfully have) even though without the device he would die.

3 Duties that Become Supererogatory?

We have investigated whether supererogation that produces a greater good may sometimes take precedence over positive and negative duties that involve lesser good. We have also considered whether refusing to supererogatorily acquiesce to an imposed personal loss may sometimes take precedence over both a duty not to harm and the greater good. Now consider whether and how what is ordinarily a duty may become supererogatory.[23]

1. Suppose one has a contractual duty as a personal lifeguard to rescue one's client from drowning even to the extent of giving up one's life. It would be supererogatory to save even one hundred nonclients at such a personal cost or even at a considerably lower cost. Suppose a conflict arises between saving one's client (at low cost) and saving one hundred nonclients at moderate supererogatory cost. It may be permissible to save the hundred on the ground that the greater good is enough to justify abandoning the client. This could be so even though by the efforts test saving the client is a highly stringent duty in comparison to saving the one hundred. (Call this the Personal Lifeguard Case.) So where "P" is "avoiding personal costs," "GG" is "greater good," and "D(p)" is "personal positive duty", P>GG, GG>D(p) even though -(P>D(p)).

But now suppose that it is possible for the lifeguard to save both the hundred and his client. In one case of this type (to be called Lifeguard Saving Both), saving the hundred raises the cost of saving his client from low to the upper limit on his duty—forfeiting his life. In this case, unlike those in Sects. 1 and 2, achieving the greater good does not require abandoning one's personal duty; it merely requires abandoning one's significant personal interest in staying alive. The question is whether (i) the personal lifeguard has a choice of either (a) not doing the supererogatory act of saving the one hundred and saving his client at low cost or (b) doing the supererogatory act and meeting the requirement of duty to save his client at the cost of his life, but (ii) the personal lifeguard lacks a choice of abandoning his client and just saving the one hundred at moderate supererogatory cost.

Here is one proposal: He may still abandon his client because it is his doing an act that can override his duty to his client that greatly raises the cost of at the same time saving the client. If this proposal were correct, it would turn what was a duty (to give up his life for his client) into supererogation. When there is already a justification for abandoning the duty and this justified act is what makes the personal cost to save one's client go up greatly, one has no duty to pay the high personal costs. It is only when there is no such factor that already justifies abandoning the duty to one's client, and there is a simple conflict between the high personal costs and one's client's welfare, that one must pay the cost.

Suppose there was only a small (rather than moderate supererogatory) cost needed to save either the hundred or the client. Then a conflict of (perfect and imperfect) duties arises between saving the client and saving the hundred. If saving the hundred justifies not saving the client and causes the cost to also save the client to rise to the

---

[23] This section is based on my "Duties that Become Supererogatory or Forbidden?" in *Principles and Persons: The Legacy of Derek Parfit* (Oxford: Oxford University Press, 2021).

limit, it may also become supererogatory to save the client. It is not the supererogatory cost of saving the hundred that is crucial to making it supererogatory rather than dutiful to also save the client.

Similar issues can arise in foregoing a negative rather than a positive duty. For example, suppose saving five people (a greater good) can justify redirecting the trolley though it will kill another person. In addition, suppose one needn't give up one's life to save the five but ordinarily one might have to give up one's life rather than kill someone else. Now suppose one could both save the five and avoid killing the other person by sacrificing oneself (i.e., by turning the trolley on oneself) instead.[24] Must one do this if one wants to save the five or else not save the five, with no option to save them by turning the trolley to one other person instead?

One proposal is that when the great cost to oneself required to avoid harming the one person arises because one is doing something (saving the five people) that itself would justify overriding the negative right of the one person, efforts to avoid killing someone that one would have to make if only one's personal interests were at stake need not be made. The great efforts that would ordinarily be a duty become supererogatory.[25] (This leaves it open that to avoid infringing a right even permissibly, lesser efforts that would ordinarily be supererogatory remain dutiful.) Consider an analogous case in war: A pilot goes on a supererogatory mission to bomb a munitions factory. Suppose it is permissible to cause some civilian deaths as a side effect but these deaths could be avoided if the pilot takes a different route which would cause his own death. Ordinarily, a military pilot might have a duty to sacrifice himself rather than cause the death of civilians when their deaths were not justified by some greater good. But when the deaths would result from an end (bombing the munitions factory) that (it is here assumed) justifies those deaths (i.e., GG>D(n)), then he need not take the route that would kill him instead. What would ordinarily be a duty to avoid killing civilians at the cost of his life becomes supererogatory.

This version of the Lifeguard Saving Both type of case should be contrasted with two other versions: (a) The lifeguard saved the one hundred by a moderate supererogatory effort, leaving his client to drown when he could otherwise have saved him at small cost. However, contrary to reasonable expectations, the client has not yet drowned but due to the lifeguard having saved the one hundred, saving his client now physically requires that the lifeguard give up his life. In this case, it seems that he must fulfill his duty at its maximal morally required cost. This is so even though the high cost is due to his having done something that justified his not

---

[24] In some ways this is like the Bystander Three Option Case in Thomson 2008 with the private lifeguard substituted for a bystander. Peter Unger discussed a case like Thomson's in his *Living High and Letting Die* (New York: Oxford University Press, 1996).

[25] Thomson reaches a different conclusion about this sort of case because she thinks it is impermissible to save five people when the only way to do this is by turning the trolley that will kill one other unconsenting person. Some think it is permissible to turn the trolley when it is the only way to save the five but deny it is permissible to do so when it is possible to substitute oneself. They face a version of the so-called "all or nothing" problem that we will discuss in Sect. 4, namely that to avoid doing what they think is the wrong of not substituting oneself, it is better to permissibly let the five die.

saving his client earlier. In this case it is not the actual saving of the one hundred that raises the cost of also saving his client since the saving is already complete. In general, having done something in the past that could justify abandoning a duty can raise the subsequent costs of doing our duty but paying these costs does not thereby become supererogatory.

(b) Circumstances make it necessary for the lifeguard to give up his life to save his client but this also makes it possible for him to save a hundred other people as they are in the same shark-infested waters as his client. Here the high cost to save his client is not due to his also saving the one hundred and he has a duty to pay the cost for his client. Given that there is no additional cost to saving the hundred beyond what he would have to pay anyway to save his client, he should save the one hundred as well. Indeed, it might be suggested that even though it is his dutifully giving up his life for his client that makes it possible to do the additional good for whose sake alone it would have been supererogatory for him to have given his life, saving the one hundred becomes a duty rather than supererogatory.[26]

Here is a (barely) possible objection to this view: Given that the cost would be supererogatory for the purpose of saving the hundred per se, even though it is no more than he would do anyway to save his client, the lifeguard should have the option not to save the hundred that he would lack were he saving them at a cost low enough to make that a duty. On this view, it isn't whether the cost would be incurred anyway that is important but what it is incurred for or what effect it has. When the cost is beyond what one is morally required to bear to produce a certain effect, one can object to making that effort with that effect. A case that could substantiate this view is one in which an outcome morally must be brought about at small personal cost but it would be positively wrong to bring it about at higher personal cost. Possibly saving a terrible criminal must be done at small cost (simply because he is a human being) but it would be wrong in itself to expend supererogatory efforts on saving him. When great efforts are made to fulfill a duty, that with no extra cost they could be expended to also save the criminal might be less important than that there is an actual causal relation between the great efforts and the criminal being saved.

2. What would ordinarily be a duty might also become supererogatory in the following Monet Case[27]: Ordinarily if one faces a choice between easily saving either a great Monet painting or a person, even an art lover would have a duty to save the person (assuming being saved is significantly good for the person). But suppose the Monet is in one burning building and the person is in another, each needs to be saved, and it is only possible in the time available for an art lover to go into one building. The risk to him is so great that saving either would be supererogatory (though justified). He is willing to supererogate only to save the Monet and doing this seems permissible.

---

[26] Horton discusses a case like this (in his "The All or Nothing Problem") except that the initial effort needed to save one person is supererogatory rather than dutiful. We will discuss it further in Sect. 4. Horton's case is unlike my first Lifeguard Saving Both Case in which the cost to save everyone is greater than the cost to save either the client or the hundred on its own.

[27] Shelly Kagan discussed a case like this (in his The *Limits of Morality* (New York: Oxford University Press, 1989) involving a parrot instead of a Monet.

If so, it should also be permissible for the art lover to save the Monet rather than the person when they are in different wings of one building. However, suppose that having taken the risk to save the Monet and about to make an easy escape with it, he comes upon the person in the same building. It is now possible easily to save her instead of the Monet but he cannot save both. The supererogatory risk that made it possible for him to reach the point where he faces the choice of easily saving either the painting or the person was for the sake of the Monet; it seems wrong if the use to which he wanted the risk to be put were 'hijacked' by something for which the risk was not and morally need not have been undertaken. (This goes beyond claiming that if additional supererogatory risk were needed to get either out of the building, the art lover would not be duty bound to take the risk for the person instead of for the Monet.[28]) Hence, in virtue of an initial supererogatory act, it becomes supererogatory rather than a duty to easily save the person rather than the Monet.

However, suppose that sometime after leaving the house with the Monet the agent encounters a flood that would destroy the Monet and also drown a person. He can save only one and either easily. Does the same argument concerning supererogatory risk already taken to save the Monet justify the agent saving the Monet rather than fulfilling what would otherwise be an easy duty to rescue the person? This might be denied while accepting that he may easily exit the burning building with the Monet instead of the person. This would be so if there were a morally significant distinction between (i) an easy rescue of a person made possible only because of the difficult rescue justifiably undertaken to save the Monet when that difficult rescue would otherwise have been needed to save the person alone and (ii) an easy rescue of a person that is possible without such a prior supererogatory act having been needed to save the person alone as is true when one comes upon the person in the flood.

4 Supererogation That Becomes Dutiful and Secondary Permissibility

1. Consider my Disturbed Visitor Case[29]: An out of town stranger who is prone to emotional disturbances will visit in the apartment near yours. If you stay in town rather than leave during his visit, he will be able to get needed information that you don't mind providing about the town. However, you will also then be present during an emotional disturbance and (assume) you would then have a duty to expend some extra effort to help him through it (it would be wrong not to). You very much want to avoid dealing with his emotional disturbance and you would not have a duty to do so if you were not physically near him, so you arrange to be out of town. This is permissible since you have no duty to stay merely to give him information and you

[28] Jeff McMahan draws a moral distinction between these types of cases by contrast to what I have suggested. See his "Doing Good and Doing the Best," in *The Ethics of Giving: Philosophers' Perspectives on Philanthropy* ed. P. Woodruff (Oxford: Oxford University Press, 2018).

[29] Introduced in *Morality, Mortality*, Volume 2 (New York: Oxford University Press, 1996).

have no duty to do what keeps you physically near him when he has an emotional disturbance. In this case, if you do the first supererogatory act of staying to give information you will be near him and so in a position that makes helping him in the second way a duty. Since a duty arises only on condition of doing the initial supererogatory act, the duty is a conditional duty.[30] (In the Monet Case, if one had a duty to leave a single building with the person rather than the Monet, one would also have a duty conditional on having done a supererogatory act. In that case, the suggestion was that because one's permissible supererogatory act to save the Monet led to being in a position to help the person, there was no duty to help at the cost of losing the Monet. By contrast, in the Disturbed Visitor Case helping the visitor with his emotional disturbance would not lead to the loss of something for the sake of which one did a supererogatory act.) Is it more important to help the visitor with information (to which you do not object) rather than avoid doing what is wrong in not helping him further if you stay? After all that is the Pareto superior outcome and he might prefer it to not being helped at all. Could you be relieved of your duty to help with his emotional disturbance when you are near by an agreement made prior to his arrival in order to get you to stay to provide information?

This case involves one person who could be helped to a greater or lesser degree. Might it, therefore, have elements in common with cases I have elsewhere discussed in which someone could be harmed to a greater or lesser degree? Suppose a bystander is about to permissibly turn a trolley from killing five to killing one other person when he discovers that if he topples that same one person into the trolley instead, he will only paralyze his legs. Toppling someone in front of a trolley even when that will only paralyze his legs (not kill him) would be impermissible (on a nonconsequentialist view) were it all one could do to stop the trolley. However, it becomes 'secondarily permissible' as a substitute for permissibly turning the trolley that will kill the same person, This is because both acts would be against the person's will and being paralyzed is better for him than being dead. I have argued for the Principle of Secondary Permissibility (PSP) which implies that sometimes acts that are initially impermissible can become secondarily permissible as substitutes for permissible acts we would otherwise do that have worse effects for the same person.[31] Is it then also true in the Disturbed Visitor Case that staying to only give information and not helping with the emotional disturbance is no longer wrong if it is the substitute for permissibly doing what is worse for the one person (leaving town)? But here is a disanalogy: Suppose that if one makes a small effort in how one topples, the person will only suffer a broken ankle rather than paralysis. Ordinarily, one would have a duty to make a small effort to reduce the harm one causes. That one comes to be in the position to do this by attempting to topple rather than kill the person, even if this were supererogatory, need not mean that it is permissible to omit the small effort to

[30] The duty is conditional only in the sense that one comes to be in a situation that would ordinarily give one a duty because one has done something else permissible. It is not conditional in the way that having a duty to kill gently is conditional on being about to kill. In this case, the conditional duty would ordinarily be contrary to duty and what puts one in the position of having the duty is also ordinarily contrary to duty.

[31] See *Morality, Mortality*, Volume 2.

cause only a broken ankle. This suggests that if one supererogatorily stays to give information (something less bad for the visitor than leaving town), it could still be wrong not to make the additional effort to help with his emotional disturbance.

Nevertheless, even if it remains wrong not to give the additional aid rather than just give information, might it be right to only give information rather than leave? Staying to only give information might be permissible relative to leaving town even if it is impermissible relative to also giving additional aid. How should one decide if it is justifiable to bring about a situation when one will then intentionally do something that is not justified? In the Disturbed Visitor Case, we can weigh the benefit that would be lost if one does not give information against the significance of subsequently doing a wrong act that has the same consequence for the visitor as one's leaving town but without doing something wrong. (Contrast this with a case in which if one stays to give information and does not help with his emotional disturbance, (i) the person would suffer more as a result of additional help being denied than of it being merely unavailable if one left town and (ii) this additional suffering would not be compensated for by getting information.)

My original conclusion about the Disturbed Visitor Case[32] was that (i) an agreement would not relieve you of your duty to give further help if you stay and (ii) you have only the choice of leaving town or staying and helping in both ways. Though what I have already said implies that one might accept (i) and yet reject (ii), I think in this case the 'all or nothing' conclusion is correct because the benefit gotten from information is small enough that it does not outweigh the lack of respect toward the person (even if he doesn't know about it) by refusing the additional aid when giving it is a duty. Leaving town to avoid that additional effort does not show a similar lack of respect since staying would be supererogatory and only staying would trigger the duty to make the additional effort given that the duty depends on being near.

2. Now consider a case, discussed by Joe Horton,[33] in which one makes a large supererogatory effort to save someone and then must decide whether to save another person as well with that same effort. Suppose that it is wrong to refuse to save the additional person. It seems one acquires a duty to save the other person when it would otherwise have been supererogatory to do so given the effort involved. Since there is a duty only on condition of doing the supererogatory act (of saving the first person), the duty is a conditional duty. Call this the Saving All Case. (It should not to be confused with the version of the Lifeguard Saving Both Case discussed in Sect. 3 (B) in which a *duty* to make a great sacrifice to save a client arguably leads to another duty to save others).

This case is the basis for what Horton calls the 'all or nothing problem'[34]: Suppose that we should avoid doing something wrong and always do what is at least permissible. This implies that we should either refrain from doing a supererogatory saving of the one person or save the other person along with him (both of which are at least permissible) but not go merely part of the way and save the one person without

---

32 Described in *Morality, Mortality,* Volume 2.

33 In Horton 2017.

34 Due to Horton 2017 and discussed in Pummer 2022.

saving the other (which is wrong).[35] This is strongly counterintuitive since what makes it worthwhile to save the one person is that a life is saved whereas no life is saved if we permissibly refrain from supererogatorily saving the one to avoid doing the wrong of not saving the other. Even if not as many morally right acts are done as should be done if another person is not saved, saving one seems more important than avoiding the wrong of failing to save another as well, especially when the other person would die anyway if one does nothing.

There are several differences between this case and my Disturbed Visitor Case: (i) In the Disturbed Visitor Case, one would have a duty to make an *additional* effort to which one objects if one makes the original supererogatory effort. In the Save All Case there is no additional effort required to save another person. (ii) In the Disturbed Visitor Case there is no duty to stay in town to give information or to stay in town to deal with a visitor's emotional disturbance. Hence there is no pro tanto duty that has to be overridden by a personal prerogative not to make big efforts in order for not staying to be permissible. By contrast, in the Save All Case it is the great effort required to save a life that keeps saving from being a duty to begin with. (iii) In the Disturbed Visitor Case the reason not to do the initial supererogatory act may be that one knows one will not be able to resist doing the additional right act yet one wishes to avoid it. If so, leaving town is not seen as a way to avoid doing something wrong but a way to avoid doing a burdensome second right act. (This is related to (i).) (iv) It seems more crucial to give partial aid when doing so would save a life rather than merely provide information even though the wrong of not saving an additional person at no additional cost is more serious than the wrong of not making an additional effort to provide emotional support.

One way of dealing with the Saving All Case is to deny that saving the second person becomes a duty it is wrong to omit. Here is a (barely) possible defense of this view (that we first raised in Sect. 3 (1.B): The effort that saves the additional person would be supererogatory if done to save him alone. Hence, even though it is no more than would be done anyway to save the first person, one should have the personal prerogative whether to save the additional person that one would not have if one were only saving him at an effort low enough to make doing that a duty. On this view, it isn't whether the effort would be made anyway that is important but what effect it has; when the effort is beyond what is morally required to produce a certain effect, one can object to that effort having that effect. Furthermore (as argued earlier in Sect. 1 when discussing the Chewing Gum Case) not having a good reason for acting on a personal prerogative need not eliminate the right to act on the prerogative.

A second analysis of the Save All Case based on the Trolley Problem and the PSP might also imply that it is not wrong to save only one person. Suppose it is permissible and supererogatory for you as a bystander to turn the trolley away from killing five people when it will then kill two other people who do not want this done to them. You are about to do this when you discover that if you topple one of the people into the trolley stopping it but killing him, the other person can survive unharmed.

---

35 This could not be said about the version of the Lifeguard Save Both Case in Sect. 3 (B) where there is a duty to save one's client and not doing so would be impermissible.

Ordinarily, it is impermissible to topple someone into a trolley killing him, but the PSP would imply that if it is no worse for the same person to be killed by the trolley in either instance, toppling him is permissible as a substitute for what you would otherwise permissibly do that would kill him and someone else too.[36] In this case, doing what kills one rather than two seems to be a duty. (This is so, even if turning the trolley at all is supererogatory and involves flipping a switch whereas toppling involves pushing someone). Analogously, it might be said, it is ordinarily wrong to save one person without also saving another at no extra cost as it is ordinarily wrong to topple someone into a trolley. However, if you permissibly fail to save anyone in the Save All Case this would be the worst outcome as two people would die, just as permissibly killing two people in the trolley case would be the worst outcome. If you may permissibly topple one of the two into the trolley though toppling is ordinarily wrong, could saving only one person rather than two in the Save All Case be secondarily permissible as a substitute for doing something permissible that has a worse outcome? Indeed, if (as seems true) it would be wrong not to topple the person rather than kill both, would it be wrong not to save only one person rather than leave both to die?

What makes this analogy imperfect is that in this trolley case there is no better alternative than toppling the one but in the Save All Case there is a still better option, namely to save two people at the same cost as saving one. This would be analogous to revising this trolley case so that as you are about to topple the one person as a substitute for turning the trolley and killing both, you discover that with no extra effort you could topple him in a different way only breaking his leg. It would ordinarily be your duty to do this. That you are in a position to do this only because you chose to permissibly and even supererogatorily do something (kill one by toppling him) less bad than the worst that you could permissibly have done (turn the trolley killing him and another person) doesn't make failing to do what only breaks the one's leg secondarily permissible rather than wrong.

Nevertheless, that failing to only break his leg is wrong does not imply that rather than topple this one person to his death you should kill him and another by turning the trolley. Toppling one person can be secondarily permissible relative to permissibly killing the two even while it is impermissible relative to breaking the one's leg. Similarly, in the Saving All Case saving one can be permissible (though not required) relative to saving no one even while it is impermissible relative to saving two. That saving one not only can be permissible but is permissible relative to saving none depends on comparing the significance of the benefit of saving one and the significance of the wrong of not saving two. In this case, unlike the Disturbed Visitor Case, the benefit of saving one person overrides the wrong of not saving (and seriously wronging) a second person especially since he would die anyway if one saved no one. This wrong (and wronging) makes the failure to-save the second person more significant than the mere failure to save him would be (without doing wrong or wronging) if one permissibly saved no one at supererogatory cost. However, avoiding

[36] I discuss both these types of cases (as well as an Extended Principle of Secondary Permissibility (EPSP)) in *Intricate Ethics*.

the wrong is not sufficient to outweigh the benefit of saving at least one life.[37] I think that saving the one person (by the supererogatory act) could even be more important than wrongfully not saving many additional people by the same effort, thus wronging many more people.

The cases in Sect. 4 not only involve what would ordinarily be supererogatory acts becoming dutiful. They also show that supererogation may permissibly take precedence over *wrongfully* failing in a duty. (This contrasts with cases in other sections that involved supererogation *justifying* failing in a duty, hence making it not wrong to do so.)

5 Conclusion

I have argued that contrary to what is usually thought, it is not always wrong not to do one's duty and instead do a supererogatory act when they conflict but also that supererogation is not always morally worthwhile. I have also argued that supererogation can be permissible even when someone other than the supererogator suffers the greatest losses, acquiescence to which would itself be supererogatory; it can be more important that supererogation can justify those greatest losses than that the losses to the supererogator be justified. Furthermore, an act that can justify not doing one's duty can sometimes make fulfilling the duty supererogatory even when there need be no conflict between doing both the act and the duty. Finally, I argued that sometimes supererogatory acts can become dutiful but also that doing a supererogatory act can have priority over avoiding wrongfully failing to perform the duty.[38]

[37] Horton's own solution to the Saving All Case and the all or nothing problem gives rise to his view (considered in Sect. 1) that if one is willing to make what would otherwise be a supererogatory effort, one has a duty to make it. This view implies that refraining from saving anyone is impermissible if one is willing to make the effort, and this explains why it is morally right to save one rather than refrain from saving anyone even if one ought to save two. However, if it is not true that willingness to make the large effort makes it a duty rather than supererogatory to do so, this explanation of the Saving All Case will fail. Also, given Horton's explanation: (i) might one toss a coin on whether to fail in one's duty to the first person or fail in one's duty to the second person, a duty that would only come to exist conditional on doing what is needed to save the first person? (ii) might not failing in one's duty to save the first person be outweighed by the foreseen failure to perform many duties one would come to have if there are many additional people (not just one) who one could but won't save at no additional cost.

[38] This chapter is dedicated in memory of Lily Safra who supported my work and many philanthropic projects. I thank David Heyd for his comments on an earlier version of this chapter.

Supererogation, Conditional Obligation, and the All or Nothing Problem

Joe Horton

Abstract If doing good is often beyond the call of duty, instances of the All or Nothing Problem abound. I have argued elsewhere that we should solve this problem by accepting a principle that I call Optimific Altruism, which has interesting implications both for the correct account of supererogation and for our obligations to give to charity. However, Theron Pummer and Daniel Muñoz have argued that we should instead solve this problem by rejecting an inference rule that I call Conditional Obligation. I here recap my preferred solution and argue against this alternative.

1 Introduction

There are many cases in which you could make a significant sacrifice to bring about a good outcome, and in which you could make the sacrifice either in a way that brings about a lesser good or in a way that brings about a greater good. In some of these cases, it is very plausible both that, because of the sacrifice required, it is permissible for you to bring about neither good, and that, because the greater good requires no greater sacrifice, it is wrong for you to bring about the lesser good. But it seems to follow from these judgments that, if you are not going to bring about the greater good, you ought to bring about neither good. This result strikes many people as counterintuitive. We can call this *the All or Nothing Problem.*[1]

Here is a simple, extreme illustration:

[1] I introduced this problem in my article 'The All or Nothing Problem', *The Journal of Philosophy* 114 (2017): 94–104. However, cases in which the problem arises were already being discussed for other reasons. For some of these earlier discussions, see Charles Fried and Derek Parfit, 'Correspondence', *Philosophy & Public Affairs* 8 (1979): 393–397; Derek Parfit, 'Future Generations: Further Problems', *Philosophy & Public Affairs* 11 (1982): 113–172, at 131; Shelly Kagan, *The Limits of Morality* (Oxford: Oxford University Press, 1989), 16; Derek Parfit, *On What Matters: Volume Two* (Oxford: Oxford University Press, 2011), 225; and Victor Tadros, *The Ends of Harm: The Moral Foundations of Criminal Law* (Oxford: Oxford University Press, 2011), 161–162.

J. Horton (✉)
University College London, London, UK
e-mail: joe.horton@ucl.ac.uk

D. Heyd (ed.), *Handbook of Supererogation*,
https://doi.org/10.1007/978-981-99-3633-5_4

A building is collapsing, with two children inside. If you hold open the door to your left, one child will escape, but you and the other child will be killed. If you hold open the door to your right, both children will escape, but you will be killed.

Since saving the children requires sacrificing your life, it is very plausible that:

(1) It is permissible for you not to save the children.

Since saving both children requires no greater sacrifice than saving only one, it is very plausible that:

(2) It is wrong for you to save only one child.

But these claims together seem to imply that:

(3) If you are not going to save both children, you ought to save neither.

Here is a way of bringing out the counterintuitiveness of (3). Suppose that you are willing to sacrifice your life to save one child, but you have taken a dislike to the other child, and so you are not willing to save both. It would be very odd for us to say, or even to think, "well, if you are not going to save both children, you really ought to save neither."

I argued in an earlier article that (3) is counterintuitive precisely because it is false when you are willing to save one child.[2] I argued that it is false when you are willing to save one child because (1) is false when you are willing to save one child. More precisely, I argued that (1) should be rejected in favor of:

(4) If you are not willing to save either child, then it is permissible for you to save neither, but if you are willing to save one, then you ought to save both.

If we reject (1) in favor of (4), accepting (2) does not commit us to (3). We can instead accept:

(5) If you are willing to save one child, then you ought to save both, but if you are not going to save both, then you ought to do the next best thing, which is to save one.

In Sect. 2, I briefly recap my argument for rejecting (1) in favor of (4), recap how this argument can be generalized into a solution to the All or Nothing Problem, and recap some interesting practical and theoretical implications of this solution.

However, my primary aim in this chapter is not to defend my preferred solution to the All or Nothing Problem, but rather to explain why I am not satisfied with an alternative solution, which has recently been defended by Theron Pummer and Daniel Muñoz.[3] According to this alternative solution, (3) does not follow from (1) and (2). In Sect. 3, I explain this alternative solution, and I argue that it should be rejected.

[2] Horton, 'The All or Nothing Problem'. In this earlier article, I stated (3) not as a conditional but instead as the claim that you ought to save neither child *rather than* save only one. However, I noted the conditional formulation in footnote 1.

[3] Daniel Muñoz and Theron Pummer, 'Supererogation and Conditional Obligation', *Philosophical Studies* 179 (2022): 1429–1443. For a precursor to this proposal, see Theron Pummer, 'All or Nothing, but If Not All, Next Best or Nothing', *The Journal of Philosophy* 116 (2019): 278–191. For a helpful restatement, see Theron Pummer, *The Rules of Rescue: Cost, Distance, and Effective*

2 Optimific Altruism

I argued in an earlier article that (1) should be rejected in favor of:

> (4) If you are not willing to save either child, then it is permissible for you to save neither, but if you are willing to save one, then you ought to save both.

Since what we ought to do does not normally depend on what we are willing to do, (4) might seem counterintuitive. But we can defend it as follows:

> Plausibly, if our acts are not justifiable to the people they affect, then these acts are wrong.[4] Plausibly, what matters is not simply that our acts are justifiable to these people, but also that they are justifiable in a way to which we can reasonably appeal. If you are not willing to save either child, you can reasonably appeal to having to sacrifice your life as a justification for not saving both. But if you are willing to save one child, you cannot reasonably appeal to this justification. Since there is no other adequate justification for not saving both children, there is then no adequate justification to which you can reasonably appeal. So, these plausible assumptions imply that, if you are willing to save one child, then you ought to save both.

It might be objected that, if (4) were true, you could escape the obligation to save both children simply by making yourself unwilling to save either child.[5] This would be an unwelcome result, for a moral view that incentivizes you to make yourself unwilling to save either child seems just as deficient as a moral view that implies (3).

I am sceptical that we have enough deliberative control over what we are willing to do for you to simply make yourself unwilling to save either child. But even if we do, this would not release you from the obligation to save both children. If the only reason you are not willing to sacrifice your life to save either child is that doing so would make you obligated to save both, then you cannot reasonably appeal to having to sacrifice your life as a justification for not saving both. Since there is no other adequate justification for not saving both, there is then no adequate justification to which you can reasonably appeal. So, even if you make yourself unwilling to save either child, you will remain obligated to save both.

We can generalize the above argument into a solution to the All or Nothing Problem:

Altruism (New York: Oxford University Press, 2022), Chap. 3. For other recent discussions of the All or Nothing Problem, or of the cases in which it arises, see Theron Pummer, 'Whether and Where to Give', *Philosophy & Public Affairs* 44 (2016): 77–95; Jeff McMahan, 'Doing Good and Doing the Best', in Paul Woodruff (ed.), *The Ethics of Giving* (Oxford: Oxford University Press, 2018), 78–102; Thomas Sinclair, 'Are We Conditionally Obligated to be Effective Altruists?', *Philosophy & Public Affairs* 46 (2018): 36–59; Douglas Portmore, *Opting for the Best: Oughts and Options* (Oxford: Oxford University Press, 2019), 206–215; Ralf Bader, 'Agent-Relative Prerogatives and Suboptimal Beneficence', *Oxford Studies in Normative Ethics* 9 (2019): 223–250; Tina Rulli, 'Conditional Obligations', *Social Theory and Practice* 46 (2020): 365–390; Daniel Muñoz, 'Three Paradoxes of Supererogation', *Noûs* 55 (2021): 699–716; and Christian Barry and Seth Lazar, 'Supererogation and Optimisation', *Australasian Journal of Philosophy* (forthcoming).

[4] For support for this premise, see T. M. Scanlon, *What We Owe To Each Other* (London: Harvard University Press, 1998), Chap. 5; and Derek Parfit, 'Justifiability to Each Person', *Ratio* 16 (2003): 368–390.

[5] I owe this worry to an anonymous reviewer for my earlier article.

> If there is no adequate justification of our acts to which we can reasonably appeal, then these acts are wrong. If bringing about an outcome O would require a sacrifice S, we can often appeal to S as an adequate justification for not bringing about O. But if we are willing to make a sacrifice that is not significantly smaller than S to bring about an outcome that is not significantly better than O, and we do not have adequate agent-relative reasons to favor this other outcome, then we cannot reasonably appeal to S as a justification for not bringing about O.[6] So, if there is no other adequate justification for not bringing about O, then we ought to bring about O.

The upshot of this argument is a principle that we can call

> *Optimific Altruism:* If the only adequate justification for not bringing about an outcome O is that it requires a sacrifice S, and we are willing to make a sacrifice that is not significantly smaller than S to bring about an outcome that is not significantly better than O, and we do not have adequate agent-relative reasons to favor this other outcome, then we ought to bring about O.

This principle is a little wordy, but the rough idea is simple: if we are willing to make a sacrifice, unless we have adequate agent-relative reasons to bring about a suboptimal outcome, we ought to bring about the best outcome that we can permissibly bring about by making this sacrifice. By accepting this principle, we solve the All or Nothing Problem.[7]

Optimific Altruism has two interesting implications, one theoretical and one practical.

The theoretical implication concerns our understanding of supererogation. If you make a great sacrifice to bring about a good outcome, we would normally say that your act was supererogatory, meaning that it was morally praiseworthy but not morally obligatory. However, Optimific Altruism implies that your act was obligatory. Since you chose to make this great sacrifice, you must have been willing to make it, and so Optimific Altruism implies that you were obligated to make it. Optimific Altruism thus implies that your act was not actually supererogatory, at least on the standard definition. More generally, Optimific Altruism seems to imply that many of the acts that we would normally take to be paradigm instances of supererogation are not actually instances of supererogation at all.

I argued in my earlier article that, rather than seeing this as a challenge to Optimific Altruism, we could see it as a reason to revise the standard account of supererogation. Optimific Altruism does imply that, because you were willing to make this great sacrifice, you were obligated to make it. But although you were obligated to make this sacrifice because of your willingness to make it, you were not obligated to be willing to make it. So, Optimific Altruism leaves space for us to acknowledge an important way in which you went beyond what morality required—you went beyond what morality required by being willing to make a great sacrifice even when it was permissible for you to instead give priority to your own interests. So, to make

[6] For an explanation of the clause about agent-relative reasons, see Horton, 'The All or Nothing Problem', 98–99.

[7] For a more detailed defence of Optimific Altruism, see Horton, 'The All or Nothing Problem', 97–100.

Optimific Altruism compatible with what we normally regard as paradigm instances of supererogation, we could just revise the standard account of supererogation. Rather than saying that acts are supererogatory when they are morally praiseworthy but not morally obligatory, we could say that acts are supererogatory when they are morally praiseworthy but not morally obligatory *unless we are willing to perform them.*

This revised account of supererogation has independent explanatory power. It is relatively common for people who perform apparently supererogatory acts to subsequently insist that they were just doing their duty. If we accept the standard account of supererogation, this phenomenon is difficult to explain.[8] We would have to claim either that these people are often mistaken about what morality requires, or that they are often insincere when reporting their moral beliefs. Since we take these people to be moral heroes, neither explanation is attractive. But if we accept the revised account of supererogation, we need not attribute any mistake or insincerity to these people. On the revised account, because these people were willing to perform these acts, they were obligated to perform them.

The practical implication concerns our obligations to give to charity. Most people think that giving to charity is supererogatory, at least when we have already given a significant amount. Most people also think that it is permissible for us to give to whichever charities we like. But charities differ significantly with respect to both the kind of help they deliver and the efficiency with which they deliver it. A gift to some charities will do much more good than the same gift to others. So, Optimific Altruism implies that, if we are willing to give to charity, unless we have adequate agent-relative reasons to give to a suboptimal charity, we ought to give to one of the charities that would do the most good.[9]

Of course, there might be some charities that are incomparable in terms of the good that they do. So, there might be no charity that is uniquely optimal. But there are clearly many charities that are suboptimal, in the sense that they do less good than other charities with which they are comparable. For example, there are several charities that aim to restore eyesight to people with curable blindness, and there is good data, readily available, indicating which of these charities are most efficient.[10] If a charity is not suboptimal relative to any other, we can then count it as one of the charities that would do the most good.[11]

---

8 For a discussion of this issue, see Alfred Archer and Michael Ridge, 'The Heroism Paradox: Another Paradox of Supererogation', *Philosophical Studies* 172 (2015): 1575–1592.

9 I defend this claim in greater detail in Horton, 'The All or Nothing Problem', 102–104. For a defence of the related but distinct claim that it is wrong to give to suboptimal charities, see Pummer, 'Whether and Where to Give'. For scepticism about both these claims, see Sinclair, 'Are We Conditionally Obligated to be Effective Altruists?'; and Barry and Lazar, 'Supererogation and Optimisation'. For a defence of the more radical claim that we are obligated to give to charity, see Peter Singer, 'Famine, Affluence, and Morality', *Philosophy & Public Affairs* 1 (1972): 229–243.

10 For some relevant data, see GiveWell.org, GivingWhatWeCan.org, and TheLifeYouCanSave.org.

11 For discussion of efficiency in charitable giving, see Peter Singer, *The Life You Can Save* (New York: Random House, 2009); Peter Singer, *The Most Good You Can Do* (London: Yale University Press, 2015); and William MacAskill, *Doing Good Better* (New York: Avery, 2016). For a more sceptical assessment of our ability to compare how much good different charities do, see the

3 Conditional Obligation

Pummer and Muñoz have recently defended a very different solution to the All or Nothing Problem. According to this alternative solution, (3) does not follow from (1) and (2).

In my earlier article, I briefly considered this kind of response to the problem. I argued that the inference from (1) and (2) to (3) is an application of the following principle.

> *Conditional Obligation:* If your only options are A, B, and C, and only A and B are permissible, then if you are not going to do A, then you ought to do B.[12]

This principle is very intuitive, and there are countless cases that seem to verify it. Suppose, for example, that it is permissible to say something nice, permissible to say nothing, and wrong to say something nasty. Conditional Obligation has the obviously correct implication that, if you are not going to say something nice, you ought to say nothing.

Conditional Obligation also follows from a plausible view about the meaning of the moral 'ought' combined with an influential view about the semantic role of conditionals. According to the view about the meaning of the moral 'ought', 'you morally ought to do A' is true if and only if, from all the possibilities under consideration, the possibilities in which you do least wrong are all possibilities in which you do A. According to the view about conditionals, the antecedent of a conditional restricts the set of possibilities under consideration, by excluding possibilities in which the antecedent is false.[13] It follows from the combination of these views that 'if you are not going to do A, you ought to do B' is true if and only if, from all the possibilities under consideration in which you do not do A, the possibilities in which you do least wrong are all possibilities in which you do B.

Pummer and Muñoz think that, despite its plausibility, Conditional Obligation is false. This is in part because they see the All or Nothing Problem as a collection

symposium on effective altruism in the July/August 2015 *Boston Review*; Thomas Nagel's review article, 'Ways to Help', in the November 2015 *Times Literary Supplement*; Amia Srinivasan's review article, 'Stop the Robot Apocalypse', in the September 2015 *London Review of Books*; and Amia Srinivasan, 'Effective Altruism and Its Limits', in Alex Guerrero and Elizabeth Harman (eds.), *The Norton Introduction to Ethics* (forthcoming).

[12] Horton, 'The All or Nothing Problem', 96. In this earlier article, I stated Conditional Obligation not as a claim about when you are conditionally obligated to do B but instead as a claim about when you ought to do B *rather than* A.

[13] This view about conditionals is originally from David Lewis, 'Adverbs of Quantification', in Edward Keenan (ed.), *Formal Semantics of Natural Language* (Cambridge: Cambridge University Press, 1975), 3–15. However, it received important development in Angelika Kratzer, 'The Notional Category of Modality', in Hans–Jürgen Eikmeyer and Hannes Rieser (eds.), *Words, Worlds, and Contexts* (New York: Walter de Gruyter, 1981), 38–74; Angelika Kratzer, 'Conditionals', *Chicago Linguistics Society* 22 (1986): 1–15; and Angelika Kratzer, *Modals and Conditionals: New and Revised Perspectives* (Oxford: Oxford University Press, 2012). For discussion of its influence in linguistics and philosophy, see Daniel Rothschild, 'A Note on Conditionals and Restrictors', in John Hawthorne and Lee Walers (eds.), *Conditionals, Probability, and Paradox* (Oxford: Oxford University Press, 2021), 19–39.

of counterexamples to Conditional Obligation.[14] But they also think there are other, independent counterexamples to Conditional Obligation, including the following case (which was originally devised for a different purpose by Frances Kamm):[15]

> You promised to meet a friend for lunch. But on your way to the lunch, you come across a burning building with a stranger trapped inside. You can continue on to lunch, keeping your promise; enter the burning building, risking your life to save the stranger; or just head home, breaking your promise for no good reason.

Plausibly, saving the stranger is not obligatory, given the risk. But saving the stranger is the only justification for breaking your promise. Conditional Obligation thus implies that, if you are not going to keep your promise, you ought to save the stranger. Pummer and Muñoz think this result is implausibly demanding and no less counterintuitive than (3).

I noted above that Conditional Obligation seems to underlie many instances of valid reasoning, such as our inference to the conclusion that, if you are not going to say something nice, you ought to say nothing. If Conditional Obligation is false, what alternative principle is at work here?

Pummer and Muñoz rise to this challenge.[16] They begin by distinguishing between *requiring reasons*, which are reasons that "tend to make actions obligatory", and *justifying reasons*, which are reasons that "tend to make actions permissible". They then propose that an act A is permissible if and only if, for every alternative B under consideration, the requiring reason to do B (rather than A) does not outweigh the justifying reason to do A (rather than B). They also propose that conditional antecedents of the form 'if you are not going to do A' restrict the set of alternatives under consideration, by excluding A. If we combine these two proposals, we arrive at the following principle.

> *Conditional Permission:* It is permissible for you to do A conditional on you not doing B if and only if, for any alternative C (compatible with you not doing B), the requiring reason to do C (rather than A) does not outweigh the justifying reason to do A (rather than C).[17]

This principle, Pummer and Muñoz claim, is what underlies those instances of valid reasoning. Return to the earlier example. Conditional Permission implies that it is permissible to say nothing conditional on you not saying something nice, because the only alternative is saying something nasty, and there is no requiring reason to say something nasty. Conditional Permission also implies that it is *not* permissible to say something nasty conditional on you not saying something nice, because there

[14] Muñoz and Pummer, 'Supererogation and Conditional Obligation', 1431.

[15] Muñoz and Pummer, 'Supererogation and Conditional Obligation', 1434. For the original case, which differs is some presentational details, see Frances Kamm, 'Supererogation and Obligation', *The Journal of Philosophy* 82 (1985): 118–138, at 119–120; and Frances Kamm, *Morality, Mortality: Volume 2* (Oxford: Oxford University Press, 1996), at 313–314.

[16] Muñoz and Pummer, 'Supererogation and Conditional Obligation', 1437–1440.

[17] Pummer and Muñoz refer to this principle as 'Conditional J&R'. Pummer also defends this principle in *The Rules of Rescue*, Chap. 3. For a precursor to this principle, see Pummer, 'All or Nothing'.

is requiring reason to say nothing rather than something nasty, and there is no justifying reason to say something nasty. So, like Conditional Obligation, Conditional Permission has the obviously correct implication that, if you are not going to say something nice, you ought to say nothing.

Furthermore, Conditional Permission does not generate the All or Nothing Problem. Return to the case in which you could save the two children. Conditional Permission implies that it is permissible for you to save only one child conditional on you not saving both, because the only alternative is saving neither, and there is no requiring reason to save neither rather than saving only one. So, rather than implying (3), Conditional Permission implies:

(6) If you are not going to save both children, it is permissible for you to save only one.

Pummer and Muñoz think this result is much more plausible.

I agree that (6) initially seems much more plausible than (3). But, on reflection, (6) seems to me unacceptable. If (6) were true, it would be coherent to say things like:

(7) You ought to save both children or save neither, but if you are not going to save both, then it is permissible for you to save only one.

(8) Your only permissible options are saving both children and saving neither, but if you are not going to save both, then it is permissible for you to save only one.

These claims strike me as incoherent. The assertion in the first clause seems to be immediately undermined by the conditional in the second clause.

Notice that the corresponding claim licensed by Conditional Obligation seems perfectly coherent, even if the conditional in the second clause is counterintuitive:

(9) You ought to save both children or save neither, so if you are not going to save both, then you ought to save neither.[18]

I think there are also some clear counterexamples to Conditional Permission. Suppose that your only options are doing nothing, kicking me in the shin, and killing yourself. Presumably, any requiring reason for killing yourself rather than kicking me in the shin does not outweigh the justifying reason for kicking me in the shin rather than killing yourself. So, Conditional Permission seems to imply that, if you are not going to do nothing, it is permissible for you to kick me in the shin. But that seems clearly false.

I think the apparent incoherence of (7) and (8), together with this apparent counterexample, is strong evidence against Conditional Permission. So, where does the rationale for Conditional Permission go wrong?

---

[18] Pummer thinks that conditional permissions should be interpreted as having the 'dyadic' form 'it is permissible to (do A, given that you are not going to do B)'. See his *The Rules of Rescue*, Chap. 3. This interpretation is necessary, he argues, to explain why the consequents of these conditionals do not detach via *modus ponens*. He has suggested to me, in conversation, that conditionals like (7) and (8) seem incoherent only because we fail to interpret them in this dyadic way. However, if these conditionals should be interpreted in this dyadic way, then so too should (9), and yet (9) does not seem incoherent at all.

I think the key mistake Pummer and Muñoz make is assuming that conditional antecedents of the form 'if you are not going to do A' restrict the set of alternative actions under consideration, by excluding A. This assumption initially looks like an application of the influential view that conditional antecedents restrict the set of possibilities under consideration.[19] But restricting the set of *possibilities* under consideration does not entail restricting the set of *alternative actions* under consideration. For example, when we consider all the possibilities in which you do not save both children, we still include possibilities in which saving both children is an action that you could have chosen to perform, and we evaluate the wrongness of alternative actions in part by how they compare to it.

We can now return to the purported counterexample to Conditional Obligation. Recall that you can keep your promise to meet a friend for lunch, risk your life to save a stranger from a burning building, or just head home. Conditional Obligation implies that:

(10) If you are not going to keep your promise, then you ought to save the stranger.

Pummer and Muñoz think this result is implausibly demanding.

It might seem that my counterexample to Conditional Permission also serves as a further counterexample to Conditional Obligation. Recall that you can do nothing, kick me in the shin, or kill yourself. Plausibly, it is permissible to do nothing and permissible to kill yourself, but wrong to kick me in the shin. Conditional Obligation then implies that:

(11) If you are not going to do nothing, then you ought to kill yourself.

This result might seem implausibly extreme.

I think (10) and (11) are both correct. (10) is correct because saving the stranger is your least-wrong option compatible with you not keeping your promise. (11) is correct because killing yourself is your least-wrong option compatible with you not doing nothing. If these claims are counterintuitive, I think that is for reasons independent of their truth. For example, (10) might be counterintuitive because it would be weird to assert it, and it would be weird to assert it because we would not expect the potential wrongness of breaking your promise to have any effect on your motivation to save the stranger; (11) might be counterintuitive simply because we hope that you will not kill yourself.

Of course, similar things might be said to explain the counterintuitiveness of (3). But, as I indicated in the introduction, it seems to me that (3) is counterintuitive at least in part because it seems false when you are willing to save one child. If my arguments in the previous section are correct, that is indeed the case. If instead those arguments are wrong, we must then explain the counterintuitiveness of (3) in some other way. But for the reasons presented here, it seems to me unlikely that Conditional Obligation is the issue.

---

[19] Pummer and Muñoz draw this connection explicitly. See 'Supererogation and Conditional Obligation', 1430, footnote 1, and 1440–1441.

4 Summary

If doing good is often beyond the call of duty, instances of the All or Nothing Problem abound. I have argued that we should solve this problem by accepting Optimific Altruism, which has interesting implications both for the correct account of supererogation and for our obligations to give to charity. Pummer and Muñoz argue that we should instead solve this problem by rejecting Conditional Obligation and accepting Conditional Permission. But Conditional Obligation is extremely plausible, and it is supported by a plausible and influential view about the semantic role of conditionals. Conditional Permission is not supported by this view about conditionals, and it has some implications that seem incoherent or false. There may be other ways to solve the All or Nothing Problem, but it seems to me unlikely that the best solution involves rejecting Conditional Obligation.

References

Archer, A., & Ridge, M. (2015). The heroism paradox: Another paradox of supererogation. *Philosophical Studies, 172*, 1575–1592.

Bader, R. (2019). Agent-relative prerogatives and suboptimal beneficence. *Oxford Studies in Normative Ethics, 9*, 223–250.

Barry, C., & Lazar, S., Supererogation and optimisation. *Australasian Journal of Philosophy* (forthcoming).

Fried, C., & Parfit, D. (1979). Correspondence. *Philosophy & Public Affairs, 8*, 393–397.

Horton, J. (2017). The all or nothing problem. *The Journal of Philosophy, 114*, 94–104.

Kagan, S. (1989). *The limits of morality*. Oxford University Press.

Kamm, F. (1996). *Morality, mortality* (Vol. 2). Oxford University Press.

Kamm, F. (1985). Supererogation and obligation. *The Journal of Philosophy, 82*, 118–138.

Kratzer, A. (1986). Conditionals. *Chicago Linguistics Society, 22*, 1–15.

Kratzer, A. (1981). The notional category of modality. In H.-J. Eikmeyer & H. Rieser (Eds.), *Words, Worlds, and Contexts* (pp. 38–74). Walter de Gruyter.

Kratzer, A. (2012). *Modals and conditionals: new and revised perspectives*. Oxford University Press.

Lewis, D. (1975). Adverbs of quantification. In E. Keenan (Ed.), *Formal Semantics of Natural Language* (pp. 3–15). Cambridge University Press.

McMahan, J. (2018). Doing good and doing the best. In P. Woodruff (Ed.), *The Ethics of Giving* (pp. 78–102). Oxford University Press.

MacAskill, W. (2016). *Doing good better*. Avery.

Muñoz, D. (2021). Three Paradoxes of Supererogation. *Noûs, 55*, 699–716.

Muñoz, D., & Pummer, T. (2022). Supererogation and conditional obligation. *Philosophical Studies, 179*, 1429–1443.

Nagel, T. (2015). Ways to help. *Times Literary Supplement*.

Parfit, D. (1982). Future generations: Further problems. *Philosophy & Public Affairs, 11*, 113–172.

Parfit, D. (2003). Justifiability to each person. *Ratio, 16*, 368–390.

Parfit, D. (2011). *On what matters* (Vol. Two). Oxford University Press.

Portmore, D. (2019). *Opting for the best: Oughts and options*. Oxford University Press.

Pummer, T. (2019). All or nothing, but if not all, next best or nothing. *The Journal of Philosophy, 116*, 278–191.

Pummer, T. (2022). *The rules of Rescue: Cost, distance, and effective altruism*. Oxford University Press.
Pummer, T. (2016). Whether and where to give. *Philosophy & Public Affairs, 44*, 77–95.
Rothschild, D. (2021). A note on conditionals and restrictors. In J. Hawthorne & L. Walers (Eds.), *Conditionals, Probability, and Paradox* (pp. 19–39). Oxford University Press.
Rulli, T. (2020). Conditional obligations. *Social Theory and Practice, 46*, 365–390.
Scanlon, T. M. (1998). *What we owe to each other*. Harvard University Press.
Sinclair, T. (2018). Are we conditionally obligated to be effective altruists? *Philosophy & Public Affairs, 46*, 36–59.
Singer, P. (1972). Famine, affluence, and morality. *Philosophy & Public Affairs, 1*, 229–243.
Singer, P. (2009). *The life you can save*. Random House.
Singer, P. (2015). *The most good you can do*. Yale University Press.
Srinivasan, A., Effective altruism and its limits. In *Alex Guerrero and Elizabeth Harman* (eds.), *The Norton Introduction to Ethics* (forthcoming).
Srinivasan, A. (2015). Stop the robot apocalypse. *London Review of Books*.
Tadros, V. (2011). *The ends of harm: The moral foundations of criminal law*. Oxford University Press.

Latitude, Supererogation, and Imperfect Duties

Douglas W. Portmore

Abstract In this chapter, I seek a better understanding of both supererogation and imperfect duties in the hopes of coming up with an account of what it is to go above and beyond the call of an imperfect duty. I argue that we can go above and beyond the call of duty, not only by performing actions but also by forming attitudes. And I argue that what's constitutive of fulfilling an imperfect duty is forming certain attitudes. I conclude, therefore, that we can go above and beyond the call of an imperfect duty by forming attitudes that are morally better than those that we must at a minimum form simply to count as having the end that it requires us to have. Along the way, I show that imperfect duties provide us with neither too much nor too little latitude in choosing which ends to pursue as well as when, and to what extent, to pursue them.

Keywords Beneficence · Discretionary ends · Duty to rescue · Imperfect duties · Indeterminate ends · Kant · Latitude · Maximalism · Moral options · Moral vagueness · Obligatory ends · Perfect duties · Supererogation

1 Introduction

Moral theories should accommodate both supererogation and imperfect duties, as both are deeply entrenched aspects of our commonsense morality. Supererogation consists in going above and beyond the call of duty—e.g., doing more than you are required to do to help those in need. And imperfect duties are duties to have certain ends—e.g., the duty to have helping those in need as "a serious, major, continually relevant, life-shaping end" (Hill, 2002, 206). But any theory that accommodates both will need to account for what it is to go above and beyond the call of an imperfect duty. For if supererogation consists in going above and beyond the call of duty and some duties are imperfect, then supererogation must involve going above and beyond the call of both perfect *and* imperfect duty. The problem is that it's unclear what it would mean to go above and beyond the call of a duty to have an end. Take the imperfect

D. W. Portmore (✉)
Arizona State University, Tempe, AZ, USA
e-mail: dwportmore@gmail.com; douglas.portmore@asu.edu

D. Heyd (ed.), *Handbook of Supererogation*,
https://doi.org/10.1007/978-981-99-3633-5_5

duty to have helping those in need as a serious, major, continually relevant, and life-shaping end. This doesn't seem to be the kind of duty that you could dispatch, let alone go beyond dispatching. For you dispatch a duty if and only if you bring it about that you no longer have it, and there seems to be no way to rid yourself of a duty to have an end. After all, it's not like the duty to pay a debt, which you can dispatch by paying what's owed and go beyond dispatching by paying more than what's owed (Guevara, 1999, 600; Hanser, 2014, 305). Also, some suggest that only acts can be supererogatory.[1] Yet, as we'll soon see, having an end doesn't equate to performing any acts. So, how can an *act* go above and beyond the call of a duty to have an *end*?

In this chapter, I seek a better understanding of both supererogation and imperfect duties in the hopes of coming up with an account of what it is to go above and beyond the call of an imperfect duty. I argue that we can go above and beyond the call of duty, not only by performing acts but also by forming attitudes. And I argue that what's constitutive of fulfilling an imperfect duty is forming certain attitudes. I conclude, therefore, that we can go above and beyond the call of an imperfect duty by forming attitudes that are morally better than those that we must at a minimum form simply to count as having the end that it requires us to have. Along the way, I show that imperfect duties provide us with neither too much nor too little latitude in choosing which ends to pursue as well as when, and to what extent, to pursue them.

2 A Broadly Kantian Account of the Distinction Between Perfect and Imperfect Duties

On the broadly Kantian account that I favor, imperfect duties prescribe (or proscribe) ends. They "command [one] to make [or to refrain from making] a certain object of choice [one's] end" (Kant MM 6:419).[2] To have an end is not merely to want something to happen (Hill, 2002, 249). I want the San Diego Padres to win more games, but it's not one of my ends. For it's not something that I endeavor to make happen. To have an end, then, is not only to want it to happen but also to be committed to promoting it—at least, insofar as certain opportunities arise. Thus, what's constitutive of having an end is making plans, adopting policies, forming desires, and having dispositions with respect to the promotion of that end.[3]

---

[1] For instance, David Heyd suggests both that "'supererogation' is recognized only as a class of acts" (1980, 313) and that "since Kant sometimes defines imperfect duties as duties to adopt ends (rather than engage in particular acts), supererogation and imperfect duty do not belong to the same level of discourse" (2019).

[2] Throughout this chapter, 'MM' will refer to Kant's *Metaphysics of Morals* (1797), and 'G' will refer to Kant's *Groundwork of the Metaphysics of Morals* (1785). In each case, the numbers that follow refer to the relevant volume and page numbers.

[3] These all consist in forming certain attitudes. As Michael Bratman explains, plans are intentions that "typically concern relatively specific courses of action extended over time" (2007, 27). For instance, one might plan to spend the morning grading exams and the afternoon playing golf. And

To illustrate, consider someone who has helping those in need as a serious, major, continually relevant, and life-shaping end (hereafter, simply "a life-shaping end"). They will feel compassion for those in need. They will hope, and be on the lookout, for favorable opportunities to help them. They will make plans to help them—perhaps, planning to join the Peace Corps after college. And they will adopt policies for helping them—perhaps, the policy of both donating blood every other month and volunteering every Sunday at a nearby soup kitchen. Of course, having this end doesn't necessitate making these particular plans or adopting these particular policies. That's because what's required to count as having an end depends on what your other ends are and how they all vie for your limited time, resources, and opportunities for promoting them. Thus, if you could best promote your legitimate ends by spending Sundays with your family rather than at a nearby soup kitchen, you would rightly adopt a different policy: perhaps, one where you donate more money, but less time, to charity.

Although having the disposition to promote something is partly constitutive of having that something as an end, actually promoting it isn't. For you may never have a decent opportunity to do so.[4] Perhaps, you never get the opportunity to promote it at all. Or, perhaps, you get some opportunities, but they're all too costly in terms of your other ends. Thus, even if you fail to take advantage of certain opportunities to help those in need, it wouldn't follow that you don't have helping those in need as an end. What's constitutive of having an end, then, is not performing certain actions, but having certain attitudes, such as the intention to regularly take favorable opportunities to further it and a preference for doing more rather than less to further it when other relevant things are equal.

Not only is action unnecessary to count as having an end but also insufficient. For suppose that you do more over the course of your life to help those in need than is necessary to count as having helping those in need as a life-shaping end. Still, you won't count as having this as an end if you only ever do so as a means of gaining the admiration of others. In that case, helping those in need is for you only a means, not an end. So, having an end necessitates having certain motivations.[5]

Although having an end doesn't, in general, necessitate acting to promote it, it does necessitate taking advantage of certain opportunities to do so. For instance, it necessitates taking advantage of any *golden opportunities* to do so—a golden opportunity being one that is such a particularly good and exceptionally rare opportunity that you couldn't pass it up and still count as having the end in question. And this

---

policies are intentions to perform "a certain kind of action on certain kinds of potentially recurrent occasions—for example, to buckling up one's seat belt when one drives, or to having at most one beer at dinner" (2007, 27).

[4] Several philosophers (e.g., Brinkmann 2015, 72; Heyd 1982, 121; Hill 1971, 56; Stocker 1967, 508) have argued that an imperfect duty is just a *disjunctive duty*—that is, a "duty to perform (at least) one member of a non-singleton, non-empty set of actions" (Brinkmann 2015, 71). But if, following Kant, we take imperfect duties to be duties to have ends, then this can't be right given that disjunctive duties are instead duties to perform actions. As we'll see, this makes disjunctive duties perfect duties—at least, on a Kantian account of the distinction.

[5] As Kant says, "to promote according to one's means [*Vermögen*] the happiness of others in need, without hoping for something in return, is every man's duty" (MM 6:456).

explains why, in what's known as *The Drowning Child Case,* you would be required to wade into the shallow pond to save the drowning child even if this would mean ruining your expensive shoes.[6] For you are required to have helping those in need as a life-shaping end. And, given both that this is a golden opportunity and that having an end necessitates taking advantage of any such opportunities, you're required to save the child.

Although having an end necessitates taking advantage of every *golden* opportunity to promote it, it doesn't necessitate taking advantage of every *favorable* opportunity to do so. And, here, I'm using 'favorable opportunity' as a term of art meaning 'an opportunity to promote an end that would involve neither violating a moral duty nor forgoing an overall better opportunity to promote your legitimate ends', where your *legitimate ends* include all and only both the ends that you're required to have (i.e., your obligatory ends) and the permissible but non-obligatory ends that you, in fact, have (i.e., your discretionary ends).[7] Instead, having an imperfect duty necessitates only that you take advantage of a sufficient number of such favorable opportunities over time, where what counts as sufficient depends both on what your other legitimate ends are and on what limitations there are on your time, resources, and opportunities for promoting them.[8] Thus, someone who has a severely disabled child to which they're required to give near-constant attention will, other things being equal, be permitted to do less to promote their other legitimate ends than someone who has no family members with special needs. And someone who has a particularly demanding career that leaves them with little free time will, other things being equal, be permitted to dedicate less time to their other legitimate ends than someone who has a less demanding career.

So, imperfect duties leave agents with a fair bit of latitude in that they can choose which favorable opportunities to take advantage of so long as, in the end, they take advantage of a sufficient number of them.[9] As Kant put it, "if the law can prescribe only the maxim of actions [that is, the end that those actions would promote], not

---

[6] See Singer (1972, 231).

[7] To illustrate the difference between a golden opportunity and a favorable opportunity, consider *The Drowning Children Case*, which I borrow with revision from Travis Timmerman (2015, 208–9). In this case, opportunities to save children from drowning are ubiquitous rather than extremely rare, as they are in *The Drowning Child Case*. Indeed, in *The Drowning Children Case*, there is—nearby to where you live, work, and play—a vast stretch of land covered with hundreds of shallow ponds, each with a small child who will drown unless you pull them to safety. Saving any individual child would cost you only $200. But whenever a child drowns or is pulled to safety, they are replaced with some other drowning child, and each additional child that you save costs you an additional $200. Thus, whereas saving *the* drowning child is, in *The Drowning Child Case*, a golden opportunity, saving *a given* drowning child, in *The Drowning Children Case*, is only a favorable opportunity.

[8] This *perfect* duty to take advantage of a sufficient number of the favorable opportunities that arise derives from the *imperfect* duty to have helping those in need as a life-shaping end. For, as Barbara Herman points out, "imperfect duties impose requirements directly on ends, only indirectly on actions" (2022, 65, note 16).

[9] We have to be careful here, for even the perfect duty to, say, show up to a specific event at a precise time on a particular date leaves you with a fair bit of latitude, as there are many different token instances of this required act-type to choose from. There's your showing up in a black dress, and there's your showing up in a red dress. There's your showing up with a smile, and there's your

actions themselves, this is a sign that it leaves a latitude (*latitudo*) for free choice (*freie Willkür*) in following (complying with) the law, that is, that the law cannot specify precisely in what way one is to act and how much one is to do by the action for an end that is also a duty" (MM 6:390).[10] For instance, someone who has helping those in need as a life-shaping end is to some extent permitted to choose who to help, when to help, and even how much help to provide. But this latitude is not unlimited. For instance, one would not be permitted to choose to provide less rather than more help for those in need if other relevant things were equal (Pummer, 2016). And one would not be permitted to choose to help one person over another simply because of the color of their skin. For, interpreted plausibly, having the end of helping those in need necessitates caring about each person's needs and in proportion to their needs—at least, insofar as all relevant things are equal.[11]

The fact that an imperfect duty provides such latitude doesn't make it any less of a duty than a perfect duty is. Certainly, you're not permitted to fail to fulfill your imperfect duties any more than you're permitted to fail to fulfill your perfect duties. So, when Kant contrasted imperfect duties with perfect duties, suggesting that only the latter "permits no exception to the advantage of inclination" (G 4:421, note), he wasn't claiming that your inclinations can exempt you from having to fulfill your imperfect duties. Rather, he was claiming that, in the case of imperfect duties, your inclinations can exempt you from having to take a particular opportunity to further the end that it requires you to have. For instance, you are, with respect to your imperfect duty to have helping those in need as a life-shaping end, permitted to choose not to volunteer for Meals on Wheels today simply because you're not in the mood to drive today.[12] By contrast, if you promised to give someone a ride today and, so, have a perfect duty to do so, you can't just choose not to do so simply because you're not in the mood to drive today. Thus, perfect and imperfect duties don't differ in their deontic force. They differ only in the sort of thing that they require. Whereas imperfect duties prescribe (or proscribe) *ends*, perfect duties prescribe (or proscribe) *actions*.[13]

---

showing up with a frown, etc. See Chisholm (1963, 4) and Stocker (1967). But, whereas a perfect duty allows you to choose only which permissible instance of a required act-type to perform, an imperfect duty allows you to choose, to some extent, whether to perform more or less of a required act-type. As Kant notes about the imperfect duty of beneficence, "the duty has in it a latitude for doing more or less, and no specific limits can be assigned to what should be done" (MM 6:393).

[10] Like Marcia Baron (1987, 242), I interpret Kant to be holding that a duty to adopt a maxim of actions is the same thing as a duty to adopt the end that those actions would promote.

[11] This does not mean that one must always choose to donate to the most effective charity, for effectiveness needn't be the only thing that's relevant—see Portmore 2019a, 214–5. In any case, I'll have a lot more to say about the sort of latitude that imperfect duties provide in Sect. 4.

[12] Volunteers for Meals on Wheels deliver meals and friendly greetings to seniors in need.

[13] Perfect duties and imperfect duties are logical contraries like 'tall' and 'short', not logical contradictories like 'tall' and 'not tall'. For there may be duties that prescribe (or proscribe) something other than either acts or ends, and such duties would be neither perfect nor imperfect. For an example, see note 26 below.

A perfect duty, then, is a duty either to perform some token instance of a required act-type or to refrain from performing any token instance of a prohibited act-type.[14] For instance, your duty to save the child in *The Drowning Child Case* is a perfect duty (though one that derives from your imperfect duty to have helping those in need as a life-shaping end). But, because saving the child is a perfect duty, it isn't enough to make saving the child your end. After all, to fulfill a perfect duty to φ, you must φ and not merely make φ-ing your end. For it's the action, not the end, that it requires.

It's not that perfect duties require success, whereas imperfect duties require only an attempt at success. For an imperfect duty to ψ requires you to succeed in ψ-ing just as much as a perfect duty to φ requires you to succeed in φ-ing. It's just that, in the case of an imperfect duty to ψ, 'ψ' must stand for "having, or not having, a certain end", whereas, in the case of a perfect duty to φ, 'φ' must stand for "performing, or refraining from performing, a certain act-type". The difference, then, is that one can count as having a certain end without ever succeeding in furthering that end. Merely trying or intending to promote an end is, in certain circumstances, enough to count as having that end. But if you have a perfect duty to perform some instance of a certain act-type, merely trying or intending to perform such an instance isn't enough. Thus, you can count as having helping those in need as a life-shaping end and yet fail to save the child in *The Drowning Child Case*. Perhaps, although you had helping those in need as a life-shaping end, you also, illegitimately, had safeguarding your shoes (your prized Air Jordans) as a *life-shaping* end. And, as a result, you took the time to take off your shoes, which resulted in your getting to the child too late. Still, you will have fulfilled your imperfect duty to have helping those in need as a life-shaping end so long as you had the requisite attitudes. It's just that you violated both your perfect duty to save the child and your imperfect duty to refrain from having safeguarding your shoes as a major end. You violated the latter in that you should have safeguarding your shoes as at most a minor end—and, thus, not one that you would pursue at the risk of failing to save a drowning child.

Whereas some perfect duties prescribe actions, others proscribe them. For instance, your duty to refrain from lying is a perfect duty that proscribes your lying. To fulfill it, you must refrain from performing any token instance of lying. It's not enough to make refraining from lying an end, because it would, then, have to compete with your other legitimate ends rather than regulate your pursuit of them. And perfect duties are meant to regulate the pursuit of your ends. Thus, threshold deontologists—i.e., those deontologists who think that it's permissible to lie if enough good is at

---

[14] Some suggest that a perfect duty is a duty to perform *particular* acts (see, e.g., Igneski 2006, 445; Noggle 2009, 7; and Stohr 2011, 50). But this won't do. As Michael Stocker notes, "We fulfil our duties by performing acts. But it is never a duty to do any [particular] act" (1967, 507). Thus, we are never obligated to perform particular acts; we are only ever obligated to perform some (but no particular) instance of a given act-type. What's more, the duty to perform at least one of acts A, B, C, and D (what's known as a *disjunctive duty*) is a perfect duty even though it's not a duty to perform any particular act. It, like all perfect duties, is a duty to perform (or refrain from performing) some act-type. It's just that, in this case, the relevant act-type is the at-least-one-of-the-given-set type.

stake—must deny that there is a perfect duty to refrain from lying.[15] They must hold instead either that there is a perfect duty only to refrain from certain specific types of lies (e.g., lies that produce less than *n* amount of goodness)[16] or that the duty to refrain from lying is merely an imperfect duty: a duty to have refraining from lying as an end.

Thus, perfect and imperfect duties differ in how each impinges on the pursuit of our legitimate ends. Imperfect duties prescribe an end that must then compete with our other legitimate ends for our limited time, resources, and opportunities. Perfect duties, by contrast, constrain how we may act in the pursuit of our legitimate ends. For one, they place strict limits on the types of acts that we may perform in the pursuit of them. Thus, if I have a perfect duty to refrain from lying, then I must refrain from lying even if lying is the best way for me to achieve my legitimate ends. For another, perfect duties require us to perform certain types of acts regardless of the opportunity costs. So, if I have a perfect duty to keep my promise, then that's what I must do even if that means forgoing a better opportunity to achieve my legitimate ends. If, instead, keeping a promise (or refraining from lying) were only an obligatory end, I would be permitted to break a promise (or lie) for the sake of better achieving my legitimate ends overall. I may, then, forgo the opportunity to further one legitimate end to take advantage of a better opportunity to further another.

There are, of course, other differences between perfect and imperfect duties. One is that imperfect duties are not correlated with a right to compel the performance of certain actions. To illustrate, suppose that you've done me a big favor and that I, consequently, have an imperfect duty to have adequately expressing my gratitude to you as an end. Given that my having this end doesn't necessitate my expressing my gratitude on any particular occasion, or even within any particular timeframe, there is nothing that you can rightly compel me to do. By contrast, if I promised to pay you $50 upon receipt of a service that you've now provided, then, given that I have a perfect duty to pay you $50, you have the right to compel me to do so—perhaps, even to compel me to do so on this particular occasion.

Another difference is that whereas perfect duties are typically *directed* (that is, owed to specific people), imperfect duties are typically *undirected* (that is, not owed to any specific people). For instance, the perfect duty that I have to pay you $50 is owed to *you*. If I fail to fulfill it, *you* will have been wronged. By contrast, the imperfect duty that I have to make helping those in need a life-shaping end is not owed to any specific person. If I fail to fulfill it, there is no one who can claim to have been wronged by this failure. For no one had a right, say, to my intending to assist them. But it's important to note that this is only a difference between many *typical* perfect duties and many *typical* imperfect duties; it isn't an essential difference between the two types of duties. For consider that my imperfect duty to have adequately expressing my gratitude to you as an end seems to be a directed duty—specifically, one that's

---

[15] If there is a perfect duty to refrain from lying, then lying is never morally justified. But, even so, one could be excused from blame for lying. For instance, it seems that one wouldn't be blameworthy for lying under duress. For it seems that duress can excuse, even if not justify, lying.

[16] Frances Kamm (1996, 264–75) calls this a *specified constraint*.

directed toward you. Thus, if I fail to fulfill it, *you* will have thereby been wronged (see also Hanser, 2014, 310–11). Likewise, my duty to have my self-improvement as an end seems to be one that's directed to me. Furthermore, there seem to be perfect duties that are undirected. For, perhaps, there's a perfect duty to bring more happy people into existence. But such a duty may be undirected, for there may be no specific happy people that we're required to bring into existence.

In taking this to be only a typical difference and not an essential difference, I am taking a somewhat controversial stand on how to draw the distinction between perfect and imperfect duties. After all, some just equate the perfect/imperfect duty distinction with the directed/undirected duty distinction (see, e.g., Mill 1979, 48). What's more, I've followed Kant in taking the perfect/imperfect duty distinction to be the distinction between duties that prescribe (or proscribe) actions and duties that prescribe (or proscribe) ends.[17] Yet, the distinction has been drawn in many other, not necessarily equivalent, ways: (1) duties whose universal violation would make human life intolerable versus duties whose universal violation would not make human life intolerable (Hutcheson, 1969); (2) duties to perform actions whose maxim cannot be thought of as a universal law without contradiction versus duties to perform actions whose maxim can be thought of as—but not willed to be—a universal law without contradiction (Kant G 4:421–424); (3) duties that allow for no exceptions in favor of inclination versus duties that allow for some exceptions in favor of inclination (Kant G 4:421, note); and (4) duties that don't allow for latitude regarding on which particular occasions to perform an instance of the prescribed act-type versus duties that do allow for such latitude (Mill 1979, 48).[18]

But my goal here is not to adjudicate between the different ways that philosophers have historically drawn the perfect/imperfect duty distinction. Rather, it's to adopt a broadly Kantian (but not necessarily Kant's) account of imperfect duties as one that takes them to prescribe (or proscribe) ends as opposed to actions and to figure out how a moral theory might plausibly accommodate such duties along with supererogation. Note, then, that I'm interested in how these two can fit together, not in how we label them, and not in what Kant's actual views about them were. So, the reader shouldn't put too much stock in my 'interpretations' of Kant. I don't so much care whether the views that I take to be Kantian in spirit are in fact Kant's. I'm mostly interested in whether they're plausible. And if the reader objects to the way I use the labels 'perfect duty' and 'imperfect duty', they should just replace them with the following less misleading labels. Call the end-regulating duties that prescribe (or proscribe) ends *telic duties*—the word 'telic' deriving from the Greek word 'τέλος' (or 'télos'), meaning 'end' or 'purpose'. And call the action-regulating duties the prescribe (or proscribe) actions *praxic duties*—the word 'praxic' deriving from the Greek word

[17] I'm certainly not alone in drawing the distinction this way. See, for instance, (Betzler 2008, 16; Gregor 1963, 98; Hill 1971, 62; Igneski 2006, 445; Kant MM 6:419; Noggle 2009, 7; Stohr 2011, 50; and Stratton-Lake 2008, 108).

[18] See Rainbolt (2000) or Salam (2016) for a more complete list.

'πρᾶξ˘ις' (or 'prâxis'), meaning 'action', 'activity', or 'practice'.[19] Nevertheless, I'll stick to the more familiar labels in the remainder.

3 Supererogation and Going Above and Beyond the Call of Imperfect Duty

The word 'supererogatory' is a term of art that philosophers and theologians use in place of such colloquial expressions as 'going the extra mile' and 'going above and beyond the call of duty'. But since these expressions are merely metaphorical, philosophers often go on to provide a more precise definition. And the standard one goes as follows:

> *The Standard Definition:* For any subject S and any act φ, S's φ-ing is supererogatory if and only if there is an available alternative act, ψ, such that both of the following two conditions are met: (c_1) S's φ-ing and S's ψ-ing are both morally permissible and (c_2) S's φ-ing is morally better than S's ψ-ing.[20]

On this definition, an act is supererogatory if and only if it is both morally optional and morally better than some morally permissible alternative. Now, if this is the correct definition, then imperfect duties are irrelevant with respect to supererogation. For, given that the variable 'φ' ranges over only acts, it implies that only acts can be supererogatory. And, as we saw above, having an end doesn't equate to performing any acts. Thus, the standard definition implies that there's no way to go above and beyond the call of an imperfect duty.

But we should reject the standard definition. There are at least two problems with it. First, it wrongly assumes that only acts can go above and beyond the call of duty. Second, it wrongly assumes that the presence of an available alternative that satisfies both conditions c_1 and c_2 is sufficient for an act's being supererogatory. I'll start by explaining why the second assumption is false using the following example.

> *The Selfish Miser:* Mr. Scrooge was a wealthy businessman who lived from 1800 to 1865. In all that time, he never did a thing for anyone except on Christmas Eve 1818, which is when he found a penny on the ground and used it to buy a loaf of bread for a hungry pauper. Had he not done this, he would have instead spent that penny on a quart of beer for himself. In which case, he would have ended up never having done anything for anyone.

Mr. Scrooge was not obligated to do what he did for the pauper. He was permitted instead to buy a quart of beer for himself. After all, he was only eighteen at the time, and, so, there were going to be plenty of other opportunities for him to help those in need. He could, therefore, have fulfilled his perfect duty to take advantage of a sufficient number of favorable opportunities to help those in need without taking

[19] I borrow the term 'action-regulating' from Salam (2016, 106).

[20] See, for instance, (Archer 2018, 4–5; Benn & Bales 2020, 919; Dorsey 2013, 356; Ferry 2015, 51; and Muñoz 2021, 699).

this particular opportunity to do so.[21] Thus, what he did was morally optional. What he did was also morally better than the morally permissible alternative of buying a quart of beer for himself. For had he not bought a loaf of bread for the pauper, he would have ended up doing nothing for anyone. And doing something for someone is morally better than that. So, according to the standard definition, Mr. Scrooge acted supererogatorily in buying a loaf of bread for the pauper. His doing so was both morally optional and morally better than the morally permissible alternative of purchasing a quart of beer for himself.

Yet, it seems that he didn't even meet, let alone exceed, the minimum that morality required of him.[22] For let's assume that, given Mr. Scrooge's wealth, privilege, and other ends, he was required by the duty of beneficence to dedicate at least 15% of his time and resources to helping those in need over the course of his life.[23] But, for him, one penny amounts to far less than 1% of his resources, or so we'll assume. The problem, then, is that the standard definition allows that an act can count as supererogatory even though it doesn't even meet, let alone exceed, the minimum required by morality. For it allows that an agent can go above and beyond the call of duty simply by performing a morally optional act that only partially fulfills a duty. It allows this because it may be both that (1) they would have done nothing to fulfill this duty had they not partially fulfilled it and that (2) their only partially fulfilling it is morally better than their doing nothing to fulfill it. But only partially fulfilling a duty doesn't count as going above and beyond the call of duty. Thus, the standard definition is incorrect in assuming that performing a morally optional act that is morally better than some morally permissible alternative is sufficient for acting supererogatorily.

Here's another example.

> *The Visit*: I promised to visit my grandmother at least twice last week but visited her only once (on Wednesday) despite having been available every day of that week. And had I not visited her on Wednesday, I would have ended up not visiting her at all, because there was

[21] Of course, given that Mr. Scrooge was a *selfish* miser, we can assume that he didn't have the attitudes that are constitutive of having helping those in need as a life-shaping end. So, he thereby violated his imperfect duty to have this as a life-shaping end.

[22] I would concede that he exceeded what he was required to do *on that particular occasion*, as the only thing that he was required to do on that particular occasion was to refrain from doing anything morally bad. But I deny that this is sufficient for going above and beyond the call of duty. For, as I see it, a subject doesn't go above and beyond the call of duty by φ-ing if they only partially or merely minimally fulfill some duty by φ-ing.

[23] Note that, as I see it, the duty of beneficence includes both the imperfect duty to have helping those in need as a life-shaping end and, derivatively, the perfect duty to act throughout one's life in a way that's consistent with having this as a life-shaping end—thus, taking advantage of a sufficient number of favorable opportunities to help those in need. Note, also, that although I've chosen to formulate the duty of beneficence as a duty to help those in need, I could have chosen to formulate it as a duty to promote the happiness of others regardless of whether they're in need. Or I could have chosen to formulate it as a duty to advance the ends of others regardless of whether doing so would promote their happiness or satisfy their needs. But none of my arguments here will hinge on this choice. Lastly, the duty of beneficence is, as I see it, a duty to make helping those in need "a serious, major, continually relevant, life-shaping end" (Hill, 2002, 206). By contrast, less demanding imperfect duties require only that you adopt something as a relatively minor end.

something else that I was really looking forward to doing the rest of the week (Portmore, 2017, 287), something that I cared about more than keeping my promise.

My visiting her on Wednesday was morally optional, for I was permitted to have visited her on only Thursday and Friday. What's more, my visiting her on Wednesday was morally better than the morally permissible alternative of not visiting her on Wednesday. After all, it's stipulated that I wouldn't have visited her at all had I not visited on Wednesday, and it's morally better to have visited her once than not at all—or so we'll assume.[24] So, the standard definition implies that my visiting her on Wednesday was supererogatory. But, clearly, my doing so didn't amount to going above and beyond the call of my duty to keep my promise. For I promised to visit her at least twice. So, again, we see that the standard definition is wrong in assuming that the presence of an available alternative that is both morally optional and morally worse than what one in fact did is sufficient for what one did to count as supererogatory. And this isn't even the standard definition's only problem.[25]

Another problem is that it wrongly assumes that only acts can go above and beyond the call of duty. This assumption would be correct if the only duties were perfect duties: duties that prescribe (or proscribe) action. But there are duties that prescribe other things besides actions (Cowley, 2015). There are duties to feel contrition, affection, gratitude, admiration, compassion, etc.[26] For instance, if you were to wrong someone, you would likely be obliged to provide them with a sincere apology, and an apology must express contrition to be sincere. Thus, the obligation to provide a sincere apology entails an obligation to feel contrite—at least, so long as you don't already feel contrite (King, 2014). And feeling contrite isn't an action; it's not even a mental action. Unlike the mental act of imagining a pink elephant, you can't just feel contrite simply by deciding or intending to do so. What's more, it seems possible to go above and beyond the call of such a duty either by having certain non-obligatory feelings or by having feelings to a greater/lesser degree than you're required/permitted to have—assuming, of course, that having them (or having them to a greater/lesser degree) is morally better than not. For instance, it seems possible to go above and beyond the call of duty by feeling compassion for someone when

---

24 Someone might deny that my visiting her on Wednesday is morally better than my refraining from visiting her on Wednesday, arguing that whether my φ-ing is morally better than my ψ-ing depends, not on how what I *would* do if I were to φ compares with what I would do if I were to ψ, but rather on how what I *could* do if I were to φ compares with what I could do if I were to ψ. But this won't work if we assume that (1) the best thing that I could do if I were to visit her on Wednesday would be to visit her each day Wednesday through Saturday, that (2) the best thing that I could do if I were to refrain from visiting her on Wednesday would be to visit each day Thursday through Saturday, and that (3) the former is morally better than the latter.

25 The idea that the standard definition fails because it wrongly counts as supererogatory acts that only partially, or merely minimally, fulfill a duty is not new—see Guevara (1999) (595–97) and Mellema (1991) (167–75).

26 I call these attitude-regulating duties (i.e., duties that regulate reasons-responsive attitudes such as fear, desire, belief, intention, resentment, admiration, etc.) *noetic duties*—the word 'noetic' deriving from the Greek word 'νοητόν' (or 'noētón'), meaning 'intelligible' or 'the object of the operation of the mind/intellect'.

they are not so badly off that you are obligated to feel compassion for them.[27] And it seems possible to go above and beyond the call of duty both by feeling more affection than you're obligated to feel and by feeling less resentment than you're permitted to feel.

Also, there are imperfect duties—duties to have ends. And it seems that, here too, we can go above and beyond the call of duty. For we can have something as a more or less significant end depending on what kinds and intensities of attitudes are constitutive of having an end of a given significance. To illustrate, suppose for the sake of argument that we're all obligated to have helping those in need as a life-shaping end—that is, as a serious, major, continually relevant, and life-shaping end that's on par with our other life-shaping ends, such as pursuing a personally rewarding career and enjoying the best things that life has to offer. But suppose that although none of us are obligated to have this as our sole, or even primary, life-shaping end, someone has it as their *sole* life-shaping end. Consequently, they have the attitudes that are constitutive of this. Thus, they intend to pursue a career on Wall Street even though they know that they would find teaching primary school more personally rewarding. It's just that they prefer (a) making and donating more money to effective charities and having a less rewarding career to (b) making and donating less money to effective charities and having a more rewarding career. They're also disposed to give up vacations, fine foods, having children, and nights out at the theater so that they can donate yet more money to effective charities. And they even plan on donating one of their kidneys to a stranger in renal failure once they get the medical approval to do so. It seems, then, that their having helping those in need as their sole life-shaping end is supererogatory. For in having this as their *sole* life-shaping end, they go above and beyond the call of their duty to have helping those in need as *one* of perhaps several life-shaping ends. After all, it certainly seems morally better that they have this as their sole life-shaping end than that they have this as just one of several life-shaping ends that must, then, compete with the others.[28]

Of course, the reader may object that what's supererogatory is not *having* helping those in need as your sole life-shaping end, but rather *doing* whatever it is that results in your having this as your sole life-shaping end.[29] For the reader may think both that only that which is under your direct voluntary control can be obligatory (or supererogatory) and that you don't have direct voluntary control over whether you

---

[27] Daniel Star suggested this case to Roger Crisp as a possible instance of supererogatory feeling. See Crisp (2015) (139, note 9).

[28] I'm imagining that despite this being their sole life-shaping end, their commitment to this end is not in any way fanatical. Thus, they don't neglect their children, as they don't have any. And they don't neglect their friends, as they're all intertwined in their altruistic pursuits. Imagine, then, that they have a good human life—just one without such non-altruistic ends as mastering Kung Fu or winning Olympic gold.

[29] But we rarely do things with the intention of acquiring certain ends. For, as Thomas E. Hill, Jr. notes, "the occasions we might describe as deliberately adopting ends are relatively rare" (2002, 268). And, as Andrews Reath notes, "it is…common to 'fall into' having certain ends. You become drawn to some activity out of interest and, over time, you find that it is important to you and has become one of your ends" (2009, 201).

have the attitudes that are constitutive of having something as your sole life-shaping end. Their thought would be that you had direct voluntary control over only the actions that led you to have these attitudes. But, as many philosophers have pointed out, it's quite commonsensical to think that we have obligations with respect to things that are not under our direct voluntary control, things such as our reasons-responsive attitudes—e.g., our beliefs, desires, intentions, and feelings. Consider that we can't form the belief that Aristotle was 5′7″, the desire to exercise more, or the intention to drink a toxin tomorrow afternoon simply by deciding or intending to form these attitudes. Yet, because such attitudes are reasons-responsive, it seems that we could be required to form them and responsible for failing to do so. And, so, it seems that the sort of control that's required for something to be obligatory (or supererogatory) is not direct *voluntary control*, but rather the sort of direct *rational control* that we exert over our reasons-responsive attitudes. Now, admittedly, this is a controversial assumption, but, given the limited space that I have here, I'll have to settle for merely flagging it and providing some sources for a fuller defense in the attached note.[30]

In any case, I've argued that there are at least two problems with the standard definition: (1) it wrongly assumes that only acts can go above and beyond the call of duty, and (2) it wrongly assumes that the presence of an available alternative that is both morally optional and morally worse than what one in fact did is sufficient for what one did to count as supererogatory. We can avoid both problems by adopting the non-standard definition of 'supererogatory' that I've argued for elsewhere (Portmore, 2019a, 136). But before I can state that definition and explain how it avoids these problems, I need to introduce some conceptual apparatus.

First, there's the concept of *performance entailment*, which is the idea that performing one option can entail performing another in the sense that the agent doesn't have the option of performing the one without performing the other. To illustrate, I have both the option of walking and the option of walking to the park, and my walking to the park entails my walking, for I don't have the option of walking to the park without walking. I also have the option of walking fast to the park. And walking fast to the park entails three other options: walking, walking fast, and walking to the park. What's more, I have the option of walking fast to the park at t_1 and then sitting on a park bench at t_2. Performing this option entails walking fast to the park as well as the three other options that it entails. Now, as we described our options with greater and greater specificity, we eventually come to a maximally specific one. This is because our options are constrained by the limitations of our control. For instance, I might have the option of walking at a rate of around 4–5 miles per hour but I don't have the more specific option of walking at the precise rate of 4.63719 miles per hour. My control just isn't that fine-grained.

And this brings us to our second bit of conceptual apparatus: the distinction between maximal and non-maximal options. For when we finally come to an option that isn't entailed by any other option, we've arrived at what I call a *maximal option*. More precisely, an option X is a maximal option if and only if there is no option

[30] See, for instance, Hieronymi (2006), McHugh (2017), Smith (2015), Portmore (2019a) (Chap. 3), and Portmore (2019b).

Y such that performing Y entails performing X but not vice versa. And any option that's not maximal is what I call a *non-maximal option*. Thus, walking fast to the park at t_1 and then sitting on a park bench at t_2 won't be a maximal option if I have some more specific option, such as either walking fast to the park at t_1 and then sitting on a *north-facing* park bench at t_2 or walking fast to the park at t_1, sitting on a park bench at t_2, *and then taking a bus back home at* t_3. And even these won't be maximal options if they too are entailed by some more specific option. Perhaps, then, the best example that I can give of a maximal option is some maximally specific way of acting over the remainder of one's life—assuming, that is, that one has at present the relevant sort of control over how one will act over the remainder of one's life.

Next, we need to distinguish going above and beyond the call of duty from merely minimally fulfilling a duty. Take any option X. A subject *merely minimally fulfills* a duty by X-ing if and only if, although they fulfill this duty by X-ing, their fulfilling this duty by X-ing is no morally better than their fulfilling this duty in any other available way. To illustrate, consider *The Visit*, and let's assume that my visiting my grandma on only Wednesday and Thursday is no morally better than my fulfilling my duty to visit her at least twice in any other available way. In that case, in visiting her on only Wednesday and Thursday, I only minimally fulfill my duty. And, thus, we can distinguish between merely minimally fulfilling my duty to keep my promise by visiting her only twice and going above and beyond the call of this duty by, say, visiting her three times.[31]

We also need to distinguish between going above and beyond the call of duty and only partially fulfilling a duty. A subject *only partially fulfills* a duty by X-ing if and only if there is a Y and a Z available to them such that both (a) their Z-ing merely minimally fulfills this duty and (b) their Z-ing consists in both their X-ing and their Y-ing. To illustrate, consider again *The Visit*. As stipulated, I visit my grandma on only one day: Wednesday. Now, as we saw above, my visiting her on only Wednesday and Thursday is an option that merely minimally fulfills my duty, and my visiting her on only Wednesday and Thursday consists in both my visiting her on Wednesday and my visiting her on Thursday. Thus, my visiting her on Wednesday only partially fulfills my duty to visit her at least twice that week.

Given this conceptual apparatus, I can now state what I take to be the correct definition of 'supererogatory'. It, unlike the standard definition, holds that the criteria for assessing whether a maximal option is supererogatory differ from those for assessing whether a non-maximal option is supererogatory. It holds that although the criteria given in the standard definition (i.e., being both morally optional and morally better than a permissible alternative) are adequate for assessing whether a maximal option is supererogatory, they're inadequate for assessing whether a non-maximal option is

[31] I'm assuming that visiting her three times is morally better than visiting her only twice. But note that going above and beyond the call of duty is not simply the oversubscription of a disjunctive duty. For it could be that visiting her six times would be too much and, so, morally worse than visiting her only twice. Visiting her six times would, then, be an oversubscription of the disjunctive duty to perform at least two members of the set {visit Monday, visit Tuesday, visit Wednesday, visit Thursday, visit Friday, visit Saturday, visit Sunday}, but it would not be supererogatory.

supererogatory. Non-maximal options must meet the additional condition of neither merely minimally nor only partially fulfilling a duty.

> *The Maximalist Definition:* For any subject S and any maximal option M_x, S's M_x-ing is supererogatory if and only if there is an alternative maximal option, M_y, such that both of the following two conditions are met: (c_1) S's M_x-ing and S's M_y-ing are both morally permissible and (c_2) S's M_x-ing is morally better than S's M_y-ing. And, for any non-maximal option N_x, S's N_x-ing is supererogatory if and only if both: (a) S does not merely minimally or only partially fulfill some duty by N_x-ing, and (b) there is an alternative non-maximal option, N_y, such that both of the following conditions are met: (c_1) S's N_x-ing and S's N_y-ing are both morally permissible and (c_2) S's N_x-ing is morally better than S's N_y-ing.[32]

Unlike the standard definition, this definition gets the intuitively correct verdicts in both *The Visit* and *The Selfish Miser*. In *The Visit*, the maximalist definition implies that I go above and beyond the call of duty with respect to keeping my promise only if I visit my grandma more than twice that week. And, in *The Selfish Miser*, the maximalist definition implies that Mr. Scrooge goes above and beyond the call of duty with respect to beneficence only if he gives more than 15% of his time and resources to helping those in need over the course of his life. Thus, it, unlike the standard definition, avoids wrongly assuming that acts, such as my visiting my grandma on Wednesday and Mr. Scrooge's buying a loaf of bread for a hungry pauper, can count as supererogatory simply because they are both morally optional and morally better than some permissible alternative.

The maximalist definition also avoids wrongly assuming that only actions can go above and beyond the call of duty. For, unlike the variables in the standard definition, the variables in the maximalist definition range over all options and not just those that are actions. Thus, if we have the relevant sort of control over whether we adopt a certain end, form a certain attitude, or feel a certain emotion, then we can thereby go above and beyond the call of duty even if we don't thereby perform any action. We can, for instance, go above and beyond the call of duty either by feeling more compassion than we're required to feel or by having the attitudes that are constitutive of having helping those in need as our *sole* life-shaping end. And, so, the maximalist definition not only allows us to fix the deficiencies of the standard definition but also allows us to account for going above and beyond the call of imperfect duty by having attitudes that are morally better than those that we must have simply to count as having the end that it requires us to have. What's more, a subject can go above and beyond the call of the perfect duty that derives from an imperfect duty: e.g., the perfect duty to take advantage of a sufficient number of favorable opportunities to promote the end it requires us to have. For instance, Mr. Scrooge could go above and beyond the perfect duty that he has to dedicate at least 15% of his time and resources to helping those in need over the course of his life by taking advantage of

---

32 This definition is maximalist because it stems from the view known as *maximalism*. Maximalism contrasts with *omnisn*. Omnism—which derives from the Latin word 'omni', meaning 'all'—holds that the deontic statuses of *all* options are determined by what ultimately matters (e.g., utility), whereas maximalism holds that only the deontic statuses of maximal options are determined by what ultimately matters and that the deontic statuses of non-maximal options are instead determined by the deontic statuses of the maximal options that entail it. See Portmore (2019a).

more favorable opportunities than is necessary to meet this threshold. In both ways, an agent can go above and beyond the requirements arising from an imperfect duty.

4 Do Imperfect Duties Provide Us with the Right Amount of Latitude?

In the previous section, I explained how a moral theory can accommodate both supererogation and imperfect duties, where the latter is conceived of as duties that prescribe (or proscribe) ends. But some may worry that, despite what I claimed at the start, there's no need for a moral theory to accommodate imperfect duties so conceived. For some may worry that, on the sort of broadly Kantian conception to which I subscribe, imperfect duties are implausible in that they provide us with insufficient latitude. The worry stems from the fact that Kant held that an imperfect duty "is not to be taken as permission to make exceptions to the maxim of actions but only as permission to limit one maxim of duty by another (e.g., love of one's neighbor in general by love of one's parents)" (MM 6:390). For some interpret this to be saying that the only limits on what we're required to do for the sake of an imperfect duty are those imposed by other moral duties—perfect and imperfect. They accept the following interpretation of the limits on what we're required to do for the sake of our obligatory ends.

> *The Rigorist Interpretation:* You are morally justified in forgoing an opportunity to promote an obligatory end if and only if taking it would entail either violating a perfect duty or forgoing an overall better opportunity to promote your obligatory ends.[33]

To illustrate, consider again the imperfect duty to have helping those in need as a life-shaping end. On the rigorist interpretation, the only moral justification for forgoing an opportunity to help those in need is that taking it would entail violating some moral duty. So, assuming that I have a perfect duty to refrain from stealing, I would be morally justified in forgoing an opportunity to help those in need if taking it would necessitate my stealing. And, assuming that I have an imperfect duty to have my self-improvement as a life-shaping end, I would also be morally justified in forgoing an opportunity to help those in need if taking it would necessitate my failing to do what I must do to count as having my self-improvement as a life-shaping end. But whenever I have the opportunity to help those in need without violating any moral duty, I would be morally required to take it. What's more, it seems that I have many such opportunities. For I could donate all the money that I would otherwise spend on non-necessities for myself—such as concerts, vacations, air travel, dinners out, and subscriptions to streaming services—to effective charities without violating any moral duty. Thus, on the rigorist interpretation, I'm required to donate all the money

[33] Jens Timmermann (2005) is a proponent of the rigorist interpretation. He interprets Kant to be saying that it is never morally permissible to pursue our discretionary ends when we could instead be pursuing our obligatory ends.

that I would otherwise spend on non-necessities for myself to effective charities. And that seems quite demanding.

So, one problem with the rigorist interpretation is that it implies that morality is much more demanding than our commonsense would have it. Another problem is that the rigorist interpretation is inconsistent with other plausible things that Kant says. For, as we noted earlier, Kant suggests that imperfect duties allow for exceptions in favor of satisfying our inclinations (G 4:421, note), and it's plausible to think that morality leaves room for such exceptions. But, on the rigorist interpretation, imperfect duties allow only for exceptions in favor of satisfying other moral duties. And, given these two problems, many reject the rigorist interpretation in favor of the following alternative.

> *The Latitudinarian Interpretation:* You are morally justified in forgoing an opportunity to promote an obligatory end if and only if you are both inclined to do so and prepared to take other opportunities to promote this obligatory end.[34]

But, whereas the rigorist interpretation makes imperfect duties out to be overly demanding, the latitudinarian interpretation makes them out to be underly demanding. Brad Hooker explains.

> Suppose I am faced with two strangers who each need help, but one of whom has greater needs and can be helped a lot more than the other. According to [the latitudinarian interpretation of] the imperfect duties view, I can simply choose which to help. But that answer seems wrong. Other things being equal, I should help the needier one. The [latitudinarian interpretation of the] imperfect duties view leaves too much room here for arbitrary choice. Or suppose I saved someone's life this morning and now I can save someone else's life at no cost to myself. Is it really morally optional whether I go on to save the second person? Surely not! (2000, 161)

Thus, both interpretations conflict with our intuitions about how demanding morality is. That alone may be insufficient for thinking them incorrect. But I believe that even if we set these counterintuitive implications aside, we can still prove that they're incorrect. Let's start with the latitudinarian interpretation. The latitudinarian interpretation can't be correct given both that an imperfect duty is a duty to have an end and that having an end necessitates taking certain opportunities (specifically, golden ones) even if we're not inclined to take them. So, even if you had already saved someone this morning and are prepared to help others on future days, you would still be required to save a second person today if you could do so at no significant cost to yourself. That you simply don't feel like saving anyone else today doesn't justify your failing to do so. For this is a golden opportunity to help someone in need. Thus, contrary to the latitudinarian interpretation, your saving the second person is morally obligatory, not morally optional.

What's more, having an end necessitates having certain attitudes: the attitudes that are constitutive of having that end. And constitutive of your having helping those

---

[34] Thomas E. Hill, Jr. seems to be a proponent of the latitudinarian interpretation. He interprets Kant to be saying that "imperfect duties allow us to do what we please on some occasions even if this is not an act of a kind prescribed by moral principles and even if we could on those occasions do something of a kind that is prescribed" (1992, 155).

in need as a life-shaping end is your being inclined to help those in greater need when other relevant things are equal. So, you would violate your imperfect duty to have helping those in need as a life-shaping end if you were inclined, other things being equal, to help those in lesser need. And, so again, we see that the latitudinarian interpretation can't be correct.

Next, consider the rigorist interpretation. It also can't be correct. Admittedly, it would be correct if the only ends that we were permitted to have were our obligatory ends. But that's implausible, for it seems that we're also permitted to have various discretionary ends, such as mastering Kung Fu or winning Olympic gold. Kant certainly thought that we're permitted to have such discretionary ends (G 4:415, 4:420, 4:428). And such discretionary ends help to give our lives meaning. Now, given such discretionary ends and the fact that what we're morally required to do to promote a given (obligatory or discretionary) end is, as I've argued above, constrained both by our other legitimate ends and by the limitations on our time, resources, and opportunities for pursuing them, the rigorist interpretation cannot be correct. For, given this, we would, contrary to the rigorist interpretation, be morally justified in not taking a particular opportunity to promote an obligatory end on the grounds that taking it would be incompatible with our adequately pursuing our discretionary ends.

Thus, we should read Kant's passage, not as saying that the only limits on what we're required to do for the sake of an imperfect duty are those imposed by other moral duties, but as saying "that, whereas we may (and indeed must) restrict the number of times we are prepared to act on one maxim (e.g., to develop our talents) by adopting another maxim (e.g., to promote the happiness of others), we may not let our concern for one maxim keep us from adopting another" (Hill, 1992, 152). And if we read Kant this way, then we should accept something like the following interpretation of the limits on what we're required to do for the sake of our obligatory ends.

> *The Moderate Interpretation:* You are morally justified in forgoing an opportunity to promote an obligatory end if and only if taking it would entail either violating a perfect duty or forgoing an overall better opportunity to promote your legitimate ends.[35]

On this view, all your legitimate ends must compete for your limited time, resources, and opportunities for pursuing them. And how competitive each just depends on their relative significance, not on whether they're obligatory or discretionary. Thus, on this interpretation, we can be morally justified in pursuing our discretionary ends (e.g., the end of winning Olympic gold) at the price of doing less to promote our obligatory ends (e.g., the end of helping those in need). And this is a very attractive interpretation of imperfect duties, for it allows us to account for the fact that, although we must always answer for not helping those in need and wouldn't be justified in not helping simply because we don't feel like it (Herman, 2007, 221), we're justified in shaping "our own lives, balancing concern for others with our own

[35] Some label this the latitudinarian interpretation: "Latitudinarians believe that pursuing our own happiness and non-moral projects, whilst not itself morally obligatory, is rational and morally permissible" (Sticker & van Ackern, 2018, 409). But I think that it's important to keep this moderate position distinct from what I've called the latitudinarian interpretation.

particular projects and concerns" (Greenspan, 2010, 181). Thus, it avoids the pitfalls of both the rigorist interpretation and the latitudinarian interpretation.

Still, it may seem that, even on the moderate interpretation, imperfect duties don't leave us with enough latitude. For it may seem that, on this interpretation, I'll have the latitude to choose which of two legitimate ends to pursue on a given occasion only if I could do just as good a job of furthering my legitimate ends overall whichever of the two I choose and do a better job of this by pursuing one of these two than by doing anything else. In other words, it may seem that I'll have the moral option of choosing which of several courses of action to undertake on a given occasion only if they are all tied for first place in terms of furthering my legitimate ends overall. Yet, it may seem that such ties would be quite rare. And, so, it may seem that even on the moderate interpretation we cannot account for all the latitude that we seem to have. After all, it seems that I am, on this occasion, permitted to pursue any of the following ends: finishing this paper by continuing to work on it, helping those in need by volunteering for Oxfam, and getting some rest and relaxation by watching some TV. Yet, it may seem very unlikely that these are all tied for first place in terms of furthering my legitimate ends overall.[36]

Now, I've argued elsewhere that such ties for first place are much more common than they initially seem to be (Portmore, 2011, Chap. 6). Briefly, the idea is that the permissibility of a non-maximal option is a function of the permissibility of the maximal options that entail it. More specifically, if there's a permissible maximal option that entails a given non-maximal option, then that non-maximal option inherits its permissibility status. Thus, from both the fact that it's permissible to φ and the fact that φ-ing entails ψ-ing, it follows that ψ-ing is permissible. For instance, if it's permissible to walk fast, then it's permissible to walk given that walking fast entails walking. And this helps us to accommodate a multitude of moral options, because it often doesn't matter in what temporal order we pursue our various legitimate ends so long as we strike the right balance among them in the end. So, assuming that I must take some future opportunities to finish this paper, some future opportunities to help those in need, and some future opportunities to get some rest and relaxation if I'm to strike the right balance among them, it doesn't matter in what order I do them so long as I do each in the right proportions. Consequently, there will be permissible maximal options in which I spend this occasion working on this paper and various future occasions either helping those in need or getting some rest and relaxation. There will be permissible maximal options in which I spend this occasion helping those in need and various future occasions either working on this paper or getting some rest and relaxation. And there will be permissible maximal options in which I spend this occasion getting some rest and relaxation and various future occasions either working on this paper or helping those in need. And given that

[36] Here, I'm thinking of scalar ends. And "where an end is scalar, where more is better, it will be a matter of rationality to maximize" (Herman, 2022, 24). For instance, these three ends all seem to be scalar ends, where it makes sense to think that more is better. Admittedly, though, there are also non-scalar ends. And where an end is non-scalar, "where more is not better—friendship is a good example of such an end—the correct relation to the end may instead be to instantiate it, to give it a role in one's life, shaped by the non-scalar values it represents" (Herman, 2022, 24).

the non-maximal options that these permissible maximal options entail inherit their permissibility status, it follows that I'm permitted to do any of these three things on this occasion. Thus, each is morally optional.

This, of course, has been rather quick, but instead of rehearsing all the arguments that I've given elsewhere against the assumption that ties for first place are quite rare (Portmore, 2011, Chap. 6), it seems preferable for me to use the remaining space here to argue against the other assumption that leads some to worry that imperfect duties leave us with too little latitude: the assumption that one will have the moral option of choosing which of several courses of action to undertake on a given occasion only if they are all tied for first place in terms of furthering one's legitimate ends. I'll argue that most, if not all, of our legitimate ends are less than fully determinate and that, given this, we should think it permissible to undertake a course of action not only when it's tied for first place with various alternatives but also when there is no alternative course action that is *determinately* better than it. So, let's call an option that's tied for first place 'optimific' and reserve the term 'optimal' for any option that lacks an alternative that is determinately better than it. I'll be arguing that, to be morally optional, it's enough that an option be optimal. It needn't be optimific.

Note that "we are often (indeed, arguably always) engaged in actions that stretch through extended periods of time in the pursuit of less than fully determinate ends" (Tenenbaum, 2021, vii). Here is a typical example.

> Today is Monday and I am engaged in making the house look nice for a visit from my aunt, who will be arriving some time Wednesday afternoon. My having this end leaves various things indeterminate. Should I hang paintings on the wall, or put flowers at her bedside, or both? At some point, I might form a more determinate conception of what counts as making the house nice for my aunt, but it seems that I can have the end of making the house look nice for my aunt without having settled this question. Similarly, what counts as achieving this end can be vague in many ways. How nice is nice enough for my aunt? If there is a small corner of my bedroom that has some dust in it, does it count as being nice enough? How much dust in the house will be enough to determine that the house is not nice? At precisely what time should everything be ready? (Tenenbaum, 2021, 1)

Like my end of making the house look nice, the ends that agents pursue are almost always indeterminate in various ways. For instance, my end of writing an important book is indeterminate given that I haven't fully specified (even in my own mind) what counts as important. My daughter's end of finishing her college education while still young is indeterminate given that she hasn't fully specified what counts as being young. And my wife's end of increasing her physical fitness is indeterminate in that she hasn't fully specified either by when or by how much. What's more, such indeterminate ends are typically fluid, such that what counts as pursuing and/or achieving them changes over time as the agent gets a better sense of what's involved in pursuing and/or achieving them, re-evaluates their commitments to pursuing and/or achieving them, and, as a result, further specifies them as time goes on. For instance, my conception of what counts as writing an important book might change as I encounter various unforeseen setbacks or opportunities. And, so, we can determine whether an agent is adequately pursuing and/or achieving their legitimate ends only by looking at both how they act over time and how their ends develop over that time. Of course, all this is incredibly messy. But, as W. D. Ross noted, "it is more important that our theory fit the facts than that it be simple" (1930, 19).

It's not only our discretionary ends that are fluid and indeterminate. The same holds for our obligatory ends. Take, for instance, our duty to have helping others as "a serious, major, continually relevant, life-shaping end" (Hill, 2002, 206). What counts as serious? What counts as major? And what counts as life-shaping? Morality, I believe, leaves this less than fully determinate.[37] Consequently, it won't always be determinate whether an agent had helping those in need as such an end.[38] Of course, it's determinate that Mr. Scrooge didn't. But other cases will be borderline. To illustrate, consider Mr. Borderline. Assume that his actions and attitudes over his lifetime are such that it's borderline whether he had helping those in need as a life-shaping end.

What should we say about whether he fulfilled his imperfect duty to have helping those in need as a life-shaping end? It seems that we must choose between the following two alternative views. On the More Demanding View, an agent fulfills their duty to have helping those in need as a life-shaping end if and only if it's determinate that they had this as a life-shaping end. By contrast, on the Less Demanding View, an agent fulfills this duty if and only it's not determinate that they didn't have this as a life-shaping end.[39] The two agree that Mr. Scrooge failed to fulfill his duty to have helping those in need as a life-shaping end, but they disagree on whether Mr. Borderline so failed. Whereas Mr. Borderline counts as having fulfilled his duty on the Less Demanding View, he doesn't on the More Demanding View. Now, I think that we should accept the Less Demanding View on the grounds that it would be unfair to blame someone for not φ-ing unless it's determinate that they didn't φ. Thus, an option needn't be optimific, but only optimal, to count as morally optional. And, so, we should accept the following interpretation of imperfect duties.

> *The Refined Moderate Interpretation:* You are morally justified in forgoing an opportunity to promote an obligatory end if and only if taking it would entail either violating a perfect duty or forgoing *what is determinately* an overall better opportunity to promote your legitimate ends.

On this interpretation, when there are several courses of action at the top, none of which is determinately better than the others, we'll have the moral option of doing whichever one we're most inclined to do. And given that our legitimate ends are almost always (if not always) fluid and indeterminate, there will often be many

---

37 The thought here is not one about language and the vagueness of certain words. Rather, the thought is that morality sometimes fails to give any precise specification of what counts as pursuing/achieving our obligatory ends and that agents sometimes fail to give any precise specification of what counts as pursuing/achieving their discretionary ends.

38 To hold that, for some proposition *p*, it's indeterminate whether *p* is to hold that it's neither true nor false that *p*.

39 Actually, there's a third view: it's indeterminate whether he fulfilled his imperfect duty to have helping those in need as a life-shaping end. For why we should avoid coming to this conclusion, see Dougherty 2016 and Portmore forthcoming (Sect. 2.3).

different maximal options that all optimally achieve our legitimate ends. Thus, these maximal options will all be morally permissible, and the non-maximal options that are entailed by them will inherit their permissibility status. Thus, imperfect duties will leave us with quite a bit of latitude in terms of the non-maximal options that we're permitted to perform on a given occasion.

5 Conclusion

On a broadly Kantian conception, imperfect duties are duties to have ends. And I've argued that attitudes, not actions, are constitutive of having ends. What's more, I've argued that we can go above and beyond the call of duty by forming attitudes and not just by performing actions. Thus, we can go above and beyond the call of an imperfect duty by forming attitudes that are morally better than those that we must at a minimum form simply to count as having the end that it requires us to have. Lastly, I've argued that imperfect duties provide us with neither too little nor too much latitude. For I've argued both that we are permitted to perform any optimal option and that our legitimate ends are often fluid and indeterminate such that there are often many different optimal ways of balancing them over time. Admittedly, this account has been rather schematic at times, but I plan on developing it further in future work.[40]

References

Archer, A. (2018). Supererogation. *Philosophy Compass, 13*, 1–9.

Baron, M. (1987). Kantian ethics and supererogation. *The Journal of Philosophy, 84*, 237–262.

Benn, C., & Bales, A. (2020). The rationally supererogatory. *Mind, 129*, 917–938.

Betzler, M. (2008). Kant's ethics of virtue: An introduction. In her (ed.) *Kant's Ethics of Virtue*. Berlin: Walter de Gruyter.

Bratman, M. E. (2007). *Structures of agency*. Oxford University Press.

Brinkmann, M. (2015). Disjunctive duties and supererogatory sets of actions. *Royal Institute of Philosophy Supplement, 77*, 67–86.

Chisholm, R. M. (1963). Supererogation and offence: A conceptual scheme for ethics. *Ratio, 5*, 1–14.

Cowley, C. (2015). Introduction: The agents, acts and attitudes of supererogation. *Royal Institute of Philosophy Supplement, 77*, 1–23.

Crisp, R. (2015). *The cosmos of duty: Henry Sidgwick's methods of ethics*. Oxford University Press.

Dorsey, D. (2013). The supererogatory, and how to accommodate it. *Utilitas, 25*, 355–382.

Dougherty, T. (2016). Moral indeterminacy, normative powers and convention. *Ratio, 29*, 448–465.

Ferry, M. (2015). Beyond obligation: Reasons and supererogation. *Royal Institute of Philosophy Supplement, 77*, 49–65.

[40] I plan on doing so in a book tentatively entitled *Kantsequentialism: A Morality of Ends*. For helpful comments on earlier drafts of this chapter, I thank Joseph Bowen, Peter Graham, David Heyd, and Theron Pummer.

Guevara, D. (1999). The impossibility of supererogation in Kant's moral theory. *Philosophy and Phenomenological Research, 59*, 593–624.
Greenspan, P. (2010). Making room for options: moral reasons, imperfect duties, and choice. *Social Philosophy & Policy, 27*, 181–205.
Gregor, M. (1963). *Laws of freedom: A study of Kant's method of applying the categorical imperative in the 'Metaphysik der Sitten.'* Blackwell.
Hanser, M. (2014). Imperfect aiding. In S. Luper (ed.), *The Cambridge Companion to Life and Death*. Cambridge: Cambridge University Press.
Herman, B. (2022). *Kantian commitments: Essays on moral theory and practice*. Oxford University Press.
Herman, B. (2007). *Moral literacy*. Harvard University Press.
Heyd, D. (2019). Supererogation. In E. N. Zalta (ed.), *The Stanford Encyclopedia of Philosophy (Winter 2019 Edition)*. https://plato.stanford.edu/archives/win2019/entries/supererogation/.
Heyd, D. (1982). *Supererogation: Its Status in Ethical Theory*. Cambridge University Press.
Heyd, D. (1980). Beyond the call of duty in Kant's ethics. *Kant-Studien, 71*, 308–324.
Hieronymi, P. (2006). Controlling attitudes. *Pacific Philosophical Quarterly, 87*, 45–74.
Hill, T. E., Jr. (2002). *Human welfare and moral worth*. Oxford University Press.
Hill, T. E., Jr. (1992). *Dignity and practical reason in Kant's moral theory*. Cornell University Press.
Hill, T. E., Jr. (1971). Kant on imperfect duty and supererogation. *Kant-Studien, 62*, 55–76.
Hooker, B. (2000). *Ideal code, real world*. Oxford University Press.
Hutcheson, F. (1969). An inquiry concerning moral good and evil. In D. D. Raphael (Ed.), *British Moralists* (pp. 1650–1880). Clarendon Press.
Igneski, V. (2006). Perfect and imperfect duties to aid. *Social Theory and Practice, 32*, 439–466.
Kamm, F. M. (1996). *Morality, mortality. Vol. 2, Rights, Duties, and Status*. New York: OxfordzUniversity Press.
Kant, I. (1785). *Groundwork of the metaphysics of morals*. Translated by Marry Gregor. Cambridge: Cambridge University Press.
Kant, I. (1797). *The metaphysics of morals*. Translated by Mary Gregor. Cambridge: Cambridge University Press.
King, A. (2014). Actions that we ought, but can't. *Ratio, 27*, 316–327.
McHugh, C. (2017). Attitudinal control. *Synthese, 194*, 2745–2762.
Mellema, G. (1991). *Beyond the call of duty: Supererogation, obligation and offense*. State University of New York Press.
Mill, S. J. (1979). In S. George, (Ed.), *Utilitarianism*. Hackett Publishing Company. ISBN: 0915144417, 9780915144419
Muñoz, D. (2021). Three Paradoxes of supererogation. *Noûs, 55*, 699–716.
Noggle, R. (2009). Give Till it hurts? Beneficence, imperfect duties, and a moderate response to the aid question. *Journal of Social Philosophy, 40*, 1–16.
Portmore, D. W. (forthcoming). Consequentializing agent-Centered restrictions: A Kantsequentialist approach. *Analytic Philosophy*.
Portmore, D. W. (2019a). *Opting for the best: Oughts and options*. Oxford University Press.
Portmore, D. W. (2019b). Control, attitudes, and accountability. *Oxford Studies in Agency and Responsibility, 6*, 7–32.
Portmore, D. W. (2017). Transitivity, moral latitude, and supererogation. *Utilitas, 29*, 286–298.
Portmore, D. W. (2011). *Commonsense consequentialism: Wherein morality meets rationality*. Oxford University Press.
Pummer, T. (2016). Whether and where to give. *Philosophy & Public Affairs, 44*, 77–95.
Rainbolt, G. (2000). Perfect and imperfect obligations. *Philosophical Studies, 98*, 233–256.
Reath, A. (2009). Setting ends for oneself through reason. In S. Robertson (ed.), *Spheres of Reason: New Essays in the Philosophy of Normativity*. Oxford: Oxford University Press.
Ross, W. D. (1930). *The right and the good*. Oxford University Press.
Salam, A. (2016). *Perfect and imperfect rights, duties and obligations: From Hugo Grotius to immanuel Kant*. Ph.D. thesis, University of Oxford.

Singer, P. (1972). Famine, affluence, and morality. *Philosophy and Public Affairs, 1*, 229–243.
Smith, A. M. (2015). Attitudes, tracing, and control. *Journal of Applied Philosophy, 32*, 115–132.
Sticker, M., & van Ackeren, M. (2018). The demandingness of beneficence and Kant's System of duties. *Social Theory and Practice, 44*, 405–436.
Stocker, M. (1967). Acts, perfect duties, and imperfect duties. *The Review of Metaphysics, 20*, 507–517.
Stohr, K. (2011). Kantian Beneficence and the problem of obligatory aid. *Journal of Moral Philosophy, 8*, 45–67.
Stratton-Lake, P. (2008). Being virtuous and the virtues: Two aspects of Kant's doctrine of virtue. In M. Betzler (ed.) *Kant's Ethics of Virtue.* Berlin: Walter de Gruyter.
Tenenbaum, S. (2021). *Rational powers in action: Instrumental rationality and extended agency.* Oxford University Press.
Timmerman, T. (2015). Sometimes there is nothing wrong with letting a child drown. *Analysis, 75*, 204–212.
Timmermann, J. (2005). Good but not required?—assessing the demands of Kantian ethics. *Journal of Moral Philosophy, 2*, 9–27.

The Staircase Scene: Supererogation and Moral Attunement

Dale Dorsey

Abstract This paper considers a pair of mutually puzzling first-order intuitions: a case in which it seems both supererogatory for an agent to perform a specified act, and also seems as though were that act not performed, this would have been a failure of moral obligations. I argue that these intuitive reactions are difficult to dislodge and resist accommodation by standard accounts of supererogation. I then argue that this puzzle motivates a new form of supererogatory action: action that, though morally required, is responding to moral circumstances or facts that the ordinary upright agent would typically overlook.

Keywords Supererogation · Attunement · Collective obligation · Consequentialism

In a particularly tense scene in the gangster film *The Untouchables*, Elliot Ness and his men are waiting for a key witness to the nefarious activities of Al Capone to board a train at Chicago's Union Station. Standing at the top of a grand staircase, Ness spies a mother struggling to get her crying child's baby carriage, along with her luggage, up the stairs. After watching her struggle for several moments, and watching several other people ignore her, he intervenes to help.

Now, the scene in the movie goes off the rails a bit, but ignoring the bloody shootout, a few things might very well be said about this case. First, it seems right that, were no one to help her, the beleaguered mother would have a moral complaint. She was clearly in distress, it would have taken a stranger approximately 20 seconds to give her a hand, and so on. Second, and somewhat more puzzling, when Elliott Ness helps the mother with her child up the staircase, she expresses *profound* gratitude. And while she doesn't exactly say this in so many words, it's clear that her attitude is something of the form of 'you didn't have to do this!'. And it doesn't seem like this reaction is inappropriate or ill-fitting. For Ness to have helped the mother up

D. Dorsey (✉)
University of Kansas, Lawrence, KS, USA
e-mail: ddorsey@ku.edu

D. Heyd (ed.), *Handbook of Supererogation*,
https://doi.org/10.1007/978-981-99-3633-5_6

the stairs does seem to be the sort of thing that he didn't *have* to do, but that was especially praiseworthy or meritorious. In short, *supererogatory*.

But herein lies a puzzle. It seems right to say that Ness's action was supererogatory. But it also seems right to say that the total absence of help would have given rise to a moral complaint—the mother could surely form a negative reactive attitude of the form that typically signals that she was on the receiving end of a moral wrong. But how could this be? How could no one be obligated to help the mother, when in so failing she is entitled to moral indignation?

In this paper, I argue that this puzzle resists easy solution, but tells in favor of a form of supererogation that, to my knowledge, has as yet been unexplored. I argue that it is appropriate to describe actions as supererogatory not *only* in cases in which they are especially morally good but nevertheless not required but also when the consideration of the relevant moral factors would be beyond the typical *attention* of agents that have developed appropriate psychological dispositions, given the complex balance of moral life.

1 The Puzzle

The introduction to this paper outlined a puzzle, but it would do to make it a little more precise here. So take two versions of the case:

> *Ness Helps*: A mother is struggling to get her baby and luggage up an ornate marble staircase after a long train ride. After trying unsuccessfully for several moments, and being ignored by several strangers, Elliot Ness intervenes and helps her and her baby to the top of the staircase. "Sir, thank you so much! You didn't have to do this," she says.

Now compare *Ness Helps* to:

> *Nobody Helps*: A mother is struggling to get her baby and luggage up an ornate marble staircase after a long train ride. Several strangers pass by her, but none stop to help. After several minutes of very awkward wrangling, she finally gets her baby to the top of the stairs. She looks back at the gathering of strangers who failed to help her and feels anger and indignation that no one bothered to lend her a hand.

I submit, and shall be operating on the assumption, that the attitudes on display by the mother in both instances of this case are appropriate or fitting. In other words, they seem to point to what are plausibly the proper verdicts about the action or lack of action involved in the cases. It seems right to say that, in *Ness Helps*, Elliot Ness acted in a way that went above and beyond, that he did not have to act, that was supererogatory. Indeed, consider the potential reaction of a morally sensitive bystander, seeing Ness help this struggling mother. We might imagine this bystander holding Ness in high regard, saying 'what a nice guy', perhaps going over to shake Ness's hand or stand him a drink. None of this seems like an over-the-top reaction.

But in *Nobody Helps*, it also seems right to say that the mother's reaction is tracking the moral facts. Perhaps her indignation isn't directed to any *single* person, but she rightly feels that *someone* should have helped her, that perhaps there was a kind of moral failure that was generalized over a group (the group of bystanders that did nothing). Indeed, one might imagine once again a bystander, unable to help, watching what is happening. "Why won't someone stop and help this poor woman?" one might imagine her saying, with a touch of indignation. And this indignation, or so it seems to me, would be perfectly appropriate in the circumstances.

Now, one point should be noted straightaway. We very often refuse to take first-person testimony concerning the status of supererogatory acts seriously (cf. Urmson, 1958). A soldier who lays down his life for his comrades, for instance, may sincerely declare that he was only doing his duty. But of course, these reactions seem inappropriate. But the current cases aren't instances of this kind. The key here is not simply that the mother *had* the reactions in the various cases, but rather than the mother's reactions *seem, in reflection, justified.*

But if that's right, here's the puzzle: how could Ness's act in *Ness Helps* be supererogatory while the mother's (and, indeed, bystander's) indignation in *Nobody Helps* be appropriate? One might put the puzzle in semi-formal terms in the following way:

1. In *Ness Helps*, Elliot Ness' helping was supererogatory.
2. If an act is supererogatory, it is not morally required.
3. Hence, Ness' helping was not morally required.
4. If Ness' helping was not morally required, then no one was morally required to help.
5. If no one was morally required to help, moral indignation at not being helped would be inappropriate.
6. Hence, moral indignation at not being helped would be inappropriate.
7. But, in *Nobody Helps*, moral indignation at not being helped is appropriate.

(1) is a simple statement of the intuition elicited by *Ness Helps*. (2) is implied by the standard conceptual account of the supererogatory. Most hold that supererogatory acts are morally optional, despite being morally praiseworthy. And hence, if an act token is supererogatory for a person at a time, that act could not have been morally required of that person at that time. But there is nothing *special* about Elliot Ness in this case. Anyone who behaved as Ness did would surely have acted in a supererogatory way, justifying (4). And (5) is a simple claim about moral indignation.[1] If someone acted in a way that was morally justified, i.e., violated no moral obligations, then in most cases, indignation seems out of place.[2] But given that no one

---

[1] Darwall (2006), 292.

[2] I say "in most cases" because, while many argue that the appropriateness of moral indignation waits on moral failure in *all* cases, I have argued that this is false (Dorsey, 2018). But the argument I offer to this end has no bearing on the case in question. The argument I offer suggests that when a person is engaged in a specific sort of norm-governed practice, even if this practice is not normatively weighty, this person can be subject to reactive attitudes such as indignation without also succumbing to normative error. But nothing like that seems to be occurring in this case; walking down a staircase in the middle of a train station does not seem to be the sort of norm-governed practice that can justify indignation without normative significance.

violated any moral obligations in not helping, it would appear that there is nowhere to properly direct moral indignation.

But of course (6) is just a denial of the considered judgment elicited in *Nobody Helps*, outlined in (7). It seems quite wrong to say, in that case, that moral indignation is inappropriate. Rather, both the mother and a bystander can feel moral indignation quite fittingly. So we have a problem here. How to explain our divergent judgments? One possibility, of course, is that we simply revise our considered judgments in either *Ness Helps* or *Nobody Helps*. But that, I think, would be a drastic revision of our common moral sentiments. Better, I think, to scrutinize the individual premises of the argument above to see if we can find some weak spots.

2 Supererogation and Moral Requirement

Since (1) is a simple statement of considered judgment in *Ness Helps*, let's start with (2). This premise claims that, if an act is supererogatory, then it is *not* morally required. Indeed, this claim stands behind much contemporary (post-Urmson) theorizing about the structure of the supererogatory. Consider, for instance, the classic examples of supererogatory action. A soldier jumping on a grenade to save his comrades is surely doing something that is especially morally good; it is a profound act of self-sacrifice. But of course, no one would have argued that he was required to do so. This seems a paradigmatic example of "going beyond the call of duty," which the category of the supererogatory is meant to capture.

For instance, in introducing the notion of the supererogatory, Gregory Mellema writes this:

> The concept of supererogation has so far been characterized in terms of satisfying three conditions. According to this characterization, the performance of an act qualifies as supererogatory if and only if: (1) The performance of the act fulfills no moral duty or obligation; (2) The performance of the act is morally praiseworthy; and (3) The omission of the act is not morally blameworthy (Mellema, 1991, 13).

Along the same lines, Douglas Portmore writes: "A supererogatory act is, as the name suggests, an act that goes above and beyond what is morally required," (Portmore, 2011, 91). The same thoughts are echoed in Urmson (1958), Rawls (1971), Dreier (2004), Horgan and Timmons (2010), and a wide range of other thinkers. Indeed, contemporary accounts of the underlying mechanism of supererogatory action presume that such action is not morally required. For instance, Portmore proposes that non-moral reasons, such as prudential reasons, can have *moral justifying force* in comparison to moral reasons, strictly so-called. So it may be, for instance, that all the strictly moral reasons favor my performance of an action ϕ, but that ϕ-ing would entail a prudential sacrifice, rendering some alternative, ψ, morally justified, and thus rendering ϕ-ing supererogatory (Portmore, 2011, 91–97). This analysis presumes that supererogatory actions are not morally required. Indeed, the so-called 'puzzle' of the supererogatory—how could an action that is morally favored not be morally required?—seems only to arise if we assume that supererogatory acts are not morally required (cf. also Archer, 2016).

And of course, it's not at all hard to see why the analysis trends in this direction. After all, if we fail to conform to our moral requirements, we fail our moral duty. But if supererogation goes *beyond* the call of duty, then surely failing to perform a supererogatory act is not failing to conform to duty. And hence supererogation cannot be required. If that weren't so, it would seem strange to hear the mother in *Ness Helps* suggest that Ness 'didn't have to do that', and so on.

3 Indignation and Collective Obligations

So let's move on to premise (4). This premise states that if Ness was not morally required to help the mother carry her child and luggage up the stairs, then neither were any other bystanders, similarly placed.

On the face of it, this premise seems quite plausible. After all, there is nothing *special* about Elliot Ness in this case. He just happens to be one among many other bystanders, each of whom could have helped. If some *other* bystander would have helped, then this other person surely would have been deserving of the same plaudits by the mother and anyone who happened to see what was happening: a supererogatory act. So, if Elliot Ness did not face a moral requirement to act, it seems hard to see why anyone else would, either.

I think this reasoning is convincing. But if this is correct, then the next premise (i.e., (5)) also seems to hold. After all, if the mother reaches the top of the stairs with no help, and *no one* failed in any moral obligation to help her, then surely indignation is not appropriate in this case. As noted already, in typical cases the appropriateness of moral indignation waits on moral failure. But if no one failed, how could indignation be appropriate in the current case?[3]

Well, perhaps there is one way. One might argue that though no *particular* person failed in their moral obligations to help the mother in question, nevertheless there may have been a kind of *collective* moral obligation in place: an obligation faced by the eligible bystanders *as a collective* (cf. Wringe, 2016). One might find some motivation for this hypothesis in the (supposed) reaction of the mother in *Nobody Helps*. She is morally indignant, and appropriately so. But *toward whom* is her indignation directed? Not at any one particular person. Rather, her indignation is directed at the *group* of bystanders who, collectively, did nothing to help.

Let's examine this possibility. As suggested by Schwenkenbecker:

> Collective obligations are not a novel type of obligation but moral obligations held in a collective *mode*: I can individually hold an obligation to do x or we (for instance, you and I) can jointly hold an obligation to do x (for instance, where x is only collectively feasible). To jointly hold an obligation is a plural predicate—it can only meaningfully apply to two or

[3] See note 2.

more agents, very much like other (non-moral) plural predicates such as 'playing a duet' or 'walking past one another'. (Schwenkenbecker, 2021, 10)

Immediately, however, this account of collective obligations seems like an ill fit for the cases under consideration. After all, it is not as if helping the mother up the staircase is something the bystanders do *collectively*. Rather, it is something they do, if, in fact, they do it, individually. Indeed, the paradigmatic case that Schwenkenbecker (2021, 7) uses to illustrate the concept of a collective obligation seems not to describe the case under consideration:

> Ten passersby witness a car accident in which a motorcyclist gets trapped underneath a car, which has caught fire on one side. Somebody has to act very quickly to pull him out the other side and in order to do so the car will need to be ever so slightly lifted. None of the passersby can lift the car on their own and pull the man out, but together they can (without taking any undue risks to their own health and safety). As it happened, the people manage to lift the car and save the motorcyclist's life.

In this case, it is plausible to say that the individuals in question face a collective obligation to save the biker—none of them can do it on their own; they each have to pitch in to help. But this isn't the case in either of the cases under consideration. Ness needs no help at all to assist the mother. It is something he can accomplish all on his own.

But perhaps there can exist collective obligations that apply to a group that only *one* member of the group can carry out. Note that many analyses of collective obligations entail that not all members of the collective need to act in any particular way for them to be part of a collective that faces an obligation. If, for instance, a table needs to be moved by a group of three movers, but only two movers can work the table around the difficult corners, then the group faces a collective obligation to move the table that does not imply obligations of a particular action for each member of the group (cf. Schwenkenbecker, 2021, 10) But even if we accept that this applies in the current case, it doesn't go any distance toward solving the puzzle. This is because, *were it the case* that the collective faced a moral obligation that was thereby carried out by Ness, it doesn't seem plausible to hold that Ness' action, in this case, was supererogatory. Ness' action would simply be the fulfillment of a collective obligation to help. And while it is true that he is the *only one* who helped, we could hardly say that Ness went 'above and beyond' the call of duty in such a case, given that his action was in fact the fulfillment *of* a duty that he had in virtue of being part of the collective, viz., to see to it that this mother was helped up the stairs.

Another possibility on this score should be considered here.[4] Let's say that the group of bystanders (of which Elliot Ness is a member) has a collective obligation to assist the mother. Now, given that this sort of collective obligation is of the kind that only a single member can fulfill, one might wonder what, in an ideal scenario, the method by which to carry out that obligation ought to be. Perhaps they deliberate about who would be burdened least, perhaps they would draw straws, or use some other fair method of assigning the requisite group obligation to a single member

[4] Thanks to David Heyd for this excellent suggestion.

of the group. But there's no time for that here—ideal circumstances clearly do not apply; the mother is struggling and someone has to help. So Ness, in volunteering, is essentially acting, in a way that goes beyond duty, to carry out the group's moral obligation. His supererogatory action is in agreeing, without further deliberation, to take on the task required of the group. Put another way: this is supererogatory, not when it comes to *helping the mother*, but rather when it comes to *helping the group* fulfill its moral obligations.

I think there's something deeply right about this proposal. I think it's quite true that Ness does the group a favor by helping the mother, unburdening the remaining bystanders. (One might imagine another bystander seeing the mother and feeling relief that Ness decided to lend a hand where they didn't, and perhaps thanking Ness for intervening.) But I confess to believing that it cannot *really* solve the problem at issue. The key is that the *mother* feels gratitude, and this attitude seems appropriate. But she doesn't feel gratitude *on behalf of the other bystanders*. She feels gratitude *on her behalf*—she thanks Ness for helping *her*. And if this attitude is appropriate, there must be something supererogatory about his helping her, rather than his volunteering on behalf of the group.

4 Two Moralities

Thus far, I have argued that we have good reason to accept each of the premises in the argument offered in §1. But of course, because this argument leads to a contradiction, something has to go, despite its surface plausibility.

One possibility is that there is a kind of equivocation happening in the reasoning on offer. Perhaps the notion of 'moral' shifts from premise (2) to premise (7). To see this, consider the following suggestion by James Dreier:

> Maybe there isn't just one 'moral point of view.' Maybe there are (at least) two. To borrow from virtue theory, one point of view we can adopt is the point of view of the perfectly virtuous agent (one of Susan Wolf's moral saints, perhaps); or, less ambitiously, just the beneficent agent. From this perspective, there is everything to be said in favor of [the supererogatory act] and nothing to be said against it...But we can also adopt a less ambitious perspective—that of the just person (maybe 'dutiful' would be a better word). From the point of view of justice, there isn't anything to be said in favor of going to all that trouble. (Dreier, 2004, 149)

Here Dreier suggests that there are actually two 'moral perspectives', as it were.[5] The first perspective is that of the beneficent, virtuous, and so forth, agent. That agent would surely be motivated to help the mother up the staircase. But we can rightly say that this act goes 'beyond' the call of duty because the notion of duty is determined by a different sort of scale—the scale of 'justice', or what one genuinely must do. How would this proposal solve the problem? By suggesting, first, that the sense in

[5] As a historical aside, Dreier's view is clearly anticipated in the dual theories of virtue and justice proposed by Henry Home, Lord Kames as part of a criticism of the moral sense theorists, such as Shaftesbury and Hutcheson. Cf. Kames and Home (1729), 31–33.

which the act was especially praiseworthy was in the first, merely beneficent, sense, and also suggesting that the sense in which the act was morally required was in the second, 'duty'-based, sense.

However, when put like this proposal clearly cannot work. This is because this equivocation does not occur. If the action undertaken by Ness is supererogatory, this *implies* that it is not obligatory in the sense determined by the second 'justice' perspective. Hence, if there is a failure of moral obligations somewhere to be had, it must be that the actions in question were not supererogatory. The puzzle remains.

5 What Is Duty?

In this section, I'm going to focus on the possibility that the errant premise is (2). To see this, consider that much of the discourse surrounding supererogatory action involves the notion of one's 'duty'. But what *is* this, precisely? (2) relies on the assumption that 'duty', at least when understanding the supererogatory, refers to *moral* duty in particular. But notice that 'duty' is clearly ambiguous. One can clearly face moral obligations. But one can also clearly face, e.g., obligations of one's professional or personal life (i.e., one's 'duty as a king' or 'as a professor' or 'as a father') (Crisp, 2013). One can also face duties of prudence; duties 'to oneself' (Dorsey, 2021). One can also, importantly, face obligations of *all things considered* practical reason.

What does the latter concept refer to? Take a simple example. Imagine that I face a moral obligation not to, say, burn my neighbor's house down. But imagine that were I to do so I would have a much better view of the natural beauty lying just beyond. Imagine also that I could 'get away with it' in the sense that no further, downstream negative consequences would befall me. This would seem to establish a prudential requirement—assume for the purposes of argument—to burn. So I appear to have two conflicting requirements. But what should I do *all things considered*? The answer to this question will surely be determined by the relative *practical* or *normative* weight of the obligations and considerations involved. In this case, or so it seems to me, the moral obligation is far more normatively weighty. And so despite it being imprudent not to burn my neighbor's house down, I face a *duty* (all things considered) to refrain.

In previous work (Dorsey, 2016), I have argued that supererogation is best understood as action that goes beyond, not one's *moral* duty per se, but rather one's *all things considered* or *practical* obligations. It is, as it were, morally better than one is all-things-considered required to perform. Could this strategy help us in the present case?

I'm inclined to think not. While I continue to believe in my earlier diagnosis of the nature of supererogatory actions,[6] this diagnosis doesn't really assist in understanding the current puzzle. This is for a number of reasons. First, while such an

[6] *Pace* excellent criticisms by, e.g., Archer (2016).

analysis would allow us to say that the act in question is supererogatory *despite* being morally required (i.e., sometimes it may be possible that to obey a moral requirement is 'beyond the call' of what one *all things considered* must perform given, e.g., certain non-moral reasons that tell against the action), it's hard to see how that could be in operation here. For starters, the non-moral reasons against helping seem exceptionally weak in this case. Assuming that helping wouldn't make one late for one's train (in which case it would fail, plausibly, to be one's moral obligation), the act of helping the mother would be of very little cost. So the reasons telling against helping from, e.g., the prudential angle seem quite weak, certainly insufficient to render a moral obligation (helping the mother) optional from the all-things-considered normative perspective.

But even if this were false, and the non-moral reasons telling against the action were sufficient to make helping all-things-considered optional, we continue to face the puzzling fact that the mother's indignation at *not* being helped seems fitting. This is because, if we understand one's 'duty' as one's 'all-things-considered' obligations, then it appears we should[7] treat indignation and other reactive attitudes not as sensitive to moral duty *in particular*, but rather sensitive to obligations that we face in an 'all-things-considered' sense. Only in the *latter* sense do we maintain obligations to which we can properly be held to account (Dorsey, 2016). But, once again, this seems to contradict our considered judgments in *Nobody Helps*. The mother's indignation is justified. But if none of the people who refused to help were, as a matter of practical normativity, required to intervene, then this indignation seems, on the face of it, out of place.

6 Supererogatory Moral Salience

In this section, I'm going to try my hand at solving the present dilemma. My solution will, I hope, preserve some of the key insights of the previous proposals, but will hopefully go some further distance toward actually resolving the tension between wanting to say that Ness' action was supererogatory, but that, were no one to help, moral indignation is appropriate.

6.1 Attunement, Salience, and the Upright Agent

I begin my attempt to explain the puzzle by noting that human agents are psychologically limited. I take this to be uncontroversial. We are limited in our ability to possess, process, and explain information. We are limited in our attention.[8] We are distractible. We are prone to exhaustion, burn-out. Now, our evolutionary history has

[7] Absent, for instance, special practices in which one is engaged, cf. note 2.

[8] Schwartz (2000).

provided us a set of rather ingenious methods by which to cope with these limitations. We have certain 'heuristics' that render certain things rather than others salient to us—allowing us, as it were, to sort out the wheat from the chaff.[9] In cases, in which concentration is required, we can enter 'flow states' to avoid various forms of distraction.[10] But of course, these coping tools—innocuous in and of themselves—can lead us into mistakes in particular cases,[11] as well as allowing us to more efficiently navigate our environment on a broad scale.

These cognitive challenges operate no less in a normative context than they do in other contexts. Given our cognitive challenges, we must develop certain sorts of psycho/behavioral dispositions that help us to make decisions with imperfect information, imperfect attention, exhaustion, and so forth. Indeed, this fact has long been recognized by act-utilitarians. Utilitarianism tells us that we act in a morally permissible manner if and only if we maximize aggregate well-being. But it has also long been known that directing our attention toward that goal will lead us into failure (Hare, 1981, ch. 3; Railton, 1984, 148–156). So good utilitarians will seek to develop psychological propensities that, *on the whole*, lead one to do as much good as one can, even if developing those dispositions will lead one to act *not* to maximize aggregate happiness in particular cases. The key here is that moral agents will need to develop certain psychological propensities that by and large direct them to the right moral answers.

Crucial among these psychological propensities are what I'll call *dispositions of attunement*. This is just a fancy way of saying *what we pay attention to*. We can't, of course, pay attention to everything.[12] Doing so would lead us into exhaustion, inaction, distraction, and so forth. So we pay attention to those things that seem most significant *given* the goal, on a utilitarian view, of maximizing aggregate happiness.

But notice that utilitarianism is not the only normative scheme that requires us to develop efficient psychological dispositions and heuristics. Consider, for instance, the possibility of agent-centered prerogatives. On any view that accepts agent-centered prerogatives, I will have permission to favor my family members in contrast to the overall good, at least to some degree. Let's say I choose to do so. As with utilitarianism, directing my rational attention toward "favoring my family members to degree *d*, as permitted by my agent-relative prerogative" is apt to go astray at doing so. This person will likely be bogged down by constant calculation and pathologic forms of deliberation ("is my daughter's dental appointment *really* as important as a stranger's flat tire?", and so on). So this person will develop certain sorts of dispositions and heuristics, including dispositions of attunement, when it comes to their family, the overall good, and other sorts of moral considerations. And this is also the case when it comes to, e.g., personal prerogatives to act in accordance with their own projects, goals, ends, and aims.

---

[9] Cf. most importantly Kahneman (2011).

[10] Csikszentmihalyi et al. (2014).

[11] Darley and Batson (1973).

[12] Cf. Schwartz (2000), 79–80.

With this in mind, I'd now like to introduce the notion of what I shall call an 'upright' individual. The upright individual, as I understand it, will have a generally *permissible* set of *psychological dispositions*. Put another way, the upright individual is a person who develops, among other psychological propensities, dispositions of attunement in accordance with the normative facts. If act-utilitarianism is true, the upright individual will be one who develops such dispositions that, overall, lead to the greater promotion of the good. If we have personal prerogatives, the upright individual will develop dispositions of attunement that generally lead to them acting in accordance with those prerogatives, but not by so large a degree as to ignore other very significant moral circumstances. And so forth. Notice that the upright individual will *on occasion* act out of accordance with the moral facts. Because their attention is not directed *at* those moral facts (for good reason, given the underlying normative circumstances) they will on occasion, say, favor their families too strongly, or act in some other way that goes beyond what would otherwise be permitted. This is especially clear when it comes to dispositions of attunement. In any particular situation, whether we are attuned to any particular set of considerations (whether moral or otherwise) cannot be determined *by* those considerations. In other words, whether I pay attention to moral fact *f* in circumstance *c* cannot be determined *by the presence of f in c*. Quite obviously: *f* will only influence my behavior or deliberation if I am attuned to it, and if I am attuned to it, it is not its presence that explains *why* I am attuned to it.[13] I must have a preexisting disposition to be attuned to the relevant moral fact in question.

However, practical reason (whether this is captured by morality or not) permits of certain forms of agent-centered concerns. These can, in principle, take the form of prerogatives or constraints.[14] But at the very least, it is justified for us to seek to, e.g., promote our own projects, look after our own families and friends, and so forth. But if this is right, then it would be natural for the upright agent to adopt dispositions of attunement that mark a certain 'balance' between the attention paid to morally significant other-regarding concerns and the salience granted to self-involving concerns, such as, e.g., promoting the success of one's projects and promoting the success of one's family or children. (Indeed, this would be essential in the case of associative constraints, of the form defended by Brink (2001).) Developing these dispositions of attunement will, on occasion, lead one to fail to be attuned to morally salient facts, including morally salient facts that give rise to moral obligations. But this does not entail that the person in question does not maintain attitudes that are in keeping with normatively justified concerns, just as the failure, in a particular case, to maximize utility does not entail that the agent in question fails to have attitudes that are in keeping with utilitarianism.

---

13 Of course, this is not to say that we cannot train ourselves over time, given exposure to moral facts or other sets of facts to may more attention to them. Surely this is possible. But whether in any *particular* case I am attuned to *f* will not be determined by the presence of *f* in that case—it will be because I either am, or am not, attuned to it.

14 Scheffler (1994), Brink (2001), Jeske (2008), Dorsey (2016).

But notice the need to develop general dispositions of attunement go well beyond simply trying to adopt a general balance between attention paid to self- and other-regarding considerations. Virtually any commonsense moral theory will consist in a number of broad moral principles (such as Ross' *prima facie* duties, 1930). But these principles need to be weighed against each other, given our imperfect information. We thus need to develop certain dispositions of attunement: what do I pay most attention to when, e.g., I could help some stranger at the expense of another? What do I treat as salient when I put myself at some minor risk for the sake of another? And so on. The upright agent will develop dispositions of attunement that, in general, provide solid (if not inerrant) guidance in such cases.

6.2 Going Beyond Upright Attunement

With all this in mind, it's time to introduce my solution to the puzzle with which this paper is concerned. To begin, I think we should say that the moral indignation on the part of the mother at not being helped is appropriate. There was, in fact, a failure of moral obligations in *Nobody Helps*.

However, it's not at all clear that, in failing their moral obligations, the bystanders were not *upright*. (Of course, I am not committed to saying that they are, in fact, upright. Some of them may prioritize their own minor inconvenience to moral considerations of any kind. Rather, I mean simply to say that their behavior was *compatible* with being upright.) Because we have some license to prioritize our own interests, circumstances, and so forth, it seems plausible to hold that the upright agent will miss moral circumstances of this kind, especially when in a hurry to catch a train.

Assuming that this is correct, I think it may be worthwhile positing a category of supererogatory act that runs as follows: an act that is reflective of attunement to facts or considerations of moral significance to which an upright agent would not ordinarily be attuned. Such acts reflect an especially praiseworthy *attention* to moral considerations. Now, of course, because these moral considerations are genuine and normatively significant, it may very well be that this form of supererogatory attunement just is, in many cases, special attunement to moral obligations that we, in fact, have. But in such cases this form of supererogation just is the fulfillment of moral obligations that the upright agent wouldn't typically fulfill. And this is enough, I think, to explain and justify the mother's reaction to Ness's beneficence in *Ness Helps*.

Thus if we posit such a category, this would explain the equivocation in the argument offered in Sect. 2. To say that Ness' helping in *Ness Helps* is supererogatory is to say that his act is reflective of attunement to moral considerations (specifically other-regarding moral considerations) that goes beyond the permissible dispositions of attunement maintained by the upright agent. To say that there are moral failures in *Nobody Helps* is to say that the individuals in *Nobody Helps* fail in their moral obligations. But these claims are, of course, compatible. Recall that upright agents can fail to conform to moral obligations given their dispositions of attunement, dispositions that develop in response to, say, personal prerogatives and so forth. And hence to display a supererogatory form of attunement to moral considerations is not necessarily to behave in a way that is not morally required.

Above I suggested that the mothers' distress might not be treated as salient by an upright agent. But why? Why think that the upright agent will miss the mother's distress in this case? While there may be other reasons as well, I'd like to highlight two here. One disposition upright agents have a tendency to develop, in recognition not only of their own personal prerogatives but also those of others, is to, as it were, "mind one's own business," where this is typically understood as a broad set of dispositions such as:

- not intervening in others' affairs unless asked or unless there is risk of serious danger;
- not to engage in behavior that would show lack of respect for others' control of their affairs;
- to appreciate that there may be more at play that meets the eye, when others are involved;
- not to step in circumstances that might jeopardize one's own interests;
- and so on.

Here's an example of how this set of dispositions might work. Imagine that you happen to board a bus with a perfectly sprightly-looking young man sitting in a seat designated for elderly and disabled travelers. You also happen to notice that there are a number of very tired, beleaguered bus riders that are standing on a moving bus struggling to keep their balance. One might be tempted, then, to intervene and to tell the young man he needs to move to accommodate others. But a number of things might be in play here: it could be that this particular young man may have a disability that isn't immediately obvious, and which would cause deep embarrassment were he have to defend himself. It could be that the other bus riders are perfectly comfortable where they are, and would resent a stranger pressing their case in this manner. So it would be perfectly compatible with being an upright agent in this case to "mind one's own business"—not to act unless, say, there were some sort of egregious activity on the young man's part, or unless one of the other passengers asked for one's assistance. Of course, adopting this disposition can lead one into immorality: it may very well be that this young man was a complete jerk with no regard for others, and the best thing overall to do in that case would be to insist that he stand for the sake of other bus riders. But this doesn't mean that failing to do so, and hence failing *morally*, was inconsistent with the dispositions cultivated by *upright* agents. Importantly, my claim here should not be inflated—it is surely not the case that minding one's own business is always compatible with being an upright agent. Far from it! Saying or doing nothing in the face of, e.g., child abuse, racial animus, intent to harm, obvious medical distress, and so forth, are not just immoral but are surely not compatible with the dispositions of attunement that would be developed by an upright agent. But the disposition to mind one's own business in cases that are more subtle, given the complexities and infelicities of moral life, seems on the whole permissible (though, of course, upright agents will continually update their dispositions when possible to refine their attunement to the morally significant in ways that better balance the existence of personal demands and prerogatives, lack of information, lack of time, and so on).

Second, I think there's something else going on in *Ness Helps* that is worth mention. This argument is somewhat more tentative but it also seems to me worth mentioning as a potential explanatory factor in the cases under consideration. As already noted, given our propensity for distraction, human beings can often to a remarkable job of keeping our focus on the goal at hand, i.e., keeping our "eyes on the prize." And this sort of disposition seems like one compatible with uprightness. In other words, once a goal or end has been set, its efficient promotion requires one to try to tune out distractions, to "go with the flow," and to focus on getting oneself to the finish line, especially when the task at hand is arduous or otherwise stressful. And it's not hard to imagine that the bystanders in the cases at hand are doing precisely this: traveling, even if one isn't faced with strong time pressure, can often be stressful, and it's sensible for individuals who are trying to catch a train to focus on that task at hand and, to the best of their ability, avoid being distracted (by, for instance, time-consuming advertisements, or conversations with folks one might enjoy, etc.). But if this is right (and once again I'm not positing it with *total* confidence), it would further help to explain why Ness' action in *Ness Helps* displays a form of salience that goes beyond what the typical upright agent will display. Not only do those upright agents "mind their own business," but they also follow the flow of traffic. They seem not even to *notice* the mother and her plight. And this is understandable—they are trying to reach a train on time; perfectly normal moral agents will, under such conditions, not notice morally significant things going on around them.

But Ness does. So not only does Ness not simply mind his own business in a way that is morally appropriate but also avoids simply being blinded to his own affairs in circumstances in which we would ordinarily think otherwise upright agents would be so blind, or at least would *often* be so blind. Rather, he perceives as salient certain moral facts, viz., that the mother is clearly in distress, that she doesn't ask for help, not because she doesn't need it, but because of, e.g., embarrassment. He treats those facts as salient and hence goes beyond the other-regarding dispositional attitudes displayed by the other bystanders. He, in other words, conforms to his moral obligations in a way that the other bystanders do not, and that, in *Nobody Helps* no one does.

7 Objections

My thesis so far: there are (at least) two categories of supererogatory actions. The first category is the familiar category: action that seems morally better than one's obligations, and is thereby not part of one's obligations. The second, proposed here, is action that is responsive to moral features of the world to which the upright agent would not typically respond. The second category is that into which Ness' action in *Ness Helps* falls, and because this sort of supererogation is compatible with obligation, the mother is perfectly justified in expressing her indignation in *Nobody Helps*.

But this proposal is subject to objections. I consider two here.

7.1 Indignation not Appropriate

I have argued that Ness' action in *Ness Helps* is supererogatory because it is action that is motivated by responsiveness to moral circumstances to which upright agents—those who balance attunement to specifically other-regarding moral concerns and one's own agent-centered concerns and prerogatives—will typically not be attuned. But if this is right, why think that, in *Nobody Helps*, the mother's indignation is appropriate? After all, these are (or so we assume) upright agents. They maintain psychological dispositions that seem to represent a responsible balance between self- and other-concern. But why, then, should moral indignation be an appropriate response? And if I haven't explained why the mother's moral indignation is appropriate, I haven't explained the paradox.

But there are, I think, two answers to this. First, it seems plausible to hold that indignation is appropriate in this case because the bystanders acted immorally. The fact that they were not attuned to the various facts that explain their immorality does not entail that it wasn't immoral, nor does the fact that the failure of such attunement was compatible with agential uprightness entail that their action wasn't immoral. After all, as already stipulated, helping the mother would have been greatly benefiting someone with essentially no cost. The fact that they were not attuned to that fact does not entail that fact fails to constitute a moral reason for action, indeed a (to my mind) decisive such reason.

This argument permits of a response. If the bystanders were genuinely upright agents, then one might suggest that, while they may have behaved immorally, they nevertheless possess an *excuse*[15]: that they simply didn't notice or weren't attuned to the mother's distress given the circumstances of the case. But if they had such an excuse, surely indignation wouldn't be appropriate. However, I have two responses to this proposal. First, it is not enough to render indignation inappropriate to simply have an excuse. Surely, in *Nobody Helps*, for indignation to be rendered inappropriate, the relevant bystanders needed to have made some attempt at moral repair; an apology, a "very sorry about that I didn't see you there"-type response. And if this is correct, then the appropriateness of the mother's indignation when such repair isn't made seems to indicate that there is some sort of moral wrong. This response leads directly into my second, however. Even if indignation is not appropriate in this instance (perhaps indignation tracks, say, dispositional uprightness or some other such thing, or perhaps such moral repair is made, and so on), this does not mean that there does not remain an interesting paradox that requires solution. In other words, it is not the indignation

[15] Thanks once again to David Heyd for this response.

itself that is paradoxical in this case, but instead, how Ness' action in *Ness Helps* could be supererogatory while having *failed* that action would be *immoral.* And for that, one needs to posit another account of supererogatory action: one that arises principally when agents are attuned to moral facts that upright agents, agents who in this case are simply "minding their own business" tend to ignore.

7.2 *Is This Really Supererogation?*

Supererogation is *super*—it is 'above' something. But above 'what'? Colloquially, it is 'above and beyond' the call of *duty*, the call of *obligation*, what we *must do.* But if supererogation (stressing the 'super') is beyond the call of duty, then any act that is required would not be supererogatory. It would be, as it were, *merely erogatory* (Dorsey, 2013). Whether or not the upright agent would have done it, or would have noticed or paid attention to the moral factors involved, if it's Ness' obligation to help in *Ness Helps*, then his action in that case cannot be supererogatory, just as a matter of concept. It's one thing, or so the objector objects, to hold that supererogatory action can be morally required (but rationally permitted). It's quite another thing to hold that supererogatory actions are required not just in the moral sense, but in the all-things-considered sense, as well.

Frankly, I find this objection uncompelling. Or, at the very least, uninteresting. If one wishes to press this line (which seems to me just a terminological choice), then my thesis can be expressed in slightly different terms, with no loss of interest. First, one might say that instead of Ness' *action* being supererogatory, that his *dispositions of attunement* are supererogatory. It's not that he didn't have to act this way, but rather that he didn't (as an upright agent) have to be attuned to the plight of the mother and her child; he could have, as it were, minded his own business in a perfectly upright manner. His *dispositions* go beyond what he would otherwise be required *given* the presence of agent-centered prerogatives and other normative permissions to favor one's own circumstances.

Second, one might simply jettison talk of supererogation. Perhaps Ness' action was not supererogatory per se, but nevertheless, however, we understand it, it seems to display features that mark out our reactions to it *as* supererogatory—effusive thanks, special praiseworthiness, and so on. These all seem appropriate in Ness' case, but do not seem appropriate in cases of run-of-the-mill satisfaction of one's moral obligations. Ness' action here is clearly something that goes beyond what we would ordinarily expect of upright agents, whether we call that supererogation or not. But it is this fact, or so I argue, that my proposal captures: it shows that Ness, but not the other bystanders, was attuned to morally significant considerations that other agents, even upright agents, were not. For my money, this counts as a *kind* of supererogation. But mine may not be everybody's money.

8 Conclusion

I began this paper with a puzzle. Elliot Ness' helping of the beleaguered mother seemed supererogatory, to go above and beyond. But were no one to have helped, the mother seems to maintain a justified, appropriate moral complaint.

To explain this, I argue, we should introduce a new category of supererogation—action that results given attunement to moral features of the world in excess of that to which the upright agent is or can be attuned. Ness' action displays this in, I think, two different ways. Not only does he not mind his own business but also avoids "going with the flow"—he allows morally significant considerations to enter his deliberation in ways that other agents, even other agents who have a set of perfectly appropriate and upright dispositions of attunement, will not. This, I think, is worthy of our special praise, and the special praise heaped on actions of this kind.[16]

References

Archer, A. (2016). The supererogatory, and how not to accommodate it: A reply to Dorsey. *Utilitas*, 28.

Brink, D. O. (2001). Impartiality and associative duties. *Utilitas*, 13.

Crisp, R. (2013). Supererogation and virtue. In Timmons (Ed.), *Oxford studies in normative ethics*. Oxford University Press.

Csikszentmihalyi, M., Abuhamdeh, S., & Nakamura, J. (2014). Flow. In *Flow and the foundations of positive psychology*. Springer.

Darley, J. M., & Batson, C. D. (1973). From Jerusalem to Jericho: A study of situational and dispositional variables in helping behavior. *Journal of Personality and Social Psychology*, 27.

Darwall, S. (2006). Morality and practical reason: A Kantian approach. In D. Copp (Ed.), *Oxford handbook of ethical theory*. Oxford University Press.

Dorsey, D. (2013). The supererogatory and how to accommodate it. *Utilitas*, 25.

Dorsey, D. (2018). Respecting the game: Blame and practice failure. *Philosophy and Phenomenological Research*, 101.

Dorsey, D. (2016). *The limits of moral authority*. Oxford University Press.

Dorsey, D. (2021). *A theory of prudence*. Oxford University Press.

Dreier, J. (2004). Why ethical satisficing makes sense by rational satisficing doesn't. In M. Byron (Ed.), *Satisficing and maximizing: Moral theorists on practical reason*. Cambridge University Press.

Hare, R. M. (1981). *Moral thinking*. Oxford University Press.

Horgan, T., & Timmons, M. (2010). Untangling a knot from the inside out: Reflections on the 'Paradox' of supererogation. *Social Philosophy and Policy*, 27.

Jeske, D. (2008). *Rationality and moral theory: How intimacy generates reasons*. Routledge Press.

Kahneman, D. (2011). *Thinking, fast and slow*. Farrar, Straus and Giroux.

Kames, L., & Home, H. (1729). *Essays on the principles of morality and natural religion*. Liberty Fund.

Mellema, G. (1991). *Beyond the call of duty*. State University of New York Press.

Portmore, D (2011). *Commonsense consequentialism: Wherein morality meets rationality*. Oxford University Press.

[16] I'd like to thank Brad Cokelet, David Heyd, Jason Raibley, and Nancy Snow for helpful comments on these ideas.

Railton, P. (1984). Alienation, consequentialism, and the demands of morality. *Philosophy and Public Affairs*, 13.
Rawls, J. (1971). *A theory of justice*. Harvard University Press.
Ross, W. D. (1930). *The right and the good*. Oxford University Press.
Scheffler, S. (1994). *Families, nations, strangers*. University of Kansas.
Schwartz, B. (2000). Self-determination: The tyranny of freedom. *American Psychologist*, 55.
Schwenkenbecker, A. (2021). *Getting our act together*. Routledge Press.
Urmson, J. O. (1958). Saints and heroes. In A. I. Melden (Ed.), *Essays in moral philosophy*. University of Washington Press.
Wringe, B. (2016). Collective obligations: Their existence, their explanatory power, and their supervenience on the obligations of individuals. *European Journal of Philosophy*, 24.

Must Virtue Be Heroic? Virtue Ethics and the Possibility of Supererogation

Rebecca Stangl

Abstract When Urmson first challenged moral philosophers to account for the phenomenon of supererogation, contemporary virtue ethics was just in its infancy. So, virtue ethicists were understandably delayed in taking up that challenge, and thus the relationship between the two remained opaque. What little discussion of virtue and supererogation there was focused on the ancients rather than their contemporary intellectual heirs and tended to be skeptical about the compatibility of supererogation and virtue ethics. Lately, this has begun to change. A number of philosophers have developed and defended sophisticated virtue-ethical accounts of supererogation. But doubts linger. Recently, important proponents of virtue ethics have argued that the intuitions underlying Urmson's challenge can and should be accommodated without embracing supererogation, or by giving a highly revisionary account of it. In this essay, I confront these lingering doubts. In doing so, I hope to suggest that there is a good reason why virtue-ethical accounts of supererogation have lately proliferated. There are still some intuitions about moral heroism, and moral goodness more generally, that resist capture in a virtue ethics that deflates or eliminates the category of supererogation altogether.

Keywords Supererogation · Virtue ethics · Moral heroism · Hursthouse · Annas

1 Introduction

It is well known that the discussion of supererogation in contemporary ethics can be traced back to Urmson's seminal paper, 'Saints and Heroes' (Urmson, 1958). In the same year, Anscombe published 'Modern Moral Philosophy' (Anscombe, 1958), inaugurating the modern revival of virtue ethics. This concurrent rise of interest in virtue ethics and the supererogatory does not seem merely coincidental. Both are driven, at least in part, by their adherents' rejection of the perceived faults in

R. Stangl (✉)
The University of Virginia, Charlottesville, USA
e-mail: rls5ef@virginia.edu

D. Heyd (ed.), *Handbook of Supererogation*,
https://doi.org/10.1007/978-981-99-3633-5_7

dominant modes of modern moral philosophy, including their tendency toward oversimplification. Urmson himself later characterized the point of 'Saints and Heroes' not merely as establishing the existence of the category of supererogatory actions, but as making room for whole ranges of moral life—"actions we might call kind, considerate, chivalrous, charitable, neighborly, sporting, decent, or acts of self-denial and self-abnegation" (Urmson, 1988, 168)—that are not concerned with duty or obligation, each of which are as different from each other as they are from duty or obligation.[1] And Anscombe, in a similar vein, claimed that it, "would be a great improvement if, instead of 'morally wrong' one always named a genus such as 'untruthful,' 'unchaste,' 'unjust'" (Anscombe, 1958, 8–9).

And yet the relationship between supererogation and virtue ethics is far from clear. One can easily agree in rejecting a certain kind of moral philosophy—a philosophy one judges to be overly abstract and constrained in various ways—and yet propose differing, or even incompatible, diagnoses of the flaw at issue or prescriptions for its remedy. And as it turns out, little work on the relationship between virtue ethics and supererogation was produced in the years following Urmson's piece. Those who immediately took up Urmson's challenge tended to be working from within one of two traditions: Kantianism or consequentialism. Some aimed to undermine his case for supererogation, and some aimed to modify their moral theories to accommodate it. But until recently, virtue ethicists were on the sideline of this debate.

This is far from surprising. Kantians and consequentialists had, at the time of Urmson's article, well-developed and various accounts of right action. The main question Urmson raised was whether these accounts of right action could accommodate, or be modified to accommodate, a distinction between the obligatory and the supererogatory. But virtue ethics was itself a new movement in analytic ethics and had yet to develop a robust account of the relationship between virtue and right action. Indeed, there were serious disagreements among virtue ethicists about whether it was possible or necessary to give an account of right action at all, much less whether that account of right action should distinguish the obligatory from the supererogatory.

Add to this the centrality of Aristotle to the work of contemporary virtue ethics and the centrality to Aristotle of the claim that virtue is a mean between extremes. The virtuous person on the Aristotelian account represents an ideal of balance, in which she does things in the right way, in the right amount, at the right time, and for the right reason. To exceed the virtuous ideal is not to be 'extra-virtuous'; it is to fall into the vice of excess. So, it was not clear what supererogation could be for the Aristotelian (for versions of this argument, see Heyd, 1982; Slote, 2007).

But in recent years, things have begun to change. A number of self-identified virtue ethicists have taken up the question of the relationship between supererogation and virtue ethics. And a number of them have developed and defended virtue-ethical accounts of supererogation. Thus, we find proposals for Neo-Aristotelian accounts

---

[1] He also seemed to go further and express some regret at grouping these various actions together under the overarching term 'supererogatory'. I agree with Heyd (2015) in rejecting this stronger claim. Various types of actions such as the ones he lists might well be usefully distinguished and their differences noted, while nonetheless sharing the character of being morally good but not required, which is precisely what one is saying in grouping them together as supererogatory.

of supererogation (Stangl, 2016, 2020); exemplarist virtue theoretical accounts of supererogation (Zagzebski, 2017); virtuous observer theories of supererogation (Kawall, 2009); agent-based virtue sentimentalist accounts of supererogation (Slote, 2007); and target-centered accounts of supererogation (Kawall, 2009, developing a version of Swanton, 2003). And work has been done to show that the doctrine of the mean, at least on some reasonable interpretations, is compatible with the existence of supererogation after all (Curzer, 2012; Stangl, 2016, 2020).

But even if it is not impossible to give a virtue-ethical account of supererogation and even if such an account can be made compatible with the doctrine of the mean, questions linger about the necessity or desirability of doing so. Recently, important proponents of virtue ethics have argued that the intuitions underlying Urmson's challenge can and should be accommodated without embracing supererogation, or by giving a highly revisionary account of it. The purpose of this essay is to confront these lingering doubts. In doing so, I hope to suggest that there is a good reason why virtue-ethical accounts of supererogation have lately proliferated. There are still some intuitions about moral heroism, and moral goodness more generally, that resist capture in a virtue ethics that deflates or eliminates the category of supererogation altogether.

2 Hursthouse's Revisionary Account of Supererogation

Rosalind Hursthouse's (1999) book *On Virtue Ethics* is rightly considered a seminal text. In it, Hursthouse offered one of the first, and most sophisticated, virtue-ethical accounts of right action, defending the following qualified agent account of right action:

> An action is right iff it is what a virtuous agent would characteristically (i.e., acting in character) do in the circumstances (Hursthouse, 1999, 28).[2]

This account of right action, Hursthouse makes clear, is intended to capture something more than merely permissible action; to say of someone that they behaved as the courageous or generous person would, for example, is to say that they acted not merely permissibly but courageously or generously and therefore well (Hursthouse, 1999, 69–71). Moreover, some of these actions are not only instances of acting well but also 'absolutely required'. Indeed, she claims, not only justice but also charity sometimes implies that we are 'absolutely required' to perform a particular action such as helping a wounded stranger (Hursthouse, 1999, 6). But though the account includes the absolutely required, not all of the actions it identifies as right are required.

[2] Note that Hursthouse later modifies this definition to take account of moral dilemmas in the following way: "An action is right iff it is what a virtuous agent would, characteristically, do in the circumstances, except for tragic dilemmas, in which a decision is right iff it is what such an agent could decide, but the action decided upon may be too terrible to be called 'right' or 'good'. (And a tragic dilemma is one from which a virtuous agent cannot emerge with her life unmarred)" (Hursthouse, 1999, 79).

In at least some cases, Hursthouse thinks, two equally virtuous agents can do two different actions in the same circumstance, and yet both act well. But if this is the case, then neither of the two actions in question can be strictly speaking obligatory: otherwise, each of the agents (in doing one rather than the other) would fail to act well. This is why Hursthouse stipulates that right action is what 'a' virtuous agent would do, and not what 'the' virtuous agent would do.

So what we have in Hursthouse's (1999) account seems to be an account of right action that encompasses both obligatory and non-obligatory well-doing. What we lack is a way to distinguish between the two kinds of right action and how the latter category is or is not related to the category of supererogation. Indeed, the terms 'supererogation' and 'supererogatory' never appear in *On Virtue Ethics*.

But in a little commented-upon paper published 7 years later, and in response to being pushed on the question by Julia Driver, Hursthouse explicitly claims that her account of right action encompasses both the obligatory and the supererogatory and proposes an account of the distinction between them.[3] Such a distinction, she says, is one "no adequate account of the right can ignore" (Driver, 2006; Hursthouse, 2006, 111). The discussion is brief and somewhat difficult to interpret. But given the centrality of Hursthouse's account of right action to the modern revival of virtues ethics, it is well worth considering.

Hursthouse models her account of supererogation on a very famous point made by Philippa Foot about the moral worth of actions. Foot had claimed, "we both are and are not inclined to think that the harder a man finds it to act virtuously the more virtue he shows if he does act well" (Foot, 1978, 10). Fastening on the first intuition, we get something like Kant's theory of moral worth: a person demonstrates the most virtue when she does her duty even in the face of conflicting inclinations. Fastening on the second intuition, we get something like the traditional interpretation of Aristotle: the person of perfect virtue is the one who not only does what the virtuous person does but also does it easily and with pleasure. Foot's solution to this seeming clash of intuitions is to draw a distinction between two different reasons why a particular person might find it difficult to act virtuously. If what makes it difficult to act well is the circumstances in which she finds herself, then it is true to say that acting virtuously in the face of those circumstances demonstrates more virtue than had the circumstances been easier. But if the source of the difficulty in acting well is some flaw internal to the agent's character, then the difficulty in acting well shows that an agent is lacking in virtue.

This distinction, Hursthouse argues, can also be put to work in explaining the difference between the obligatory and the supererogatory.[4] Suppose an agent sees someone drop a full purse and restores the purse without hesitation. There are at least two different kinds of circumstances in which she might do so: one in which she is

[3] I thank Jeremy Reid for bringing this paper to my attention.

[4] Hursthouse seems to suggest that not only she, but also Foot, relies on this distinction in giving an account of the supererogatory. I am not clear on why she suggests this. Nowhere in 'Virtues and Vices' does Foot claim to be giving an account of supererogation. If Foot makes this connection somewhere else, I am not aware of it, and Hursthouse does not give a reference.

comfortably well-off, and one is which she is poor. In the former case, returning the purse is easy and reflects no great virtue on her part. But in the latter case, doing so is made difficult by the circumstances; when she does so without hesitation, she demonstrates great virtue and what she does is therefore ‘strikingly admirable’. This distinction, Hursthouse tells us, *just is* the virtue ethics distinction between the obligatory and the supererogatory. Both agents do what is right and good, but the former, whose circumstances make virtuous action easy, does only what is obligatory. The latter, whose circumstances make virtuous action difficult and therefore particularly admirable, does what is supererogatory.

It must be said that there is something quite puzzling about this account of supererogation. The case Hursthouse uses to illustrate her account is taken directly from Foot’s discussion of the relation of virtue to moral worth. But Foot describes that situation as one in which an agent “has an opportunity to steal, in circumstances where stealing is not morally permissible, but… refrains” (Foot, 1978, 11). And there is no reason to doubt that Hursthouse’s use of this case is intended to mirror Foot’s in this regard. For presumably keeping the purse would be dishonest. And, on Hursthouse’s view, each vice generates a prohibition, “do not do what is dishonest, uncharitable, mean” (Hursthouse, 1999, 36). So, it would be wrong to keep the purse, in either case. But if this is the correct reading of Hursthouse, it implies that both the agent who does what Hursthouse calls supererogatory and the agent who does what Hursthouse calls obligatory refrain from stealing when stealing is ‘not morally permissible.’ And thus both the agent who does what Hursthouse calls supererogatory and the agent who does what Hursthouse calls obligatory do what is required and whose omission would be wrong. But it is generally taken to be definitive of supererogatory action that it is not required, and its omission is morally permissible.

Of course, Hursthouse is quite right that, even among those actions that it would be morally wrong not to perform, some are more difficult to do and thus more admirable when done. Urmson himself emphasizes this point. Indeed, he addresses just the sort of case Hursthouse discusses here. But he does so in order to distinguish such cases from true cases of supererogation. Thus, Urmson points out that we can and do call people heroic or saintly when they do their duty in circumstances which make that duty difficult. A doctor who stands by his patients even during the plague, Urmson says, is heroic in that way. But his action is not heroic in the particular sense that Urmson is interested in, and which the concept of supererogation is intended to capture. Thus, Urmson writes,

> We have considered the, certainly heroic action of, the doctor who does his duty by sticking to his patients in a plague-ridden city; we have now to consider the case of the doctor who, no differently situated than countless other doctors in other places, volunteers to join the depleted medical forces in that city (Urmson, 1958, 201–202).

The first doctor’s action is heroic just insofar as he performs his duty in a situation where fear for their safety would cause many to fail in their duty. But his action is not supererogatory just insofar as he has a duty to his patients to stay. The second doctor, on Urmson’s view, has no such duty; he therefore not only performs an action we might call heroic but also an action that Urmson calls ‘heroic par excellence’ and which is supererogatory. It is the latter kind of case that is the target of analysis in his paper.

But surely Hursthouse's case mirrors the first doctor rather than the second doctor. Someone who finds a lost wallet and returns it even though she is poor does something particularly morally admirable. But insofar as it would be dishonest to keep the wallet, she would fail in her duty if she were not to return it. So the action is not supererogatory. To carry through the parallel with Urmson's discussion, we could perhaps imagine a third-party bystander who witnesses the loss of the wallet and perhaps noticing the hardship this places on the one who has lost it, generously replaces the missing money even though she herself is poor. Such a person would perform an action that would almost certainly be supererogatory. But that is not the sort of case Hursthouse, or Foot, seems to be discussing.

At this point, one might begin to wonder why Hursthouse chooses to give such a revisionary account. The answer, I think, comes in an earlier passage in the paper, where Hursthouse writes:

> We need not deny that 'virtue', in modern parlance, is a threshold concept (Swanton, 2003, 24). (Perhaps even 'excellence' is. After all, we say A– [or perhaps A] to A+ on our students' essays means 'Excellent'.) If someone gets over the threshold then they certainly are not vicious. But grammatically, 'virtue,' and the terms for the individual virtues, accept a whole range of qualifications—"quite V, admirably V, for his age/for her time/in his society/given her disadvantages"—where the qualifications enable us both to give credit where credit is due but also *to register the point that the ideal standard has not yet been met* (Hursthouse, 2006, 105, emphasis added).

What Hursthouse seems to want to say here is that any departure from perfect virtue is not, strictly speaking, virtue. Virtue requires that the ideal standard be met. But if this is right, then when Hursthouse says that an action is right if and only if it is what a virtuous person would characteristically do in the circumstances, she is committed to the view that an action is right if and only if it is what an ideally virtuous person would do in the circumstances. And presumably, an ideally virtuous person would not only do those actions that Urmson thought were required by duty but also all those actions that Urmson thought exceeded the requirements of duty. An ideally virtuous person doctor would not merely fulfill her fiduciary trust to her own patients but would also volunteer to treat patients in a plague-ridden city not her own. (Assuming, of course, that such a thing would not violate some special responsibilities she might have to, for example, her children. As always, the circumstances will matter in determining whether it is virtuous for any particular doctor to volunteer in such a way.) If someone merely fulfilled her fiduciary duties and chose not to volunteer, she would therefore fail to do what is right. So we cannot say of her that she merely did her duty; rather, she failed to what was right in any sense. Thus, Hursthouse cannot make the distinction between the 'supererogatory' and the 'obligatory' in the way in which Urmson suggests is central to the doctrine.

On a strictly literal reading, therefore, it looks like Hursthouse is not so much giving a virtue ethics account of the supererogatory as changing the subject. But if that is the case, why not just say so? Why claim to give an account of supererogation at all? She could instead have done what Crisp (2013) does and simply denied that virtue ethics is compatible with supererogation. Crisp argues that virtue not only can require a particular action but also always requires one to act on its dictates. There is thus no virtuous action whose performance would be good but whose omission is simply permissible.[5] He thus concludes that a virtue-ethical account of supererogation is impossible. Hursthouse, it seems, might well have done the same.

Here's a possible justification for sticking with the terminology of 'supererogation' even if the conceptual apparatus underlying it is highly revisionary: If the account correctly identifies all and only those actions that Urmson called supererogatory as supererogatory, Hursthouse might argue that Urmson had identified an important class of actions—a class of actions that no adequate account of right action could ignore—while being mistaken about their nature. Where Urmson had thought such actions were those that went beyond duty, Hursthouse might argue that they are in fact those actions that are difficult to perform.

So does Hursthouse's account succeed on that ground? It does not. Hursthouse's account does identify some of those actions that Urmson took to be paradigmatically supererogatory as supererogatory. The soldier who jumps on a grenade, for example, does something courageous in a circumstance where courage is objectively difficult to show. His action thus demonstrates more virtue than someone who shows courage in, e.g., facing her fear of a difficult conversation. Likewise, the doctor who volunteers to treat patients in a plague-ridden city not his own shows generosity and compassion in a situation in which generosity and compassion are objectively difficult to show and thus demonstrates more generosity than, e.g., the person who donates funds for the doctor's medical clinic.

But while Hursthouse's view correctly identifies the doctor who volunteers in a plague-ridden city not his own as performing a supererogatory action, it also identifies the doctor who remains in her own city to treat patients when plague arrives as performing a supererogatory action. And, it does so for the exactly same reasons. For the doctor who treats her own patients faces the same risk of infection as the doctor who treats patients in a foreign city, and thus shows generosity and compassion in a situation in which generosity and compassion are objectively difficult to show. She also demonstrates more generosity and compassion than, e.g., the person who donates funds for either doctor's clinic. Thus, Hursthouse's account of supererogation appears to be overinclusive.

But it is also arguably underinclusive. For Urmson, saintly and heroic instances of supererogation were the most conspicuous instances of supererogation. But they were not the only such instances, nor indeed were they the most common such instances. Many supererogatory actions are more mundane than that. As Urmson wrote,

> It should indeed be noted that heroic and saintly actions are not the sole, but merely conspicuous, cases of actions that exceed the basic demands of duty; there can be cases of disinterested kindness and generosity, for example, that are clearly more than basic duty requires and yet hardly ask for the high titles, "saintly" and "heroic" ...It is possible to

[5] Crisp grants that in certain circumstances there might be two actions, x and y, each of which are equally virtuous. But in that case virtue requires one to do 'x or y' and neither x nor y will count as supererogatory.

> go just beyond one's duty by being a little more generous, forbearing, helpful, or forgiving than fair dealing demands (Urmson, 1958, 205).

But such mundane actions are highly unlikely to count as supererogatory on Hursthouse's account. There is no reason to assume that being a little more generous than required, for example, or doing a small kindness for a stranger, will always be objectively difficult. Quite the contrary. And since Hursthouse requires that an action count as objectively difficult in order to qualify as supererogatory, she will be unable to make sense of how such actions might be supererogatory.

Of course, given that such mundane acts of supererogation are not nearly as impressive as saintly and heroic acts of supererogation, Hursthouse could simply grant this point and deny that anything other than saintly and heroic actions can qualify as supererogatory. And it must be said that she would not be alone in so doing. There is much less controversy among philosophers about the supererogatory nature of saintly and heroic action than about the supererogatory nature of small acts of kindness or generosity.

I think this may be a mistake. It is true that the philosophical literature has paid significant, perhaps even disproportionate, attention to the examples of moral saints and heroes. But Terry Horgan and Mark Timmons have recently suggested that we would in fact do better to focus at least as much on the more mundane cases of supererogation. After all, very few of us are moral saints or heroes, but many of us give and receive small kindnesses, and forgive and are forgiven small mistakes. If one follows Aristotle in holding that experience is often key to the development of practical wisdom, one might reasonably conclude that we have at least as much moral and epistemic insight into the nature of the latter as the former. And once we do focus on these mundane acts of supererogation, we find that they play an important role in morally significant relationships. We often value small kindnesses and acts of generosity precisely insofar as they are not morally required. It is their very optionality that allows them to express a kind of care that can trigger a valuable sense of connection between persons without which our relationships of love and friendship would be deeply impoverished (Horgan & Timmons, 2010).

But even if this is wrong, the main problem with Hursthouse's account remains: it does not correctly identify all and only those saintly and heroic acts that are supererogatory as supererogatory. I thus conclude that Hursthouse's account of supererogation is not only highly revisionary in the way it conceives of supererogation and obligation but also fails to identify the same class of actions that we normally refer to in using that term. It is thus doubly revisionary. Given this, one might wonder whether Hursthouse would have been better off—or, at least, more straightforward—to jettison the concept of supererogation altogether. And more recently, indeed, Julia Annas has suggested that virtue ethicists should do just that. It is thus to her arguments that I now turn.

3 Annas's Rejection of Supererogation

In her recent Lindley Lecture, 'Virtue and Heroism,' Julia Annas argues that virtue ethics—or at least, the (neo)Aristotelian version of virtue ethics which she defends—is not, after all, compatible with supererogation as traditionally conceived. But this fact, she thinks, is not to be lamented. It is not to be lamented because the concept of supererogation, she argues, is not an intuitively compelling phenomenon that all ethical theories should try to accommodate. It is, instead, an attempt to account for a general moral phenomenon that is intuitively compelling: moral heroism. Urmson rightly saw that moral heroism could not be accommodated within the dominant moral theories of the time, but Annas argues that he was wrong to see supererogation as a solution to the problem.[6] Virtue ethics, on the other hand, can explain moral heroism without appeal to supererogation, thus avoiding the paradoxes associated with the concept while still accounting for the underlying moral phenomenon that motivates its defenders (Annas, 2016).

Annas's argument begins by asking us to consider not only Urmson's famous case of the solider who jumps on a grenade and thereby saves his companions from certain death but also real-life moral heroes such as Nelson Mandela and the residents of the small French town of Le Chambon. The former led a revolutionary movement for social justice, survived unjust imprisonment without succumbing to bitterness, and eventually led South Africa through a peaceful transition to a more just future. The latter, at great peril to themselves, offered shelter to Jews fleeing Nazi persecution during WWII, saving the lives of a great many people, particularly children.

Once we focus on these actual moral heroes, a point that Urmson himself recognized comes to the fore: Namely, many such individuals claim that they in fact did only what they ought to have done, and what any decent person would have done in their circumstance. But if the traditional account of supererogation is correct, then these moral heroes are mistaken: they did not merely do what they ought to have done; they did more than they were required to do. Urmson solves this problem by postulating that moral heroes are perhaps possessed of a modesty "so excessive as to appear false" (Urmson, 1958, 203). Annas finds this solution wanting. We admire moral heroes and take them as moral models; surely we should also be able to trust them as ethical authorities. But on this view, we cannot even trust what they say about their own actions.[7]

---

[6] Annas writes that "Urmson does not use the actual term supererogatory, but his article is rightly regarded as the beginning of serious attempts to focus on the issue." (Annas, 2016, 19, footnote 5) I agree with the latter half of this claim, but Annas need not have conceded the former half of it. Urmson does in fact use the term 'supererogatory action' on page 214 and uses it in the standard way in which it is employed in the debate following (Urmson, 1958, 214).

[7] This is indeed a difficult problem for defenders of supererogation; Archer and Ridge may be right to refer to it as "another paradox of supererogation" (Archer & Ridge, 2015). Nonetheless, I have argued elsewhere that this problem can be addressed, and that we can learn from moral heroes even if they are sometimes mistaken about the deontic status of their own actions (Stangl, 2020, 77–88). Archer and Ridge (2015) also offer valuable responses to the problem. Because my focus in this

So supererogation, Annas suggests, fails to make sense of moral heroism. But simply denying supererogation, she thinks, threatens to render moral heroism unintelligible as well. The most common way of denying supererogation is by claiming that there is some ethically best thing to do, which everyone is required to do, and that moral heroes are simply the ones who succeed in doing so. The implication of this claim, though, is that any of us who are not moral heroes are thereby moral failures. But this is an impossible theory to live by: it declares almost all of us failures and gives no obvious plan for how we might become like moral heroes, who are utterly unlike us.

How, then, should we make sense of moral heroism? Annas suggests that virtue ethicists, at least, ought to reject both supererogation as well as the idea that there is some ethically best action that moral heroes perform and which the rest of us count as moral failures for not performing. In place of this view, they ought to accept a picture according to which moral heroes are not people of exceptional character, or people who do what is 'ethically best' or 'more virtuous' than ordinary people do. Rather, they are ordinary persons doing the very thing that ordinary virtue demands but doing so in extraordinarily difficult circumstances. As she writes,

> They [moral heroes] are acting well and responding well to a situation they haven't chosen, and wouldn't choose, to be in. They illustrate for us particularly well the situation of people of ordinary virtue who are faced with an extreme situation and do the best they can, in circumstances where most people don't do what they do, and so don't even rise to the level of ordinary virtue. Heroes... challenge us not to become outstanding, but to stick by the level of virtuous development we have when the situation becomes challenging (Annas, 2016, 15–16).

This account of moral heroism allows us to do justice to the self-reported verdict of the moral heroes that they were only doing what was required; ordinary virtue in fact does demand that one shelter victims of violence. But Annas thinks it also allows us to make sense of striving to be like moral heroes. In so striving, we are not aiming to become more virtuous than we are, or to live radically different lives. We are instead aiming to strengthen the stability of our own virtue, lived in our own concrete lives, with the hope that it would sustain us even if difficult circumstances were to come.

Annas thus ends up with an account of moral heroism that is in some ways quite like the revisionary account of supererogation that Hursthouse defends. Hursthouse claims that supererogation just is doing what virtue requires in difficult circumstances. Annas claims that moral heroism is doing what virtue requires in difficult circumstances that the morally heroic haven't chosen, and that neither they nor we would choose. For the reasons given above, I am inclined to think Annas's choice to defend the latter as an account of moral heroism is more perspicuous and defensible than Hursthouse's choice to defend the former as an account of supererogation. But just as I argued that Hursthouse's account could not capture all and only those instances of supererogation that we recognize as supererogatory, I will now argue that Annas's account cannot capture all those instances of moral heroism that we

---

essay is on Annas's positive account of moral heroism and whether it can account for our intuitions without appealing to the concept of supererogation, I will not repeat these arguments here.

recognize as moral heroism. It thus fails in its ambition to eliminate the motivation for giving an account of supererogation.

Before proceeding directly to that argument, however, let me say something briefly about how Annas's account of moral heroism relates to Urmson's discussion of more mundane acts of supererogation. Like Hursthouse's account of supererogation, Annas's account of moral heroism is unlikely to explain anything about the moral goodness of being a little more generous than required, for example, or doing a small kindness for a stranger. For such actions will almost never be objectively difficult in the way heroism requires, nor do they happen in circumstances that neither we nor moral heroes would choose. Indeed, just the opposite. They both happen in the ordinary course of our lives and in contributing to the good of intimate relationships are part of what make those lives choice worthy. Or at least, drawing on the work of Horgan and Timmons, so I argued.

But while this may present a problem for Hursthouse's ambition to give an account of supererogation, it is less clear if it could present a problem for Annas's argumentative aims. Annas aims to undermine the attraction of supererogation by giving an account of moral heroism that renders it unnecessary. There is a stronger and weaker way of reading this aim. On the stronger interpretation of her aim, Annas's account of moral heroism is intended to undermine the case for supererogation, full stop. On the weaker interpretation, Annas's account of moral heroism is intended to undermine any case for supererogation that bases itself on the phenomenon of moral heroism. Given that her focus is resolutely on the phenomenon of moral heroism, Annas never directly confronts the question of whether supererogation is necessary to explain any other kinds of moral goodness. For this reason, it may be most charitable to interpret her argument along the lines suggested by the weaker reading.

Therefore, let us turn our focus squarely to the question of whether Annas's account of moral heroism does adequately account for the phenomenon without appeal to something like supererogation. It seems clear to me that Annas has given an insightful account of some kinds of moral heroism. Annas is certainly right to draw our attention to moral heroes who face difficult circumstances that neither we nor they would choose, and who manage to do what virtue requires, even when so many others fail. I think she is also right to single out many of those who rescued Jews escaping persecution during WWII as illustrative of just such moral heroism. It took immense bravery, compassion, and wisdom to have saved the lives of refugees in such circumstances. But it might well have been a failure of bravery, compassion, and wisdom had one turned away the refugee at the door. Such rescuers are arguably right that they did only what virtue required of them. And if they are right about that, then such instances of moral heroism neither require nor are compatible with supererogation for their explanation.

The question I want to press is whether all moral heroism is of this kind. I do not think it is. Consider again Urmson's two heroic doctors. The doctor who treats her own patients in the face of an outbreak of plague fits Annas's picture of moral heroism very well. She finds herself in a situation which neither she nor we would choose, and bravely and compassionately continues with her work. But were she to abandon her patients, she would fail due to her cowardice or selfishness. Here, Annas seems right to say this is a case of living up to the demands of virtue under difficult circumstances which no one would choose and that no account of supererogation is required to make sense of it. (Of course, Urmson agrees on this point.)

But what of the second doctor, who volunteers to treat patients in a foreign city beset by plague? Does she also fit Annas's picture of moral heroism? I don't think she does. Consider an updated version of the case, with some more details filled in: Imagine a doctor has been treating patients for the last ten years and doing so virtuously. She is a competent and compassionate health provider. She treats her patients justly—never overcharging them or discriminating against anyone—and even offers a certain amount of free care to the uninsured in her community. One day, we can imagine, she begins to wonder if she might do more. She realizes that, as much as she enjoys treating her patients and as much as they benefit from her care, there is no shortage of qualified medical professionals in her area. Most of her patients could easily find that care elsewhere. But many people are not so lucky. They live in developing countries, far from urban centers, without access to reliable care of any kind. She begins to wonder whether she might join Doctors without Borders, at least for a time, to serve these patients. Such a life would involve a good deal of sacrifice, perhaps even physical hardship, but she would do immense good and find that good meaningful and fulfilling. She joins Doctors without Borders and fulfills her aims by treating many such patients in the years to follow.[8]

I think such a doctor could reasonably be considered a moral hero. But she doesn't fit Annas's model of moral heroism. While serving with Doctor without Borders she might well find herself in difficult situations, situations in which many people's virtues might fail them. It is not everyone who can be as generous, patient, and determined as such work requires. But she will not find herself in a situation she hasn't chosen, and wouldn't choose, to be in. Quite the opposite. She has freely chosen the life of a doctor working in difficult circumstances, treating patients to whom she had no standing obligation. She could easily have continued the life she had, and none of us would have judged her vicious or morally deficient. Had our doctor never treated patients during her service with Doctors without Borders, she would not have failed in virtue. She would have simply continued her life as a suburban doctor, which was already a life characterized by compassion, wisdom, and justice. This is a kind of moral heroism, I think, that cannot be adequately captured without appeal to the traditional concept of supererogation: by acting more generously and compassionately than virtue requires, our doctor does what is morally admirable even though refraining from so acting would have been no failure in virtue and thus not wrong.

In suggesting that such cases of moral heroism are genuine, we need not remain in the theoretical realm. Annas is quite right to push us to look at real-life moral exemplars. But when we do that, we find that moral heroes come in more than one variety. Some are exactly as Annas says: ordinary individuals who find themselves in extraordinarily demanding circumstances that neither they nor we would choose.

[8] This is a version of a case I discuss in (Stangl, 2020, 32–34) in a different context.

But not all. We might think here of Paul Farmer, the founder of Partners in Health.[9] His life was not completely unlike our imaginary doctor above. Having graduated from Duke University and Harvard Medical School, he could easily have lived a life as a prosperous suburban doctor, and none of us would have judged him vicious or morally deficient. But he instead served some of the poorest, most underserved people in the world, regardless of their ability to pay and even though a failure to do so would not have been unjust, disloyal, or vicious in some other way. For that, many rightly regard him as a moral hero. His heroism is of a kind best explained by appeal to the concept of supererogation.

As such, it is also a kind of heroism that may inspire us in a different way than the model which Annas explores. For Annas, it is a mistake to think of moral heroes as inspiring us to "to try, unrealistically, to be exceptional, still less to seek out situations in which to act in exceptional ways. Rather we are being inspired to remain stably at our level of virtue" (Annas, 2016, 17). But moral heroes such as Paul Farmer *can* and sometimes do inspire us to seek out new and even difficult situations in which we can and do act (even a bit) more generously and bravely than we do now, or than anyone else could rightly demand.

4 Conclusion

What the above considerations suggest is that there are still some intuitions about moral heroism, and moral goodness more generally, that resist capture without appeal to something like the traditional concept of supererogation. Virtue ethicists thus have some reason to try to accommodate it. And I believe that they can. On the view I defend, virtue-ethicists ought to embrace the following account of supererogation:

> An action is supererogatory *iff* it is overall virtuous and either (a) the omission of an overall virtuous action in that situation would not be overall vicious or (b) there is some overall virtuous action that is less virtuous than it and whose performance in its place would not be overall vicious. (Stangl, 2016, 2020)

I have argued elsewhere that this account is both intuitively compelling and fully consistent with a virtue-ethical approach to ethics (Stangl, 2016, 2020).

Here, I will add only that this account easily explains our intuitions about the moral difference between Urmson's two doctors. The first doctor, who treats his own patients suffering from plague, acts virtuously and well. But were he not to treat his patients, he would act unjustly and disloyally toward his patients and therefore do something overall vicious. Treating his patients is therefore good and virtuous, but not supererogatory. The second doctor, who treats plague victims in a distant city, also acts virtuously and well. But failing to treat those patients would be neither disloyal nor unjust nor vicious in any other way. Treating her patients is therefore not only good and virtuous but also supererogatory. This account thus vindicates

9 For an account of his life and work, see Kidder (2003).

key intuitions about supererogation and moral heroism that cannot be vindicated on Hursthouse's or Annas's accounts.

Of course, this alone doesn't demonstrate that virtue ethics, all things considered, should embrace this or any other account of supererogation. But if the argument of this essay is correct there will be a genuine cost to abandoning supererogation altogether. There is thus good reason why attempts to construct and defend virtue-ethical accounts of supererogation have lately proliferated.

References

Annas, J. (2016). *Virtue and heroism.* Lindley Lecture, The University of Kansas. https://kuscholarworks.ku.edu/handle/1808/19868. Accessed 1 August 2022.

Anscombe, G. E. M. (1958). Modern moral philosophy. *Philosophy, 33*(124), 1–19.

Archer, A., & Ridge, M. (2015). The heroism paradox: Another paradox of supererogation. *Philosophical Studies, 172*, 1575–1592.

Crisp, R. (2013). Supererogation and virtue. In M. Timmons (Ed.), *Oxford studies in normative ethics* (Vol. III, pp. 13–34). Oxford University Press.

Curzer, H. (2012). *Aristotle and the virtues.* Oxford University Press.

Driver, J. (2006). Virtue theory. In J. Dreier (Ed.), *Contemporary debates in moral theory* (pp. 113–123). Blackwell Publishing.

Foot, P. (1978). Virtues and vices. In P. Foot (Ed.), *Virtues and vices: And other essays in moral philosophy* (pp. 1–18). Blackwell.

Heyd, D. (1982). *Supererogation: Its status in ethical theory.* Cambridge University Press.

Heyd, D. (2015). Can virtue ethics account for supererogation? *Royal Institute of Philosophy Supplement, 77*, 25–47 (edited by C. Cowley).

Horgan, T., & Timmons, M. (2010). Untying a knot from the inside out: Reflections on the 'Paradox' of supererogation. *Social Philosophy and Policy, 27*(2), 29–63.

Hursthouse, R. (1999). *On virtue ethics.* Oxford University Press.

Hursthouse, R. (2006). Are virtues the proper starting point for morality? In J. Dreier (Ed.), *Contemporary debates in moral theory* (pp. 99–112). Blackwell Publishing.

Kawall, J. (2009). Virtue theory, ideal observers, and the supererogatory. *Philosophical Studies, 146*(2), 179–196.

Kidder, T. (2003). *Mountains beyond mountains.* Random House.

Slote, M. (2007). Famine, affluence, and virtue. In R. L. Walker & P. J. Ivanhoe (Eds.), *Working virtue* (pp. 279–296). Clarendon Press.

Stangl, R. (2016). Neo-Aristotelian supererogation. *Ethics, 126*(2), 339–365.

Stangl, R. (2020). *Neither heroes nor saints: Ordinary virtue, extraordinary virtue, and self-cultivation.* Oxford University Press.

Swanton, C. (2003). *Virtue ethics: A pluralistic view.* Oxford University Press.

Urmson, J. O. (1958). Saints and heroes. In A. I. Melden (Ed.), *Essays in moral philosophy* (pp. 198–216). University of Washington Press.

Urmson, J. O. (1988). Hare on intuitive moral thinking. In D. Seanor & N. Fotion (Eds.), *Hare and critics: essays on "Moral Thinking"* (pp. 161–169). Clarendon Press.

Zagzebski, L. (2017). *Exemplarist moral theory.* Oxford University Press.

The Expected, the Contra-Expected, the Supererogatory, and the Suberogatory

Terry Horgan and Mark Timmons

Abstract This chapter defends the claim that the space of human actions is really partitionable into five non-overlapping deontic categories: the three commonly recognized ones (the obligatory, the impermissible or wrong, and the optional), plus two additional ones labeled the expected and the contra-expected. These latter categories are typically not recognized in ethical theorizing but nonetheless they are part of everyday moral experience. The defense of these additional deontic categories appeals, via inference to the best explanation, partly to phenomenological considerations and partly to moral-normative considerations. It is further argued that this five-way partition of the deontic realm helps explain why the hybrid categories of the supererogatory and suberogatory are deontically asymmetrical.

Keywords Deontic categories · The expected · The contra-expected · The supererogatory · The suberogatory

1 Introduction

In moral philosophy, it is commonly assumed that each actual or possible human action falls under one of three mutually exclusive and jointly exhaustive deontic categories: the morally obligatory, the morally impermissible, and the morally optional.[1] But we maintain, and have recently argued in Horgan and Timmons (2022), that the space of human actions is really partitionable into five non-overlapping deontic categories: the three commonly recognized ones, plus two additional ones that we have called the *expected* and the *offensive*. (For a reason explained below, here we will

[1] The deontic category of the morally permissible encompasses both the morally obligatory and the morally optional.

T. Horgan · M. Timmons (✉)
University of Arizona, Tucson, USA
e-mail: mtimmons@arizona.edu

T. Horgan
e-mail: thorgan@email.arizona.edu

D. Heyd (ed.), *Handbook of Supererogation*,
https://doi.org/10.1007/978-981-99-3633-5_8

deploy the label 'contra-expected' for the latter category, rather than 'offensive'.[2]) Our argument primarily concerns situations in which one person has conferred some benefit upon another person when doing so is morally optional, and focuses on the deontic status of acts of gratitude expression in such situations. The argument is partly phenomenological and partly moral-normative.

The phenomenological component appeals to typical moral experiences concerning these kinds of situations, and to the moral judgments to which such experiences give rise. We argue non-demonstratively and abductively—i.e., via "inference to the best explanation"—that these moral experiences and moral judgments, including their experienced interconnections with one another, are best explained by the empirical psychological hypothesis that such experiences and judgments represent acts of gratitude expression as being neither morally obligatory nor morally optional but rather as falling under the distinct deontic category we call "morally expected."

The moral-normative component appeals primarily to the intrinsic moral value of certain non-transactional forms of human interaction—gratitude expression in particular—as a rationale for the deontic categories of the expected and (what we now are calling) the contra-expected. Here too the argument is non-demonstrative; it is an application of "wide reflective equilibrium"—the analog, for moral theorizing, of inference to the best explanation as a non-demonstrative form of reasoning about matters non-moral and empirical. The claim is that any normative-ethical theory that incorporates these two additional deontic categories will better accommodate certain commonly held, reflectively persistent, moral judgments than will a version of the given normative-ethical theory that eschews these two deontic categories.[3]

Among those in moral philosophy who acknowledge the moral categories of the supererogatory and the suberogatory, it is commonly claimed that these are hybrid notions that on one hand are partly deontic but on the other hand are partly axiological and/or partly aretaic.[4] We ourselves acknowledge these two moral categories, and we concur that they each have both deontic and axiological/aretaic aspects. But our own view about these two categories is novel and distinctive regarding their respective deontic aspects, in a way that invokes our five-way partition of deontic categories. (The term 'offensive' can be deployed in essentially (if not exactly) the same way as

[2] Under our proposed five-way partition, the category of the morally permissible would encompass not only the morally obligatory and the morally optional but also the morally expected. Moreover, it would become useful and appropriate to distinguish between *weak* moral permissibility, which also encompasses the morally contra-expected, and *strong* moral permissibility, which does not.

[3] The phenomenological component of our argument feeds directly into the moral-normative component. Suppose that inference-to-the-best-explanation methodology does indeed favor the empirical psychological claim that various commonly held, phenomenology-based, reflectively persistent, moral judgments represent certain actual or potential acts as having the deontic status of being morally expected (rather than representing these acts as being either morally required or morally optional). Then for any generic normative-ethical theory T, wide-reflective-equilibrium methodology is apt to favor a version of T that incorporates the deontic categories of the morally expected and the morally contra-expected over a version that does not incorporate them.

[4] See, e.g., Heyd (1982, Chap. 6), McNamara (2011a, 2011b) and Grigoletto (2019).

'suberogatory'.[5] That is why, since we ourselves regard the suberogatory as a hybrid category, for clarity we here deploy the coined term 'contra-expected' for the deontic category in question.)

The plan for this paper is as follows. First, we will briefly summarize the phenomenological component of our argument, in Horgan and Timmons (2022), that the deontic categories of the expected and the contra-expected are frequently operative in people's actual moral experiences and actual moral judgments. The pertinent considerations apply to numerous real-life situations other than ones involving gratitude expression. Second, we will set forth our recommended account of the hybrid categories of the supererogatory and the suberogatory—an account that treats the supererogatory as morally optional deontically and treats the suberogatory as morally contra-expected deontically (despite the symmetric nature of the five deontic categories themselves). Third, will set forth a moral-normative rationale for the deontic categories of the expected and the contra-expected, in a way that extends and generalizes the moral-normative rationale we offered in Horgan and Timmons (2022). Fourth, we will explain why, given our recommended five-way partition of deontic categories, the supererogatory and the suberogatory are deontically asymmetrical.

We intend the claims we make about the categories we discuss to apply to *prototypical* instances of these categories. This leaves open the possibility of exceptional instances, of one sort or another.

2 The Expected and the Contra-Expected

One might initially think that the very meaning of the words 'deontic' and 'deontological' precludes the possibility of mutually exclusive categories that are *genuinely* deontic, beyond the familiar categories of the morally obligatory, the morally optional, and the morally wrong or impermissible. But the terms 'deontic' and 'deontological' derive etymologically from the Greek 'deon', meaning "that which is binding." Thus, the underlying concept of moral *constraint*—moral 'bindingness'—is quasi-metaphorical, in a way that opens conceptual space for the possibility of a type of moral constraint that is weaker than outright moral obligation. Ordinary moral thought and common moral practice, we maintain, actually recognize two different kinds of prototypical moral constraint: moral obligation (with its symmetrical contrary, moral wrongness, or impermissibility), and moral expectation (with its symmetrical contrary, moral contra-expectation).[6] Ethical theorizing has largely

[5] Driver (1992) introduced the term 'suberogatory' into ethics. Chisholm (1963) introduced the term 'offence' into ethics that is sometimes taken to refer to the same category as the suberogatory. See, e.g., Mellema (1991).

[6] Moral obligation and moral impermissibility are symmetrical contraries in the following sense: it is impermissible not to perform an obligatory act, and it is obligatory not to perform an impermissible act. Likewise, for the putative categories of the morally expected and the morally contra-expected: not performing an expected act is contra-expected, and not performing a contra-expected act is expected.

failed to appreciate this fact, and instead has allowed the technical philosophical terms 'deontic' and 'deontological' to ossify in such a way that the very idea of a deontic category of moral expectation is apt to be regarded as a category mistake. But we recommend resisting this terminological tyranny. Theoretical normative ethics should acknowledge a pair of moral categories that are genuinely deontic in a non-ossified sense, that long have been an important element of real-life ethics, but that hitherto have been largely overlooked and only dimly appreciated in philosophy: the symmetrically contrary categories of the morally expected and the morally contra-expected.

So our contention, which might appear radically novel relative to received thinking in contemporary normative ethics but which we claim merely thematizes long-inherent elements of everyday moral thought, is that there are really five distinct non-overlapping deontic moral categories: the morally wrong (or impermissible), the morally contra-expected, the morally optional, the morally expected, and the morally obligatory. All these pertain to actual and potential actions, and all except moral optionality involve constraint on actions of one kind or another, as indicated in the following diagram.[7]

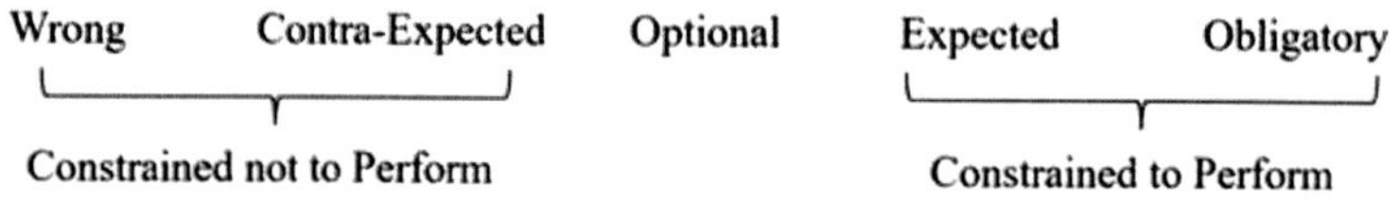

Our argument in Horgan and Timmons (2022), for acknowledging and embracing the deontic categories of the expected and the contra-expected, focuses on gratitude expression. The phenomenological component of the argument harnesses the abductive claim that typical moral experiences when contemplating whether someone who has received a morally optional benefit from another person should express gratitude, and certain widely held judgments that arise from such moral phenomenology, are best explained by the empirical psychological hypothesis that these experiences and

[7] These deontic categories can apply not only to a coarse-grained act A (i.e., an act-determinable) but also—and somewhat independently—to various fine-grained acts A_1, A_2, …, A_n that are alternative determinates of A. For example, consider being obligated to convey some bad news to a friend. Suppose one can do so either by emailing them, calling them, or meeting with them in person (an example from McNamara 2011a). Each of these acts is a determinate of one's act-determinable obligation to convey the bad news.

In some cases, it will be more natural to express the same kind of point somewhat differently. Rather than invoking the determinable/determinate distinction, distinguish instead between the deontic status of an act A itself on one hand, and on the other hand, the varying deontic statuses of different alternative *ways* of performing A. Suppose, for instance, that one has agreed to meet with a student to go over a draft paper for the class one is teaching. In discussing the paper with the student, one might, e.g., do so in an unfriendly, inpatient way that expresses one's dislike for having to meet the student as one shuffles through a pile of papers on one's desk, not making eye contact with the student, answering the student's questions in a curt manner, and all the while frowning. On the other hand, in discussing the student's paper, one might do so in a friendly way that expresses one's interest and encouragement as one makes appropriate eye contact and takes time to thoroughly answer questions and make suggestions for improving the paper.

judgments deontically categorize gratitude expression in such a situation as being neither morally required nor morally optional, but rather as being morally expected. Briefly summarized, the reasoning goes as follows. Suppose that person A has freely and willingly conferred a benefit on person B, in circumstances where doing so is morally optional. Several kinds of phenomenology-based moral judgment typically arise about such a case, strongly and with reflective persistence. First, it is commonly judged that A has no right to receive an expression of gratitude from B. Second, it is commonly judged that B nonetheless does 'owe' it to A, in some sense, to express gratitude—even though this is not a kind of 'debt' that confers upon A the moral right to receive such a response from B. Third, it is commonly judged that B would be morally blameworthy in failing to express gratitude, especially if the benefit B has received is significant—and that B would be morally blameworthy whether or not that failure would have any psychological effect on A. The simplest and most natural explanation of this combination of judgments, and of the underlying moral phenomenology that spawns them, is that gratitude expression is here being regarded neither as morally obligatory nor as morally optional, but rather as morally expected. Being regarded deontically as not obligatory explains the judgment that A has no moral right to receive gratitude expression from B; moral rights are correlative to outright moral obligations. And being regarded deontically as morally expected, and thus as morally constraining B, explains not only the judgment that B would be morally blameworthy for failing to express gratitude to A but also the judgment that B would be morally blameworthy whether or not such a failure would have—or would be likely to have—any negative psychological effect on A.[8]

Numerous other kinds of act, in addition to expressions of gratitude, are subject to pertinently similar kinds of moral assessment—a fact that it is best explained by supposing that such assessments too are deploying the deontic category of the morally expected. Suppose, for example, that person C is a vigorous young adult who is sitting in a seat in a crowded bus in which all other seats are occupied by older, frail-looking people, and that a little old person D slowly climbs onto the bus. One's moral phenomenology when contemplating this situation is as-of the following: (i) D has no moral right to C's seat, hence (ii) C is not morally *obligated* to offer C's seat to D, yet (iii) C is morally *expected* to do so in such circumstances, hence (iv) C is morally constrained by this expectation, hence (v.1) C would be morally blameworthy for failing to offer the seat to D, and moreover (v.2) C would be thus blameworthy *by virtue of* violating an operative moral constraint upon C.

Other plausible examples of acts that are morally expected or morally contra-expected are easily multiplied; they are legion in ordinary life. Consider, for example, Cheshire Calhoun's conception of what she calls 'common decencies'. These are actions "that we expect any minimally well-formed agent will *elect* to do for others

[8] Perhaps some acts that deontically are morally optional are blameworthy nonetheless, by virtue of certain morally negative axiological and/or aretaic features they instantiate—e.g., willfully insulting another person by saying something about the person that is actually true. But be that as it may, one's intuitive moral judgment that B would be morally blameworthy for failing to express gratitude to A is different; one judges that B would be blameworthy by virtue of knowingly and willingly violating an operative moral constraint.

absent any requirement to do so" (2016, 107). Besides the bus seat case, common decencies include such kindnesses as giving directions, permitting drivers to change lanes, letting someone go ahead of you in line, volunteering to take minutes at a meeting where there is an expectation that attendees will take turns doing so, and forgiving a shopper who apologizes for carelessly dropping a light package of flour on your foot. Doing these things is not morally obligatory, nor are they supererogatory, but one is morally constrained to do them. As Calhoun rightly points out, in the philosophical literature in ethics such actions "seem to occupy a shadowy territory between the obligatory and the supererogatory" (2016, 105), not recognized for the important role they have in our everyday lives nor, consequently, for the unique status they deserve in ethical theory. In some places in her article, Calhoun refers to these types of action as 'normative expectations', though she never explicitly says that moral expectedness is a distinct deontic category.

Edna Ullmann-Margalit writes about what she refers to as acts of 'considerateness'; these overlap with Calhoun's common decencies. According to Ullmann-Margalit, a considerate act in the presence of strangers is one that "benefits someone else at near-zero costs to oneself" (2011, 210).[9] Two of her many examples include: entering a department store and holding the heavy door open for someone entering behind you, and walking to your parked car in a full parking lot noticing a driver who is looking for a spot and motioning to them to follow you as you hasten your pace. As she points out, such actions are not supererogatory, nor are they obligatory, though they are owed to others out of recognition and respect. In one place, Ullmann-Margalit refers to there being a 'normative expectation' that one does the considerate thing, though like Calhoun she does not explicitly identify expectation as a distinct deontic category. Our conception of expectation includes such acts of considerateness but is more expansive in its inclusion of acts of gratitude that may involve more than doing something at near-zero cost to oneself.

3 The Supererogatory and the Suberogatory

Paradigmatic supererogatory acts have both deontic features and axiological/aretaic features. Deontically, we submit, a paradigmatically supererogatory act is morally optional: the agent is not constrained either by moral obligation or by moral expectation to perform it, and the agent also is not constrained either by moral impermissibility or by moral contra-expectation *not* to perform it. Axiologically, a paradigmatic supererogatory act is morally good in some significant respect and is not morally bad in other respects; moreover, this positive moral value constitutes the agent's motivating reason for performing the act. And aretaically, a paradigmatic supererogatory act is an expression of the agent's morally admirable 'quality of will' vis-a-vis those persons who stand to benefit from the morally good features of the act.

[9] Ullmann-Margalit also discusses the notion of considerateness as it applies to relationships within a family.

Paradigmatic suberogatory acts too have both deontic features and axiological/aretaic features. The pertinent axiological features are roughly the inverse of those exhibited by paradigmatic supererogatoriness: a paradigmatic suberogatory act is morally *bad* in some way and is not morally good in other ways; moreover, the agent is well aware of this negative moral value, and nonetheless deliberately performs the act anyway. Similarly, the pertinent aretaic aspect is roughly the inverse of that for paradigmatic supererogatoriness: a paradigmatic suberogatory act is an expression of the agent's morally *deficient* 'quality of will'—perhaps indifference—vis-a-vis those persons who stand to be harmed by the morally bad features of the act. But deontically, we submit, there is a striking and important asymmetry between the categories of the supererogatory and the suberogatory: although paradigmatic supererogatory acts are deontically *optional*, paradigmatic suberogatory acts are deontically *contra-expected*.[10,11]

4 The Moral-Normative Rationale for the Morally Expected and the Morally Contra-Expected

A familiar kind of moral-normative rationale has often been given in moral philosophy for classifying some kinds of action (or omission) as morally obligatory or as morally impermissible. We will now offer a brief articulation of that rationale, and we will then urge that there is a very similar, and quite compelling, moral-normative rationale for classifying certain other kinds of action (or omission) as morally expected or morally contra-expected.

For ease of exposition, the discussion to follow will mainly focus on actions (and not omissions), benefits (and not harms), the morally obligatory (and not the morally

[10] Like the deontic categories discussed above, the categories of the supererogatory and the suberogatory can apply not only to acts themselves but also—and somewhat independently—to various alternative determinates of a single determinable act, or to various different ways of performing a single act. (Of course, neither an act nor a manner of acting qualifies as supererogatory unless it is suitably motivated; similarly for suberogatoriness.) For instance, a determinable act A that itself is morally obligatory (and hence is not supererogatory, whatever the agent's motivation for it might be) might be performed by performing an act A_i that is a supererogatory determinate of A—where A_i itself is "above and beyond the call of duty."

[11] A by-product of our treatment of the categories of the expected, the contra-expected, the supererogatory, and the suberogatory is that there are several distinct kinds of moral reasons for performing or not performing an action. A *requiring* reason renders an action (or omission) morally obligatory (pro tanto) and renders its omission (or its performance) morally forbidden (pro tanto). A *favoring* reason, applicable to an action that is morally optional deontically, renders the action morally good (pro tanto). A *non-requiring constraining* reason renders an action (or omission) morally expected (pro tanto) and renders its omission (or its performance) contra-expected (pro tanto). (In accordance with the two preceding notes, these different kinds of moral reason also can apply, somewhat independently, to different determinates of a determinable action with a given deontic status, or to the specific way in which an action is performed.) Distinguishing multiple kinds of moral reason is theoretically important, *inter alia*, for resolving certain puzzles in moral philosophy like the putative 'paradox' of supererogation; cf. Horgan and Timmons (2010).

impermissible), and the morally expected (and not the morally contra-expected). But because moral obligation and moral impermissibility are symmetrical contraries, and likewise moral expectation and moral contra-expectation, our remarks will extend straightforwardly to omissions, harms, moral impermissibility, and moral contra-expectation.

When a particular kind of action A is regarded in a society as morally obligatory, typically certain practices will be prevalent in the society that serve to implement, maintain, and enforce general conformity to the requirement of A-performance. We will these *conformity-maintenance practices*. Familiar conformity-maintenance practices include overt moral discourse asserting that A is morally obligatory and enjoining people to conform to the obligation of performing A, forming (and often overtly expressing) negative reactive attitudes directed at those who flout the requirement of A-performance (including self-directed negative reactive attitudes, when one flouts the requirement oneself), and various kinds of sanctions (sometimes including legal sanctions) for violations of the requirement to perform A.

The conformity-maintenance practices associated with a society's regarding A as morally obligatory will tend to impinge upon the autonomy of individual agents in the society. The impingement will be largely a matter of two kinds of actual or potential cost to the individual: costs of performing A, in various actual or potential situations where not performing A would benefit the individual, and costs of not performing A, given the consequences that the individual would be likely to incur by virtue of the prevalent conformity-maintenance practices. We will call these *conformity costs* and *non-conformity costs*, respectively.

So a key question, when considering whether or not it would be morally legitimate for a society to treat a specific kind of action A as morally obligatory, is the following. When, and why, would the overall social benefits of treating A as morally obligatory be sufficient to outweigh the impingement on individual autonomy—in the form of conformity costs and non-conformity costs—that would result from the conformity-maintenance practices associated with the society's treating A as morally obligatory? Or (to rephrase this question in slogan form): When, and why, would a society's treating A as morally obligatory satisfy *the autonomy/obligation tradeoff*?

It takes quite a lot, in terms of overall social benefits, in order for a candidate moral obligation to satisfy the autonomy/obligation tradeoff. One legitimate kind of moral obligation involves justice: people are morally obligated to treat each other fairly. Another legitimate kind involves refraining from certain actions—for instance, reckless driving—that risk serious and undeserved harm to others. The bar is high, and should be, for meeting the autonomy/obligation tradeoff—because, otherwise the impingements on individual autonomy that would result from operative conformity-maintenance practices would constitute morally illegitimate *encroachments* on autonomy.

This familiar line of reasoning makes good sense, vis-a-vis the deontic category of moral obligation. But it also can be generalized and extended in a quite straightforward way, to encompass the distinct deontic category of moral expectation. The crucial points are as follows. First, there will be two distinct kinds of conformity-maintenance practices in a society—associated, respectively, with acts regarded in the society as morally obligatory, and acts regarded as morally expected. (We will call these *obligation-strength* conformity-maintenance practices and *expectation-strength* conformity-maintenance practices, respectively.) Second, expectation-strength conformity-maintenance practices will be, in general, *less intrusive* on individual autonomy—in terms of conformity costs and non-conformity costs to individuals—than obligation-strength conformity-maintenance practices. (This difference is partly constitutive of what counts for a society as morally expected, as opposed to what counts as morally obligatory.) Third, there is an autonomy/expectation tradeoff, analogous to the autonomy/obligation tradeoff. Fourth, because the conformity costs and non-conformity costs imposed by expectation-strength conformity-maintenance practices are lower than those imposed by obligation-strength conformity-maintenance practices, certain actions that fail to satisfy the autonomy/obligation tradeoff *do* satisfy the autonomy/expectation tradeoff; hence, a society can legitimately treat these latter acts as morally expected, even though treating them as morally obligatory would constitute an illegitimate encroachment on individual autonomy.

We submit that there are numerous kinds of action that qualify as morally expected, in terms of the moral-normative rationale described in the previous paragraph. In particular, the many kinds of action that are commonly classified under the rubric of *common decency* all belong to this deontic category.[12] Some pertinent facts about such actions are the following. First, widespread conformity to common decency substantially benefits society at large, in a variety of ways. (For instance, it makes everyday interactions with others much more pleasant, in general, than would be the case otherwise; it conventionally conveys respect toward others; it often enhances one's own autonomy by enabling reliance on others' helpfulness; etc.) Second, these kinds of social benefits, although substantial, do not suffice to qualify acts of common decency as morally obligatory. However, third, such social benefits do suffice to qualify these acts as morally expected—because the associated conformity and non-conformity costs to individuals are considerably lower than the conformity and non-conformity costs of moral obligations. (Often enough, the cost of conformity is minimal or null; and typically the cost of non-conformity is little more than having to endure negative reactive attitudes—perhaps including one's own—toward oneself.[13])

The upshot is that the familiar kind of moral-normative rationale for the moral obligatoriness of certain kinds of action does indeed generalize, to yield a similar moral-normative rationale for the moral expectedness of certain other kinds of action.

---

[12] Perhaps the category of the morally expected includes other kinds of acts too, besides those that typically would be regarded as acts of common decency. We leave this question open here. We certainly hold that gratitude expression is morally expected, whether or not it is aptly regarded as a matter of mere common decency. For an illuminating discussion of common decency that is very close in spirit to our own view, see Calhoun (2004) [2016] mentioned above in Sect. 2.

[13] It bears emphasis that for most people, the prospect of being the brunt of negative reactive attitudes (including toward themselves) does tend to motivate them to avoid non-conformity to moral expectations—even though non-conformity costs under expectation-strength conformity-maintenance practices normally are considerably less severe than non-conformity costs under obligation-strength conformity-maintenance practices.

Generically, there is an *autonomy/moral-constraint tradeoff*, and this genus has two distinct species: the autonomy/obligation tradeoff and the autonomy/expectation tradeoff. Acts of common decency do not satisfy the autonomy/obligation tradeoff, but they do satisfy the autonomy/expectation tradeoff. Hence, such acts belong to the deontic category of the morally expected.

5 The Deontic Asymmetry of the Supererogatory and the Suberogatory

As remarked earlier, among those in moral philosophy who acknowledge the categories of the supererogatory and the suberogatory, it is commonly agreed—and we ourselves concur—that these are hybrid categories. In addition to deontic aspects, they have axiological, aretaic, and epistemic aspects. In order for an action to be paradigmatically supererogatory it must have positive moral value with respect to its actual or likely effects on other people, the agent must know (or justifiably believe) that this is so, and the agent's motivation for performing the action must reflect a 'quality of will' involving concern for others' well-being. And in order for an action to be paradigmatically suberogatory, it must have negative moral value with respect to its actual or likely effects on other people, the agent must know (or justifiably believe) that this is so, and the agent's motivation for performing the action must reflect a morally objectionable quality of will—e.g., indifference—regarding others' well-being.

For present purposes, we need not delve into the specifics of such non-deontic features. It will be useful too to bundle the non-deontic aspects of the supererogatory under a single label, and likewise for the suberogatory. We will call an action *beneficent* if has the non-deontic features of paradigmatic supererogatory actions, and *contra-beneficent* if it has the non-deontic features of suberogatory actions.

Regarding the deontic aspects of the supererogatory and the suberogatory, this much is uncontroversial: supererogatory actions are not morally obligatory, and suberogatory actions are not morally impermissible. Accordingly, if one embraces the common assumption in moral philosophy that any action belongs to just one of the three mutually exclusive deontic categories of the morally obligatory, the morally impermissible, and the morally optional, then one if forced to conclude that deontically, both supererogatory actions and suberogatory actions are morally optional.

However, given the two additional deontic categories of the morally expected and the morally contra-expected, not only is one no longer forced to regard both the supererogatory and the suberogatory as morally optional, but also there is no longer even any prima facie presumption that this is so. On the contrary, the natural thing to say about them is this: paradigmatic supererogatory actions are those that are *beneficent and morally optional*, whereas paradigmatic suberogatory actions are those that are *contra-beneficent and morally contra-expected.*

One important reason why this is the natural construal of the supererogatory and the suberogatory is that it neatly explains a striking asymmetry about praise and blame, concerning certain actions that are neither morally obligatory nor morally impermissible—viz., the fact that in everyday moral thinking, non-obligatory actions that are merely commonly decent are considered *not praiseworthy*, whereas actions that are not impermissible and yet fail to be commonly decent, and are knowingly and willingly performed anyway, are considered *blameworthy*. The explanation, involving matters deontic, goes as follows. Performing an action that one is morally constrained to perform is not morally praiseworthy, because it merely conforms with a moral constraint; but (contra-beneficently) performing an action that one is morally constrained *not* to perform is morally blameworthy, because it violates a moral constraint. So, since people are morally constrained to act in accordance with common decency—albeit constrained by moral expectation, rather than by moral obligation—paradigmatic acts of common decency are not worthy of praise. On the other hand, since people are morally constrained not to perform acts which, despite not being morally impermissible, nevertheless fail to accord with common decency, contra-beneficent failures of common decency are worthy of blame.

According to everyday moral thinking, it is also the case that even though paradigmatic suberogatory actions are blameworthy, people do not have a *moral right* not to bear the brunt of suberogatory actions by others. This too is neatly explainable by the construal of paradigmatic suberogatory actions as those that are contra-beneficent and contra-expected. Moral rights against another person are correlative with potential actions of that person, vis-a-vis oneself, that are morally impermissible. Contra-expectation is a weaker negative constraint than impermissibility and does not undergird correlative moral rights.

Paradigmatically praiseworthy actions are beneficent actions that one is **not** morally constrained **to perform**. By contrast, paradigmatically blameworthy actions are contra-beneficent actions that one **is** morally constrained **not to perform**. Paradigmatically supererogatory actions are praiseworthy; hence, deontically they not merely are non-obligatory but also are morally optional. Paradigmatically, suberogatory actions are blameworthy; hence, although they are not morally impermissible, they are morally contra-expected.

6 Conclusion

Being as economical as possible with the range of concepts one needs in one's ethical theory is a desirable theoretical goal. However, insofar as one is interested in an ethical theory that is faithful to the nuances of the moral landscape that arguably are revealed in everyday moral thought and discourse, a more important goal is that one's theory employs enough concepts to enable an accurate reflection of that landscape. We have argued that doing so requires explicit recognition within ethical theory of the deontic categories of the expected and the contra-expected. Besides being more faithful to the nuances of common, everyday deontic judgment than is theorizing deploying only the familiar tri-partite distinction of the obligatory, the wrong (or impermissible), and the optional, the two added categories provide an illuminating construal of the asymmetry of the supererogatory and the suberogatory, as we have just explained.

References

Calhoun, C. (2004). Common decency. In C. Calhoun (Ed.), *Setting the moral compass: Essays by women moral philosophers.* Oxford University Press. Reprinted in Calhoun, C. (2016). *Moral aims: Essays on the importance of getting it right and practicing morality with others* (Ch. 4). Our references in the text are to the 2016 reprint.

Chisholm, R. (1963). Supererogation and offense: A conceptual scheme for ethics. *Ratio, 5*, 1–14.

Driver, J. (1992). The suberogatory. *Australasian Journal of Philosophy, 70*, 286–295.

Grigoletto, S. (2019). *Only through complexity: morality and the case of supererogation.* Padova University Press.

Heyd, D. (1982). *Supererogation.* Cambridge University Press.

Horgan, T., & Timmons, M. (2010). Untying a knot from the inside out: Reflections on the 'Paradox' of supererogation. *Social Philosophy and Policy, 27*, 29–63.

Horgan, T., & Timmons, M. (2022). Expressing gratitude as what's morally expected: A phenomenological approach. *Ethical Theory and Moral Practice, 25*, 139–155. https://doi.org/10.1007/s10677-021-10261-w

McNamara, P. (2011a). Supererogation inside and out: Toward an adequate scheme for common-sense morality. In M. Timmons (Ed.), *Oxford studies in normative ethics* (Vol. 1, pp. 202–235).

McNamara, P. (2011b). Praise, blame, obligation, and DWE: Toward a framework for classical supererogation and kin. *Journal of Applied Logic, 9*, 153–170.

Mellema, G. (1991). *Supererogation, obligation, and offence.* SUNY Press.

Ullmann-Margalit, E. (2011). Considerateness. *Iyyun: The Jerusalem Philosophical Quarterly, 60*, 205–244.

Supererogation and Its Conceptual Neighborhood Through a DWE Lens

Paul McNamara

Abstract I first provide an accessible overview of the DWE (Doing Well Enough) logical and semantic framework for representing going beyond the call and its family of kindred concepts in a tightly intergraded way. Next, a module, for representing some basic agent-evaluative notions is developed ("AA" for "Aretaic Assessment"), and then it is integrated with the more act-evaluative notions of DWE, thereby allowing for a representation of suberogation and supererogation (as distinct from going beyond the call) and many other combined deontic and agent-evaluative notions. I then probe more deeply into how the semantic structures for the DWE framework might themselves be generated and then offer some brief reflections on the classic supererogation puzzle, as well as that of supererogatory holes and the all or nothing problem, framing options via the preceding reflections.

Keywords Beyond the call · Supererogation · Permissible suboptimality · Suberogation · The minimum · Must · Ought · Act versus agent evaluation · Supererogatory holes · All or nothing

1 Introduction

Consider a paradigmatic case of going beyond the call. Tiny Tim is caught in a building that is on fire. Someone is running to pull the fire alarm, while a mailwoman passing by sizes up the imminent danger to the child, drops her bag, and manages to rescue the child before the fire personnel arrives. We can easily imagine that the dangers to the rescuer were transparent. We would almost all judge that the mail-woman went beyond the call in rescuing the child. She did more than she had to do, exceeding the minimum morality demands. Adding that the mailwoman had the best of intentions in doing what she did, what she did was praiseworthy (i.e., she is praise-worthy for the rescue). Philosophers would say that her action was 'supererogatory'

P. McNamara (✉)
University of New Hampshire, Durham, New Hampshire, USA
e-mail: Paul.McNamara@unh.edu

D. Heyd (ed.), *Handbook of Supererogation*,
https://doi.org/10.1007/978-981-99-3633-5_9

(a term rarely used except in philosophy).[1] Dramatic cases are not essential: a small favor, a small act of thoughtfulness/kindness, forgiveness, etc., also serve.

Why care about this notion? Well, consider moral exemplars. We typically conceive of them as often doing much more than required—often their most salient trait. Surely ethical theories then should account for this feature of moral exemplars. Yet the most famous traditional approaches in normative ethical theory (Virtue Ethics, Kantian Ethics, and Utilitarianism) have had trouble either recognizing the possibility of supererogation or giving a minimally satisfactory account of it. Supererogation is also part of a family of concepts that ethical theorists and deontic logicians have often struggled to adequately account for, often stumbling around among them, conflating members of distinct pairs with one another, and often thereby unwittingly ruling supererogation out. Representing supererogation coherently requires grappling with an enriched array of moral concepts and representing their logical relationships carefully enough to generate a coherent framework. In fact an often-touted slogan, 'the traditional deontic scheme must go!,' alleges that traditional deontic logic rules out going beyond the call. So there are a number of reasons to address supererogation.

I have developed and motivated a framework, DWE ('Doing Well Enough'), for coherently representing going beyond the call and associated concepts in its neighborhood, specifying their logical properties and relations to one another, relations validated by an interpreted semantic framework for these concepts. This chapter will approach supererogation and kin through this lens.

2 DWE and Logical Features of Supererogation and Affiliated Concepts

2.1 *The Traditional Deontic Scheme*[2]

The Traditional Definitional Scheme

The fundamental normative statuses of what I call 'the *Traditional Scheme*' are these six:

it is obligatory that (**OB**)
it is permissible that (**PE**)
it is impermissible that (**IM**)

[1] I will use 'going beyond the call' and 'supererogation' interchangeably at times for ease of exposition, but later I will use 'supererogation' for a distinct notion stronger than the first.

[2] By the 'Traditional Scheme', I mean a bit of unsystematic deontic folklore—essentially TDS plus DS and/or TTC below, along with replacement of equivalents, that is, classical modal system ED. See (McNamara, 1996a), (McNamara, 1996b), (McNamara, 2021), and (McNamara and Van De Putte 2022).

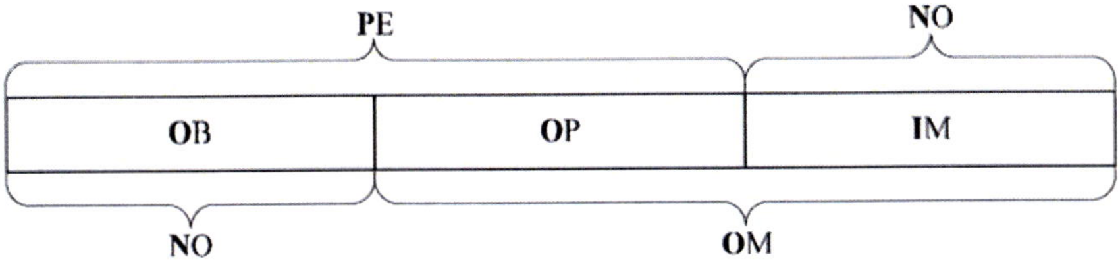

Fig. 1 The traditional threefold classification (TTC)

it is omissible that (**OM**)
it is optional that (**OP**)
it is non-optional that (**NO**).[3]

Any of the first four could be taken as basic and the other five defined via it. Taking **OB** (*deontic* necessity) as basic, we provide '*The Traditional Definitional* Scheme:'

(TDS)	$\mathbf{PE}\varphi \leftrightarrow \neg\mathbf{OB}\neg\varphi$	(Permissibility)
	$\mathbf{IM}\varphi \leftrightarrow \mathbf{OB}\neg\varphi$	(Impermissibility)
	$\mathbf{OM}\varphi \leftrightarrow \neg\mathbf{OB}\varphi$	(Omissiblity)
	$\mathbf{OP}\varphi \leftrightarrow (\neg\mathbf{OB}\varphi\ \&\ \neg\mathbf{OB}\neg\varphi)$	(Optionality)[4]
	$\mathbf{NO}\varphi \leftrightarrow (\mathbf{OB}\varphi \vee \mathbf{OB}\neg\varphi)$	(Non-optionality)[5]

These equivalences are natural enough and often employed, with the most focus on the first two, which are routinely presupposed in contexts of supererogation.

The Traditional Threefold Classification and the Deontic Square

In addition to TDS, it was assumed that the 'The Traditional Threefold Classification' (TTC) holds (Fig. 1).

The three internal categories are deemed *mutually exclusive and jointly exhaustive*: every proposition is either (overridingly) obligatory, optional, or impermissible, and no more than one of these. Let MJ($\mathbf{OB}\varphi$, $\mathbf{OP}\varphi$, $\mathbf{IM}\varphi$) stand for

---

[3] These abbreviations are mnemonics. We will be adding a number of others. Here we choose to read the basic operator as 'it is obligatory that' so that continuity with permissibility, impermissibility, and optionality is not lost as it would be with the 'it ought to be the case that' reading. A choice must be made. 'It is obligatory that' may also be read personally, but non-agentially as 'it is obligatory that I be in my office' (Krogh & Herrestad, 1996; McNamara, 2004).

[4] As we will see, it is noteworthy that we label this condition with 'OP' for 'optionality' not 'IN' for 'indifference'.

[5] Here such equivalences will be called 'definitions'.

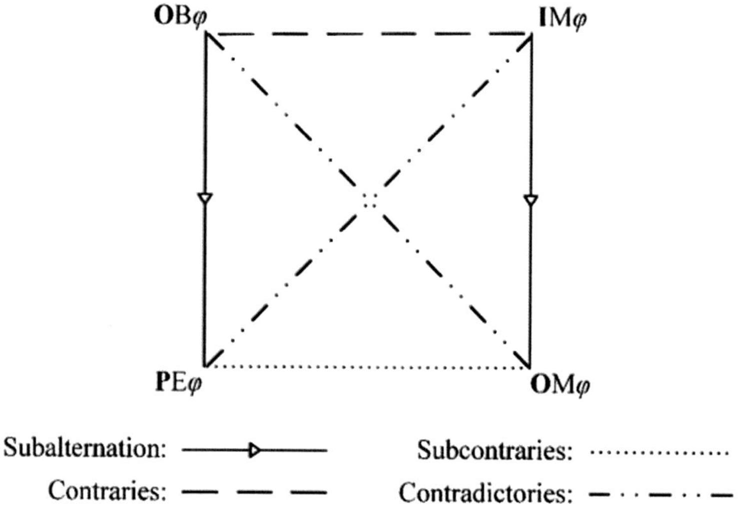

Fig. 2 The deontic square (DS)[7]

Fig. 3 The deontic hexagon (DH)

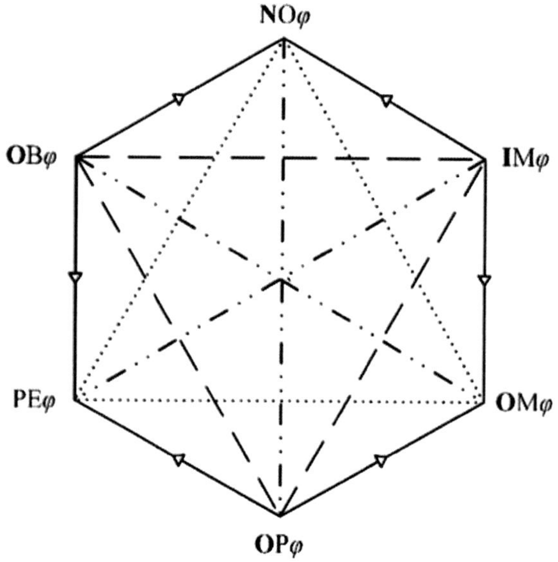

(TTC) $(\mathbf{OB}\varphi \vee \mathbf{OP}\varphi \vee \mathbf{IM}\varphi) \,\&\, [\neg(\mathbf{OB}\varphi \,\&\, \mathbf{OP}\varphi) \,\&\, (\neg\mathbf{OB}\varphi \,\&\, \mathbf{IM}\varphi) \,\&\, \neg(\mathbf{OP}\varphi \,\&\, \mathbf{IM}\varphi)]$.[6]

The Deontic Square (DS) is also part of the Traditional Scheme (Fig. 2).
DS amounts to this conjunction

---

[6] 'MJ' is chosen as a mnemonic for "mutually exclusive and jointly consistent. This function can be precisely defined (see McNamara, 2021).

[7] Recall that *contraries* cannot both be true, *subcontraries* cannot both be false, contradictories must have opposing truth values, and subalternation is entailment.

$$\text{(DS)}\quad (\neg\mathbf{OB}\varphi \leftrightarrow \neg\mathbf{OM}\varphi)\ \&\ (\mathbf{IM}\varphi \leftrightarrow \neg\mathbf{PE}\varphi)\ \&\ \neg(\mathbf{OB}\varphi\ \&\ \mathbf{IM}\varphi)$$
$$\&\ (\mathbf{PE}\varphi \vee \mathbf{OM}\varphi)\ \&\ (\mathbf{OB}\varphi \rightarrow \mathbf{PE}\varphi)\ \&\ (\mathbf{IM}\varphi \rightarrow \mathbf{OM}\varphi).$$

If we add nodes for OP (optionality) and ¬OP (non-optionality), we get a *deontic hexagon* (Fig. 3).

An important logical feature of optionality is *the indifference of optionality to negation*:

$$\text{(ION)}\quad \mathbf{OP}\varphi \leftrightarrow \mathbf{OP}\ \varphi$$

ION follows from the TDS and RE (replacement of logical equivalents) assuming a classical propositional logic.[8]

The Fundamental Presupposition Behind the Traditional Deontic Scheme

As noted, the underlying logic in the Traditional Scheme was classical propositional logic and replacement of provable equivalents

$$\text{(RE) If } \varphi \leftrightarrow \psi \text{ is a theorem, then so is } \mathbf{OB}\varphi \leftrightarrow \mathbf{OB}\psi.$$

But it turns out that TTC and DS, given the Traditional Definitional Scheme, are then each easily shown to be *tautologically* equivalent to the principle that (overriding) obligations cannot conflict,

$$\text{(NC)}\neg(\mathbf{OB}\varphi\ \&\ \mathbf{OB}\ \varphi).^{9}$$

As in much of the history of normative ethics, early deontic logicians assumed obligations could not conflict. Although it seems obvious that there can be conflicting obligations, nonetheless, in the vast majority of work in ethics on supererogation, authors assumed they were dealing with a concept of obligation for which NC held. I propose we read '**OB**' as 'it is *overridingly* obligatory that.' If it is overridingly obligatory that φ, then it is obligatory that φ and that obligation overrides all obligations it conflicts with, defeating all competitors. NC seems *analytic* so interpreted.

I now turn to a scheme that has been routinely confused with the Traditional Scheme, by both friend and foe of supererogation, and has been a source of much confusion, as well as mistaken criticisms of traditional deontic logic.

---

[8] For $(\neg\mathbf{OB}\varphi\ \&\ \neg\mathbf{OB}\neg\varphi) \leftrightarrow (\neg\mathbf{OB}\neg\varphi\ \&\ \neg\mathbf{OB}\neg\neg\varphi)$, so $\mathbf{OP}\varphi \leftrightarrow \mathbf{OP}\neg\varphi$.

[9] See (McNamara and Van De Putte 2022), Sect. 2.2.

2.2 The Traditional Scheme, Supererogation, Indifference, and Urmson's Constraint

The Conflation of Moral Indifference with Moral Optionality and Its Cost

Today, whether I put my left shoe on first or not is a matter of indifference to me, and I assume it is a matter of moral indifference too. As with optionality, we endorse 'the indifference of indifference to negation': if it was a matter of moral indifference that I put my left shoe on first, then it is a matter of indifference that I do not.

$$\text{(IIN)}\ \mathbf{IN}\varphi \text{ iff } \mathbf{IN}\neg\varphi.$$

It should also be clear that the defining condition of optionality (being neither obligatory nor impermissible) must be met by moral indifference. Thus, we endorse the 'optionality of indifference':

$$\text{(OI)}\ \mathbf{IN}\,\varphi \rightarrow \mathbf{OP}\,\varphi.$$

Moral significance can be plausibly defined via indifference:

$$\text{(SI)}\quad \mathbf{SI}\varphi \stackrel{\text{def}}{=} \neg\mathbf{IN}\,\varphi.$$

Now consider this analog of the TTC, the '*Strong* Threefold Classification' (Fig. 4).

(I will use '[#]' as a marker for rejected purported principles.) Here the operator for indifference replaces that for optionality (and that for significance replaces that for non-optionality).[10] Using our shorthand for mutual exhaustion and joint exclusion, STC amounts to this analog to TTC:

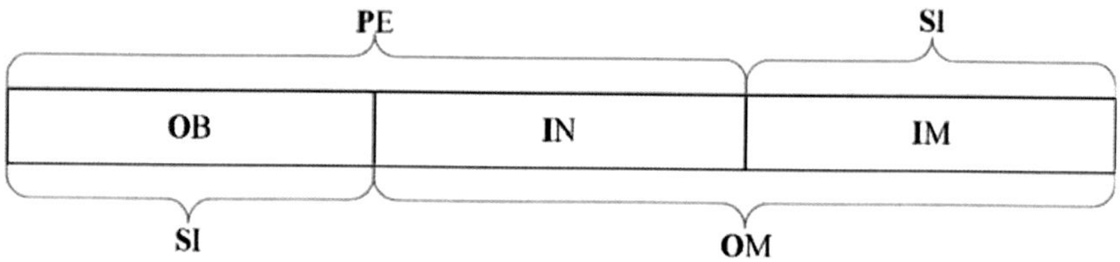

Fig. 4 The *strong* threefold classification (STC) [#]

[10] The outer labels derivatively express these dubious claims: $\mathbf{PE}\varphi \leftrightarrow (\mathbf{OB}\varphi \vee \mathbf{IN}\varphi)$ and $\mathbf{OM}\varphi \leftrightarrow (\mathbf{IN}\varphi \vee \mathbf{IM}\varphi)$.

$$\text{(STC) MJ}(\mathbf{OB}\varphi, \mathbf{IN}\varphi, \mathbf{IM}\varphi)^{11}\ [\#]$$

As with TTC, the intention is that each alternative option falls into one of the classes, but no more than one. Given the traditional scheme, STC is easily shown to be tautologically equivalent to what I call 'Moral Rigor':

$$\text{(MR) } \mathbf{OP}\varphi \leftrightarrow \mathbf{IN}\varphi.\ [\#]$$

where does supererogation fit in? Recall our mailwoman's rescue of Tiny Tim, a paradigmatic case of going beyond the call. Her rescue was neither obligatory nor impermissible. Letting '**BC**' stand for 'It is beyond the call (for Jane Doe) that,' these two features of going beyond the call can be combined into 'the optionality of going beyond the call':

$$\text{(OS)}\mathbf{BC}\varphi \rightarrow \mathbf{OP}\varphi.$$

However, our mailwoman's rescue is not exhausted by its optionality, for even though optional, her action was hardly a matter of moral indifference. Thus, we endorse 'the non-indifference of supererogation':

$$\text{(NIS)}\mathbf{BC}\varphi \rightarrow \neg\mathbf{IN}\varphi.$$

Together, the last two entailments yield 'the optional non-indifference of supererogation':

$$\text{(ONIS) } \mathbf{BC}\varphi \rightarrow (\mathbf{OP}\varphi \,\&\, \neg\mathbf{IN}\varphi).$$

Recall that we earlier saw that we must endorse the 'optionality of indifference', $\mathbf{IN}\varphi \rightarrow \mathbf{OP}\varphi$. But considering going beyond the call, it should be apparent that the converse is dubious. Let us assume that it was a matter of moral indifference that our rescuer put on her left shoe first this morning. That is hardly so for her rescuing Tiny Tim. Yet both were optional. So, replacing '**OP**' with 'IN' in the Traditional Threefold Classification to yield the Strong Threefold Classification is highly counterintuitive. STC makes morality too harsh a taskmaster. For given ONIS above, anyone committed to the possibility of going beyond the call is thereby committed to what I call 'Optionality with a Difference':

$$\text{(OWD) : } \mathbf{OP}\varphi \,\&\, \neg\mathbf{IN}\varphi \text{ is satisfiable.}$$

[11] That is, $(\mathbf{OB}\varphi \vee \mathbf{IN}\varphi \vee \mathbf{IM}\varphi) \,\&\, [\neg(\mathbf{OB}\varphi \,\&\, \mathbf{IN}\varphi) \,\&\, \neg(\mathbf{OB}\varphi \,\&\, \mathbf{IN}\varphi) \,\&\, \neg(\mathbf{IN}\varphi \,\&\, \mathbf{IM}\varphi)]$

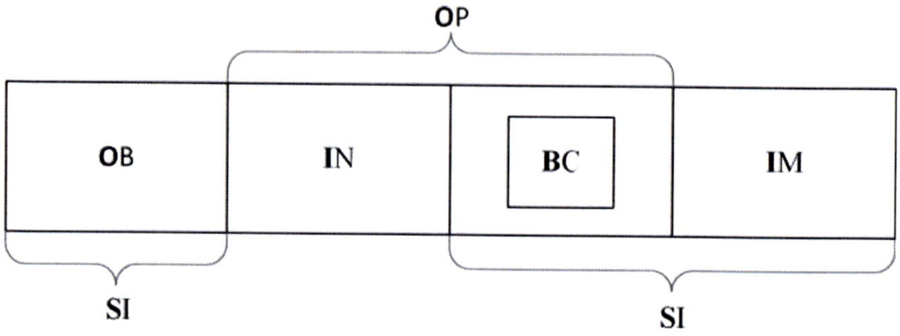

Fig. 5 The minimal five-fold classification (FC)

This is essentially Urmson's Constraint on deontic schemes.[12] STC rules out this possibility and it entails moral rigor, $\mathbf{OP}\varphi \leftrightarrow \mathbf{IN}\varphi$, thus enjoining the collapse of optionality and indifference.[13] STC and MR each exclude any possibility of going beyond the call. So when friends of supererogation cry out that 'The Traditional Deontic Scheme must go!,' they are often making the same conflation of optionality with indifference by conflating the Traditional Threefold Classification with the Strong Threefold Classification, unwittingly equating indifference with what is neither obligatory nor impermissible and so unwittingly ruling out supererogation while trying to defend it. STC is untenable. TTC is relatively innocuous.

Note also that if it is beyond the call to rescue Tiny Tim, then since it is not a matter of moral indifference to do so, it is not a matter of moral indifference to not do so either (which follows from IIN). If the rescue is significant, then so is not rescuing significant, even though both are optional.

The Minimal Five-Fold Classification

A preliminary extended minimal five-fold classificational picture naturally emerges (Fig. 5).

In symbols:

$$\text{(FC) MJ}(\mathbf{OB}\varphi, \mathbf{IN}\varphi, \mathbf{BC}\varphi, [\mathbf{OP}\varphi \,\&\, \mathbf{SI}\varphi \,\&\, \neg\mathbf{BC}\varphi], \mathbf{IM}\varphi).$$

Reflection on going beyond the call and notions in its neighborhood force this enrichment of the TTC on us, whereas neglect of such notions encourages conflations of distinct concepts, as we have already seen in the case of indifference and optionality which led to STC. These reflections also naturally lead to new questions that can generate new insights about the expanded normative positions an agent might be in. For example, you may have wondered why the diagram above partitioned the

[12] See (Urmson, 1958).

[13] We find the conflation of these two concepts in G. E. Moore, Von Wright, Prior, even in Urmson as he struggles to distinguish the two (see McNamara, 1996b), and being neither obligatory nor impermissible is still often labeled as "indifferent".

non-indifferent optional alternatives into those that are supererogatory and those that are not. For if not, we could define the supererogatory actions as *the* non-indifferent optional actions, by adding to ONIS, an endorsement of its converse 'the sufficiency of optional non-indifference for supererogation'

$$\text{(SONS)}\ (\mathbf{OP}\varphi \,\&\, \neg\mathbf{IN}\varphi) \rightarrow \mathbf{BC}\varphi\ [\#]$$

We are in a good position to see why this is unacceptable by the logical features of the concepts alone. Suppose $\mathbf{BC}\varphi$. By ONIS, $\mathrm{OP}\varphi \,\&\, \neg\mathbf{IN}\varphi$. Then by ION and IIN, $\mathbf{OP}\neg\varphi \,\&\, \neg\mathbf{N}\neg\varphi$, and then by SONS, we get $\mathbf{BC}\neg\varphi$, thereby generating 'the indifference of *supererogation* to negation':

$$\text{(ISUN)}\ \ \mathbf{BC}\varphi \rightarrow \mathbf{BC}\neg\varphi.\ [\#]$$

ISUN is surely absurd, for it says the negation of what is supererogatory is supererogatory, and it entails that for any alternative, either it is not supererogatory or *both* it and its negation are supererogatory.[14] It is plausible to think that something stronger holds: that there can be "no supererogatory conflicts":

$$\text{(NSC)}\ \ \neg(\mathbf{BC}\varphi \,\&\, \mathbf{BC}\neg\varphi)$$

For suppose that φ is supererogatory for Jane Doe and that φ consists of Jane doing A. Then her doing A while doing only permissible things must guarantee doing more than the minimum. But then, presumably, there must be permissible ways of her not performing A that do not involve doing more than the minimum. So even if she does nothing but permissible things, A's non-performance cannot assure that she has done anything beyond the minimum, and hence $\neg\varphi$, her not doing A, cannot be supererogatory. For example, surely our mailwoman's *not* rescuing the infant would not be supererogatory per se by just helping direct the fire truck to the scene (the minimum, we imagine). We must leave space for optional non-indifferent *non*-supererogatory alternatives.

So despite the fact that ethical theorists and deontic logicians have routinely, and often still do, label the condition of being neither obligatory nor impermissible as 'indifference,' this is an unwarranted and substantive conflation of two distinct and important deontic notions and one that rules out supererogation. As we will see, the conflation of distinct pairs of concepts has been one major obstacle in finding a coherent place for supererogation in deontic logic, as well as in ethical theory.

We turn now to additional concepts in the neighborhood of supererogation and how they fit in.

---

[14] Indeed, a number of analysts thought that if φ is supererogatory, then its absence, $\neg\ \varphi$, is an offense or suberogatory, and so has a *negative* valence. Although we will reject this too, it does underscore the counter-intuitiveness of SONS.

2.3 Additional Concepts in the Neighborhood and Logical Connections

Must, Ought, Can, and Cannot

I have called the claim that *must/obligation/duty* and *ought* are equivalent in moral contexts a 'Major Bipartisan Presupposition' of most of the twentieth century ethics and deontic logic. Yet

> 'Must', but not 'ought', expresses whatever ethicists and deontic logicians have virtually uniformly taken 'ought' to express: *moral or deontic necessity*. For the latter has routinely been taken to be whatever satisfies the familiar definitional equivalences involving permissibility and impermissibility. And this means that contrary to a dominant bipartisan trend this [now past] century, we can't take 'ought' as basic and then assume that what is permissible is whatever satisfies '¬ought¬,' nor that what is impermissible is whatever satisfies 'ought¬' ((McNamara, 1996c), p. 158)

The Presupposition assumes *ought* has a tight continuity with the traditional concerns with what is obligatory, permissible, and impermissible often expressed in a deontic square of opposition, but it does not (McNamara, 1996c, 2021), and in more detail, (McNamara, 1994), and Chapter 3 of (McNamara, 1990), provide a cumulative case argument that *must* properly entails *ought*, and that it is *must*, not *ought*, that has the traditional interdefinable relationships with permissibility and impermissibility. I offered the first model-theoretic account of the difference between *must* and *ought* in those works and will sketch a close cousin below, but, first, I briefly summarize some of the key points supporting the difference. First, there are *family differences* (as it were); (McNamara, 1996c) asks us to consider two groups of expressions, where the two groupings seem natural, and the members of the first group (e.g., S must do A, S is required to do A, it is imperative that S do A) seem uniformly stronger than those of the second group (S ought to do A, S should do A, it is best that S do A). Second, *ought conversationally implies* optionality. If I say, 'you ought to take this turn' rather than 'you must take this turn,' I suggest it is best to do so, but by not using 'must' I suggest that there is another acceptable (though less than best) option. Third, there are *constitutional differences*. A deadline is a time by which something must be done, not one by which something (merely) ought to be done. *Ought* is too weak here. Similarly, a job *requirement* is something that must be done, not one that merely ought to be done, as illustrated by the widespread 'Employees must wash their hands.' Fourth, there are *speech act differences*. If your supervisor says, 'you must do A,' they are ordinarily *requiring* you to do so, unlike if they say, 'you ought to do A,' thereby *recommending* you do A, but not requiring you to do so. Fifth, there are *contrastive differences*. 'You ought to, but you don't have to' (or 'You ought to, but it is not the case that you must') is perfectly apt, whereas 'You ought to, but it is not the case that you ought to' is incoherent (but shouldn't be if *must* is equivalent to *ought*). McNamara provides other considerations as well: pertaining to traditional equivalences of S *must* do A with ¬(*S can not do A*), and with *S can't not do A* (the *can* and *can't* here are of permissibility and impermissibility); pressure from asymmetric use of *must* and *ought* in non-moral contexts; and contexts, like law, where *must*, *can*, and *can't* are felicitous and *ought* is not.[15]

Now notice the impact of the pervasive conflation in the twentieth century ethical theory and deontic logic of deontic necessity with *ought*. Consider the near axiomatic claim: 'You ought to do the best you can.' By implication of the conflation, we get 'You must do the best you can,' but then how can the best you can do *ever* be supererogatory? If what you ought to do, you must do, and so you can't not do, then nothing supererogatory can ever be something you ought to do, not even returning a favor given freely. So once again, we have a conflation that makes it difficult to find a stable or coherent place for supererogation.

The Least One Can Do

This widely used expression has received virtually no attention in deontic logic or ethical theory, yet it is ripe with import. Consider *"The least you can do is call and let them know you will be very late."* The 'can' appears to be the *can* of *permissibility*.[16] The expression suggests an *ordering* quite naturally: an ordering with a *minimally acceptable level*, and lower and higher levels potentially, and with all levels lower than the least being impermissible. The latter is reflected in the aptness of this idiom when used to scold: 'The least you could have done was call' with its contextually understood, 'and you didn't even do that!' This is an extremely rich idiom and an important data point in understanding pre-theoretic moral consciousness. It is plausible that going beyond the call rules out doing the least you can do, so there is a logical relationship here, and going beyond the call seems to involve doing more than the least you can do. So this notion is vital to supererogation's conceptual neighborhood, and not one to neglect. It is also not to be confused with what is obligatory. If I committed to helping you move (you had given me a half day's help once before), then the least I can do is provide just the minimal reasonably expected help, but I can also fulfill my obligation by providing more help than I would be doing the least I could do.

'Jane *did more (good) than she had to do'* is quite colloquial speech. The 'more' (like 'least') suggests an *ordering*, and since it is more than you *had to* do, it suggests the acceptability of doing less, so *ordered acceptable options* naturally emerge. Also, it naturally suggests the possibility of doing less than you had to do. So 'the least one can do' meshes well with colloquial expressions of going beyond the call. Indeed, I propose that a plausible condition for going beyond the call is this one:

> someone does more than she would have done had she done the least she could have done.

Graded Supererogatory Options

If I am obligated to help you move, and there is some minimal amount of helping that constitutes the least I can do, and another that constitutes more, then we have

---

[15] I think the reason for this is that there is no apt ordering relation in law, and one is needed for 'ought,' but not for 'must' (McNamara, 1994) and (McNamara, 1990).

[16] It cannot be the *can* of ability or possibility plainly, and there seem no plausible candidates given the use of this but for the *can* of permissibility.

at least two permissible options, one outranking the other, and of course, there is no reason to think it must be only two-tiered. For example, if the minimal help is a half day (e.g., 8:00–noon), then presumably (set aside conflicts), I can also give you more (e.g., 8:00–2:00 pm), or still more (e.g., 8:00–4:00 pm), etc. Assume more is better here. But then, the second is beyond the call, and so is the third and it is also better than the second supererogatory act. This has sometimes been ruled out tacitly by assuming a supererogatory act is one you ought to do (the most choiceworthy) coupled with the in-tandem plausibility of the claim that what you ought to do is the best you can. But this just cannot be right if graded supererogation is possible. This is another place where one can get muddled.

Permissible Suboptimality

Consider now '*You ought to but don't have to*' and '*you can but ought not.*' It is plausible to think that what you ought to do is what is most choiceworthy, but since it need not be what you must do, it is permissible to do less than what you ought (McNamara, 2011b) argues that this is closely associated with agent-centered prerogatives to do less than the best or less than you ought. Such actions are suboptimal, but permissible, and they can also be supererogatory.

2.4 The DWE Framework

The DWE Language

Let us pull things together and regiment a bit. Here are four concepts taken as primitive in the DWE logical framework:

OB: It is *Obligatory/a Must* (for S) that φ
MIφ: The *Minimum* (for S) involves (its being the case that) φ
MAφ: The *Maximum* (for S)/What S *ought to do* involves/implies φ.
INφ: It is *Indifferent* (for S) that φ[17]

We imagine these four operators added to a language for classical propositional logic. Some defined operators and their intended readings are:

$\mathbf{PE}\varphi \overset{\text{def}}{=} \neg\mathbf{OB}\neg\varphi$	(It is *Permissible* for S that φ.) (cf. 'can')
$\mathbf{IM}\varphi \overset{\text{def}}{=} \mathbf{OB}\neg\varphi$	(It is *Impermissible* for S that φ.) (cf. 'can't')
$\mathbf{OM}\varphi \overset{\text{def}}{=} \neg\mathbf{OB}\varphi$	(It is Omissible (for S) that φ (cf. 'can not')
$\mathbf{OP}\varphi \overset{\text{deff}}{=} \neg\mathbf{OB}\varphi \,\&\, \neg\mathbf{OB}\neg\varphi$	(It is *Optional* for S that φ.)
$\mathbf{NO}\varphi \overset{\text{def}}{=} \mathbf{OB}\varphi \vee \mathbf{OB}\neg\varphi$	(It is *Non-Optional* for S that φ.)
$\mathbf{SI}\varphi \overset{\text{def}}{=} \neg\mathbf{IN}\varphi$	(It is *Significant* for S that φ.)

[17] Note that the readings are *personal* but not necessarily *agential* (McNamara, 2004).

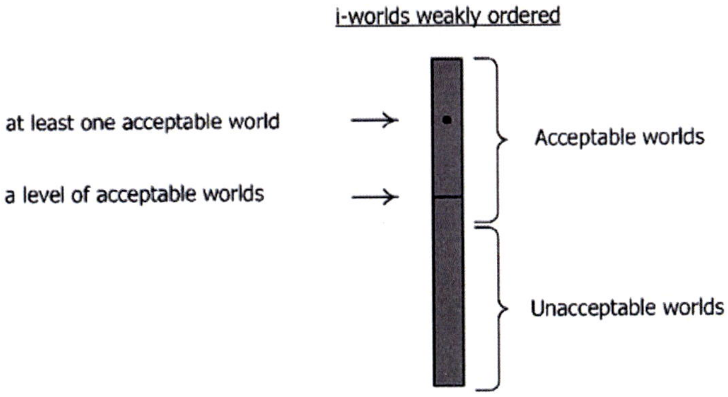

Fig. 6 DWE semantic framework

$\mathbf{BC}\varphi \overset{\text{def}}{=} \mathbf{PE}\varphi \,\&\, \mathbf{MI}\neg\varphi$ (It is *Beyond the Call* for S that φ.)

$\mathbf{PS}\varphi \overset{\text{def}}{=} \mathbf{PE}\varphi \,\&\, \mathbf{MA}\neg\varphi$ (It is *Permissibly Suboptimal* for S that φ.)

Let us set aside identifying logics (deduction systems for DWE) and other definable operators.[18]

The DWE Semantic Framework[19]

Let me sketch a semantic framework for this which meshes well with some of our reflections in Sects. 2.1–2.3. We imagine a structure like this (Fig. 6).

We assume that for an arbitrary agent, Jane Doe, and from the standpoint of a given state/world, *i*, that she is in, there is a set of realizable states/worlds (not empty) where she comports herself *acceptably* (given her conditions in *i*), as well as a set of ones where she does not. We also assume there is a moral ordering of the states, some she comports herself at least as well as in others, and some she does so better than in others.[20] Since some can be tied (different but at the same rank), there will be levels of equi-ranked states. Since we will not be concerned with groups of agents, we will leave our arbitrary Jane Doe unlabeled. Lastly, we assume that if a world *j* is

---

[18] See (McNamara, 2021; Mares and McNamara, 1997).

[19] Note that the semantic framework outlined here is a generalization that I have called DWE^G elsewhere (McNamara, 2022). In the classical DWE framework (e. g., (McNamara, 1996c, 1996a, 1996b), there is only an ordered set of acceptable worlds (per the situation in some state *i*), and there 'indifference' is parsed as essentially indifference among the permissible options, which we could easily give truth conditions for below in addition to the stronger notion of indifference at all levels of value, but we set the former notion aside here. Also, in classical DWE semantics, there is no ordering of the *i*-unacceptable worlds (or reference to them at all), nor to what we will call below, GOO, a principle to the effect that as good as ok is ok.

[20] The former relation taken as basic and as reflexive and transitive.

acceptable (per i) and a world k is ranked at least as high as j, then k too is acceptable. I call this the 'GOO principle': as good as ok is ok.

As simple as this is, we can use it to begin to model the concepts introduced in Sects. 2.1–2.3 distinguishing distinct concepts and supporting their intuitive logical properties and relationships to one another. Consider a case: Imagine that you must provide some delicate information to a colleague across campus, and for simplicity, imagine there are three ways to do this emailing, phoning, or talking in person. Lastly, suppose, not implausibly, that the permissible options are ranked according to how personal they are. So giving the info in person is ranked higher than giving it on the phone, which is ranked higher than emailing the info (Fig. 7).

Then, giving the info is something you *must* do, not giving it is *impermissible. The least you can do* is provide the info by email, but giving it in person is best, and so that is what you *ought* to do. However, giving the info by calling on the phone is not only permissible, but *beyond the call*, as is doing it in person. Furthermore, giving the info via email or by phone is *permissibly suboptimal*. Below are two diagrams

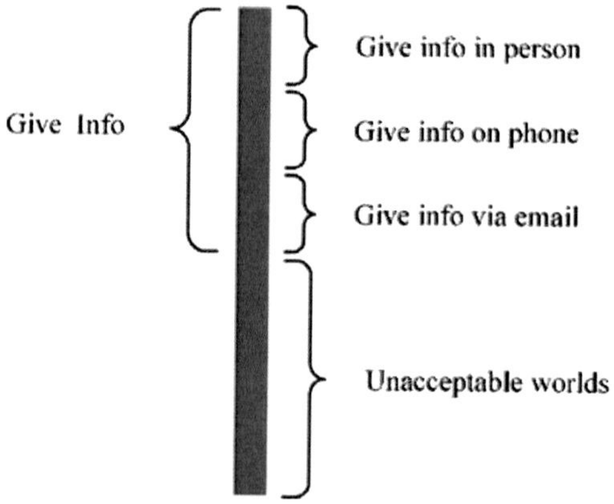

Fig. 7 DWE example

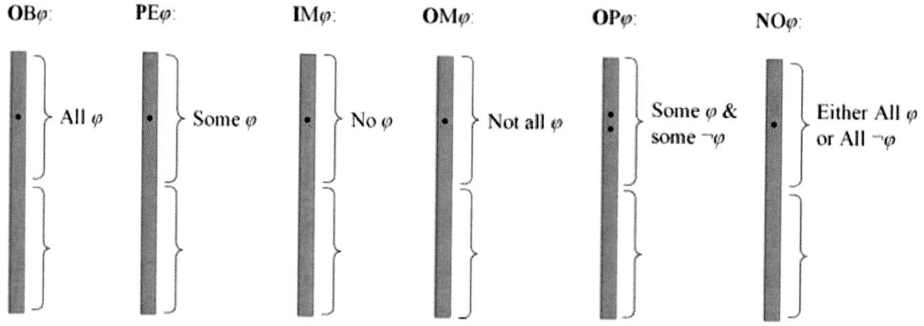

Fig. 8 Conditions for the familiar deontic operators

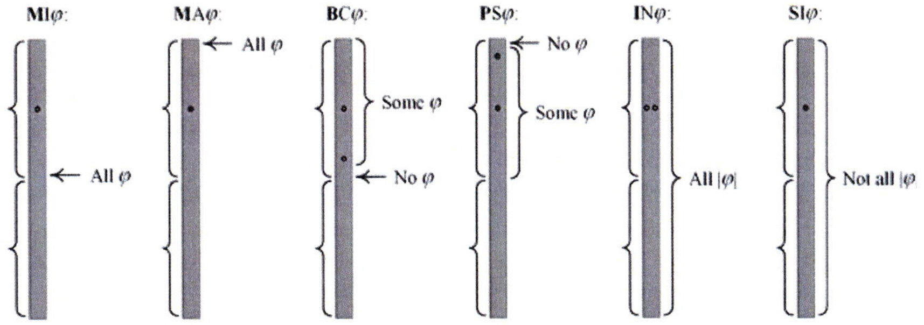

Fig. 9 Conditions for the new DWE operators

of the truth conditions for the DWE operators. Here, first are those for the familiar six deontic operators (Fig. 8).

For these, the ordering does not matter. It is *obligatory* (a *must*) for Jane Doe that φ iff φ holds in all the acceptable worlds (the worlds where she comports herself acceptably henceforth), φ is *permissible,* iff it holds in some acceptable world, φ is *impermissible,* iff φ holds in no acceptable world, φ is *omissible,* iff it does not hold in some acceptable world, φ is *optional,* iff it holds in some acceptable world, and not in some other, and *non-optional* iff either φ hold in all the acceptable worlds or in none.[21] For these, the ordering is idle. But not so for the less familiar notions in Fig. 9.

Doing the *minimum* (the *least*) (for Jane) involves φ iff φ holds in all the lowest ranked acceptable worlds, *Doing* the *maximum* (what *ought* to hold) involves φ iff φ holds in the best of the acceptable worlds, φ is *beyond the call* iff φ is precluded by the lowest ranked acceptable worlds, but φ is included in some acceptable worlds (so ones ranked above the lowest acceptable ones); φ is *permissibly suboptimal* iff it is precluded by the highest ranked acceptable worlds, but occurs somewhere among the acceptable worlds (so lower than the best level). φ is a matter of indifference iff every level of value (acceptable or not) can be achieved with or without φ (i.e., at each level, there is at least one φ world and one not-φ world, indicated by '$|\varphi|$' at a level); and φ is a matter of *significance* iff some level of value depends on either its presence or its absence throughout that level.

It is easy to see and show that all DWE's primitive and defined concepts are modeled distinctly in this semantic framework. Also, all the intuitive logical relationships between the notions discussed earlier are validated and those rejected earlier are invalidated. We illustrate informally with three. First what is indifferent is optional. **IN**$\varphi \rightarrow$ **OP**φ. Recall that in all models, there is an acceptable alternative, say *j,* to any state *i.* Now let's suppose **IN**φ holds at *i.* Then by the truth conditions for **IN**φ, at every level of worlds, there is a φ-world and a $\neg\varphi$-world, and so this must hold for the level that *j* is a part of as well; then by GOO, all worlds at that level, must be acceptable alternatives to *j*, since they are as good as *j*, which is acceptable. So the

[21] I ignore here any representation of agency, and do not maneuver to block tautologies becoming obligatory, in the hope that the picture will still be reasonably clear without the needed modifications.

truth condition for $\mathbf{OP}\varphi$ follows: there is an acceptable φ-world and an acceptable $\neg\varphi$-world. However, the converse is invalidated. Let's illustrate this in the process of showing that the optional non-indifference of supererogation (ONIS) is validated: $\mathbf{BC}\varphi \to (\mathbf{OP}\varphi \,\&\, \neg\mathbf{IN}\varphi)$. Suppose $\mathbf{BC}\varphi$ holds at *i*. Then by the truth conditions for **BC**, it follows that there is an *i*-acceptable world where φ holds (so $\mathbf{PE}\varphi$ holds) and all the lowest ranked of the *i*-acceptable worlds preclude φ (so $\mathbf{MI}\neg\varphi$ holds, and so $\mathbf{PE}\neg\varphi$ holds). So it follows that $\mathbf{OP}\varphi$; but since there is also an *i*-level with only $\neg\varphi$-worlds, $\neg\mathbf{IN}\varphi$ holds. Similarly for permissible suboptimality. In fact, we can easily show this generalization is validated: $(\mathbf{BC}\varphi \vee \mathbf{PS}\varphi) \to (\mathbf{OP}\varphi \,\&\, \neg\mathbf{IN}\varphi)$. Here is a sample of additional plausible principles that are validated, but not mentioned above:

$\mathbf{OB}\varphi \to (\mathbf{MI}\varphi \,\&\, \mathbf{MA}\varphi)$	(MA-MI SUBORDINATION)
$(\mathbf{MI}\varphi \vee \mathbf{MA}\varphi) \to \mathbf{PE}\varphi$	(PE-SUBORDINATION)
$\mathbf{BC}\varphi \to \neg\mathbf{BC}\neg\varphi$ and $\mathbf{PS}\varphi \to \neg\mathbf{PS}\neg\varphi$	(BC/PS-No Conflicts)
$\mathbf{IN}\varphi \to \neg{*}\varphi$ [* here for **OB**, **IM**, **NO**, **MA**, **MI**, **BC**, **PS**]	(IN-EXCLUSION)

Increased Expressive Power in the DWE Framework

The increase in expressive power is illustrated by contrasting the earlier Traditional Threefold Classification, with the 12-fold classification in Fig. 10.

The partition relies on the four basic DWE operators (**OB**, **MA**, **MI**, and **IN**) and shows some derivative labels in parenthesis or via curly bracket labels exterior to the main rectangle. The changes are in the enriched partitioning of the optionality sphere: where there was before just one position, optionality itself, there are now ten subclasses of optionality. This illustrates another reason reflection on supererogation is productive: it reveals subtleties in our moral conceptual scheme.

Similarly, the contrast with the deontic hexagon is pronounced. The following octodecagon in Fig. 11 only covers a portion of what a full figure would be.[22]

We now turn to some reflections on agent evaluation and classical conceptions of supererogation and suberogation.

---

[22] For example, consider just the operator **MA**. Neither $\neg\mathbf{MA}\varphi$ nor $\neg\mathbf{M}\neg\varphi$ nor $\neg\mathbf{MA}\varphi \,\&\, \neg\mathbf{M}\neg\varphi$ (optimality indifference) are listed on any of the nodes, similarly for **MI**, **PS**, and **BC**. There is *only one* subcontrary relation indicated. (There are four exterior grayed out lines that are idle—they better allow for the gestalt of the figure as 18-*sided* not just 18-pointed.).

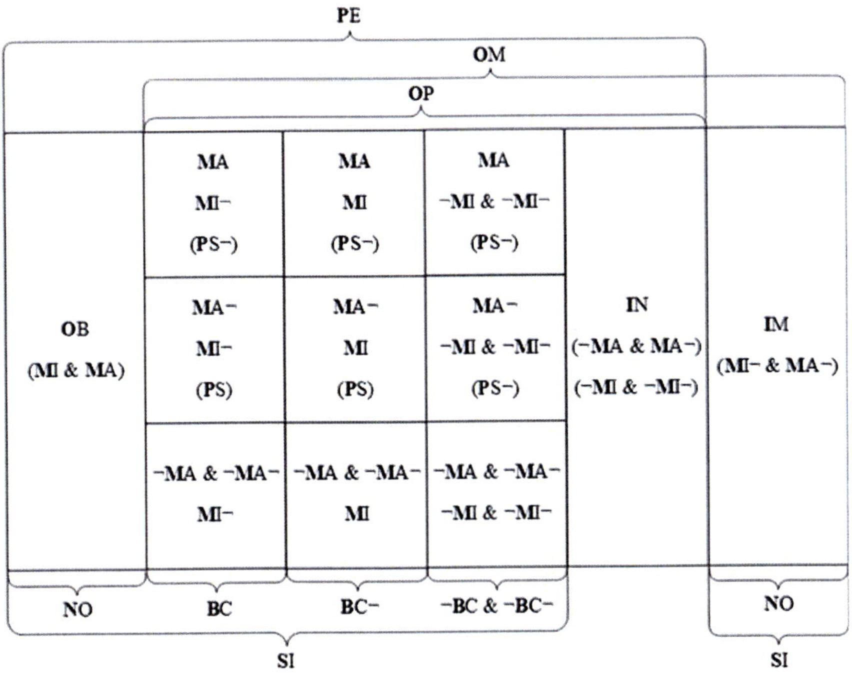

Fig. 10 The DWE twelvefold partition

3 Adding an Aretaic Module: Agent Evaluation, Supererogation, and Suberogation

3.1 A Framework for Agent-Evaluative Appraisal

3.1.1 *What of Supererogation and Offense (Suberogation)*?

There is no representation in DWE of *agent-evaluative notions* (or 'aretaic notions'—I will use these interchangeably). Yet supererogation has been widely taken to analytically entail *praiseworthiness*, a paradigmatically agent-evaluative concept. 'Supererogation' is used primarily in philosophical contexts (see Heyd, 2012) for historical discussion). Furthermore, even if some want to quibble about this, it seems clear that doing more than you must—going beyond the call, seems to have no agent-evaluative component, yet it is important to distinguish cases where this is praiseworthy. I have used '**BC**' in anticipation of using 'beyond the call' for the non-agent-evaluative notion and will use **SU** and 'supererogation' going forward for the result of adding an additional layer of agent evaluation as a requirement. The concepts are important. (McNamara, 2011a, 2011b) argues that there is a need to distinguish two concepts (at least), that one can go beyond the call without being praiseworthy

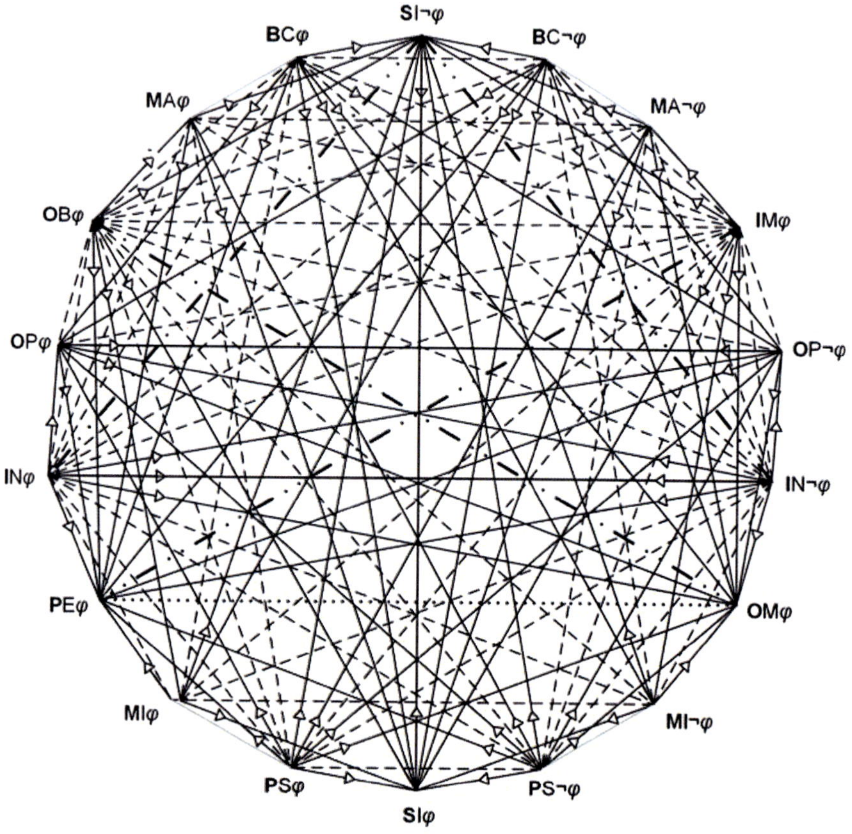

Fig. 11 A deontic octodecagon

at all, and that there are excellent reasons to use 'going beyond the call' as a marker for a more colloquial speech like that above. If I am right about 'supererogation' the operator **BC** cannot represent it, and this typically endorsed equivalence, φ is *supererogatory* iff φ is *beyond the call*, holds only if DWE fails to represent either notion. Furthermore, what of the notion of *suberogation* or *offense*? An offense, as used by philosophers in this context seems to analytically entail *blameworthiness*, but the latter is not expressible in the DWE framework, so how can offenses, the purported mirror image of supererogation, be expressed in that framework? Without agent appraisal, these concepts seem inexpressible in DWE.

Although the claim that supererogation analytically entails praiseworthiness might be disputed (Archer, 2016) and some might just assert that the italicized equivalence above is analytic, things are hardly obvious (see McNamara, 2011b), and exploring integrating deontic notions with agent-evaluative notions is clearly worthwhile. We sketch next the picture outlined in McNamara (2000), and especially (McNamara,

2011a) for adding a semantic module and extension of the language allowing for agent evaluation.

Sketch of the Framework, AA (for 'Aretaic Appraisal')

We evaluate agents for motives, intentions, traits, actions, states of affairs, etc. I will assume propositions can serve:

that Jane Doe has motive M/has intentions I/has trait T/performs action A/...

The core idea AA will invoke is that some propositions (would if realized) *reflect well* on people, others *poorly*, some *more favorably* than others, and some *neutrally*. Let's stick to *all things considered* assessments of Jane Doe for just those propositions that are *consistent with Jane Doe's abilities at a state, i:* CO_i. We imagine CO_i is derived from a standard accessibility relation, *COij*, on worlds, *W*, read as *j is a state consistent with Doe's abilities in i* as follows. CO_i will denote the set of all propositions true at some world consistent with Jane's abilities at *i*.[23] For convenience, we will think of propositions here world-theoretically, as sets of worlds where they are true. We imagine an ordering relation, so that for any propositions *X* and *Y* in CO_i,

$X \geq_i Y$ iff *X would reflect at least as well on Jane Doe as Y* (per *i*).

I will sometimes drop the qualifier 'would' in what follows, but it should be understood.[24] We imagine that $\geq_i$ is both reflexive and transitive: for propositions consistent with her ability, *X* reflects as well on her as itself, and if *X* reflects as well on her as *Y*, and *Y* as well as *Z*, then *X* reflects as well on her as *Y*. We add a corresponding basic operator, $p \geq q$, to the language of DWE-AA, and we imagine it relative to some arbitrary Jane Doe. We then can define *strong preference* and *equi-ranking* relations at the syntactic and semantic levels in standard ways:

$$X >_i Y \stackrel{\text{def}}{=} X \geq_i Y \,\&\, \neg(Y \geq_i X) \qquad p > q \stackrel{\text{def}}{-} p \geq q \,\&\, \neg(q \geq p)$$

$$X \approx_i Y \stackrel{\text{def}}{=} X \geq_i Y \,\&\, Y \geq_i X \qquad p \approx q \stackrel{\text{def}}{=} p \geq q \,\&\, q \geq p$$

We take tautological propositions to reflect neutrally on all agents.[25] This will be an anchor. We can then define these four more familiar agent-evaluative notions:

Praiseworthy (PW) propositions as ones ranked higher than a tautology
Blameworthy (BW) propositions as ones ranked lower than a tautology

---

[23] This is less strong than what is within Jane Doe's abilities, which would be analyzed as what is consistent with her abilities (at *i*) *to bring about*. It is consistent with her abilities that the sun will rise tomorrow, but it is not consistent with her abilities that she brings that about, whereas it is consistent with her abilities that she brings it about that her hand is raised. See McNamara (2000, 2011a).

[24] That is, we are interested in agent evaluation of propositions that are true but also that merely might be true consistent with her abilities, especially ones she could bring about.

[25] That is, they reflect no better nor worse on Jane than does the proposition that all cats are cats.

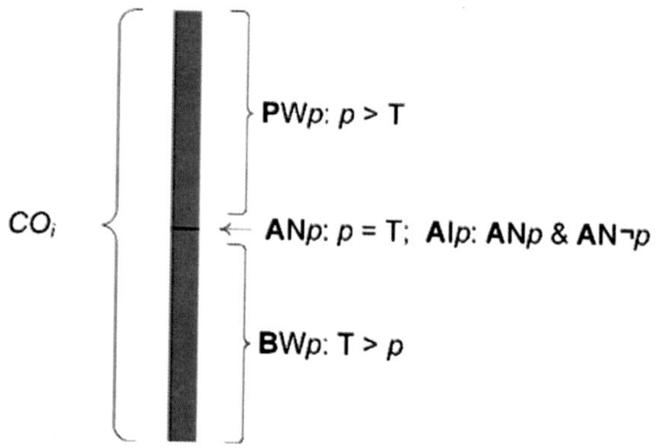

Fig. 12 Some aretaic concepts via an aretaic ordering

Aretaically Neutral (AN) propositions as propositions ranked equally with tautologies

Aretaically Indifferent (AI) propositions as neutral ones with neutral negations.

Here it is in a picture (Fig. 12).

We gather all the propositions consistent with Jane Doe's abilities together and then rank them according to how well they would reflect on Jane all-in-all if realized. Those that reflect better on Jane than tautologies, reflect favorably on her, and so she is to be evaluated as worthy of at least some degree of praise were such a proposition realized; those ranked below a tautology reflect unfavorably on her, and so she is worthy of some degree of blame were such a proposition realized; those ranked equally with a tautology are aretaically neutral for Jane, and those whose negations also are aretaically neutral for Jane are aretaically indifferent for Jane.

We need *not* endorse $\geq$ -*Connectivity* as basic,

$$\forall i \forall X \forall Y[X, Y \in CO_i \rightarrow (X \geq_i Y \vee Y \geq_i X)]$$

that for every pair of propositions, one either reflects as well on Jane as the other, or vice versa (so the diagram above simplifies things). However, connectivity generates things presupposed in what I will call the 'classical framework for supererogation and kin,' so let us assume it henceforth. Here are some consequences:

CO-COMP: $(\mathbf{CO}p \,\&\, \mathbf{CO}q) \rightarrow (p \geq q \vee q \geq p)$

AR-EXH: $\mathbf{CO}p \rightarrow (\mathbf{AN}p \vee \mathbf{PW}p \vee \mathbf{BW}p)$

AI-EQUIV: $\mathbf{AI}p \leftrightarrow (\mathbf{CO}p \,\&\, \neg\mathbf{BW}p \,\&\, \neg\mathbf{BW}\neg p \,\&\, \neg\mathbf{PW}p \,\&\, \neg\mathbf{PW}\neg p)$.

CO-Comparability says if p and q are each consistent with Jane's abilities, then either p reflects as well on her as q or vice versa. Aretaic-Exhaustion says if p is consistent with Jane's abilities, either p reflects neutrally, favorably, or unfavorably

BW	BN	BW¬

PW	PN	PW¬

Fig. 13 Two simple threefold aretaic classifications

on Jane. AI-Equivalence says p is aretaically indifferent iff p is consistent with Jane's abilities and neither praiseworthy nor blameworthy, and likewise for $\neg p$.

Do all things considered praiseworthiness and blameworthiness satisfy *No Conflicts* principles?

$$\text{PW-NC}: \neg(\mathbf{PW}p \,\&\, \mathbf{PW}\neg p) \quad \text{BW-NC}: \neg(\mathbf{BW}p \,\&\, \mathbf{BW}\neg p)$$

These are at least plausible to consider for all-in-all aretaic appraisal and they are presupposed in the classical conceptions of supererogation and offense, so let us assume they hold too. We can generate them by adding two semantic constraints:

$$\text{PW-NC}' : \forall i \forall X (X >_i \top \rightarrow \neg(\mathrm{W} - \mathrm{X} >_{\mathrm{i}} \top))$$
$$\text{BW-NC}' : \forall i \forall X (\top >_i X \rightarrow \neg(\top >_i W - X))$$

That is, if a proposition reflects better on Jane than a tautology, then not so for its negation; and if a proposition reflects worse on Jane than a tautology, then not so for its negation.

Now simple aretaic partitions of the propositions consistent with Jane's abilities for **PW** and for **BW** are derivable (Fig. 13).

Where **PN**p is read as p is *Praise Neutral* and **BN**p as p is *Blame-Neutral*, each defined as follows:

$$\mathbf{PN}p \stackrel{\text{def}}{=} \neg\mathbf{PW}p \,\&\, \neg\mathbf{PW}\neg p \quad (\text{It is } \mathit{Praise\ Neutral} \text{ that } p)$$
$$\mathbf{BN}p \stackrel{\text{def}}{=} \neg\mathbf{BW}p \,\&\, \neg\mathbf{BW}\neg p \quad (\text{It is } \mathit{Blame} - \mathit{Neutral} \text{ that } p)$$

As usual, the partitions imply mutually exclusive and jointly exhaustive categories for the propositions consistent with Jane's abilities. If we combine those two partitions, we get this sevenfold **PW-BW** aretaic partition of the propositions consistent with Jane's abilities (Fig. 14).

"Sevenfold" because the two corner shaded cells are ruled out in our framework (e.g., if p were ranked both higher and lower than a tautology (as in the top left corner), by transitivity, p would be ranked higher than p, which we've rightly ruled out).

We now have the beginning of a framework for some fundamental agent-evaluative notions. How might we integrate this with our prior more action-evaluative deontic notions?

	BW:	BN:	BW¬:
PW:	**PW & BW**	**PW & BN**	**PW & BW¬**
PN:	**PN & BW**	**AI**	**PN & BW¬**
PW¬:	**PW¬ & BW**	**PW¬ & BN**	**PW¬ & BW¬**

Fig. 14 A sevenfold **PW-BW** aretaic classification

3.2 Integrating DWE's More Act-Evaluative Notions with AA's Agent-Evaluative Notions

Weaving in the Categories of the Traditional Deontic Scheme

What happens if we blend in our earlier deontic threefold classification? (Fig. 15).

We get this twenty-one-fold aretaic-deontic partition (Fig. 16).

With this simple combination of aretaic and deontic partitions, we can begin to explore the logic of a variety of moral conditions of interest. Here is a general question:

How are deontic and aretaic conditions related to one another, and how far can deontic and aretaic valences diverge (if at all)?

Consider these bridging (or reductive) principles:

(a) No PW-IM Conflicts: ¬(**PW**p & **IM**p) [i.e.**PW**p → **PE**p]
(b) No BW-OB Conflicts: ¬(**BW**p & **OB**p) [i.e.**BW**p → **PE**¬p].

(a) Disallows all cases of impermissible praiseworthiness, and (b) disallows all cases of obligatory blameworthiness. These principles are reductive. Were we to

OB	**OP**	**IM**

Fig. 15 Traditional threefold classification

	PW & BN:	PW & BW¬:	PN & BW:	AI:	PN & BW¬:	PW¬ & BW:	PW¬ & BN:
OB:	PW & BN & OB	PW & BW¬ & OB	PN & BW & OB [elim by b)]	AI & OB	PN & BW¬ & OB	PW¬ & BW & OB [elim by a/b)]	PW¬ & BN & OB [elim by a)]
OP:	PW & BN & OP (SU*) [elim by c)]	PW & BW¬ & OP (QS) [elim by c)/d)]	PN & BW & OP (OF*) [elim by d)]	AI & OP (FI)	PN & BW¬ & OP (OF*¬) [elim by d)]	PW¬ & BW & OP (QO) [elim by c)/d)]	PW¬ & BN & OP (SU*¬) [elim by c)]
IM:	PW & BN & IM [elim by a)]	PW & BW¬ & IM [elim by a)/b)]	PN & BW & IM	AI & IM	PN & BW¬ & IM [elim by b)]	PW¬ & BW & IM	PW¬ & BN & IM

Fig. 16 Twenty-one-fold aretaic-deontic partition

add these, six possibilities in our 21-fold partition would be eliminated, as indicated in the annotation of the shaded boxes in the top and bottom rows there: (a) would eliminate the last two shaded cells of the top row, and the first two of the bottom row; (b) would eliminate the first two shaded cells of the top row and the last two shaded cells of the bottom row. Adding both principles together would leave only fifteen deontic-aretaic positions (those lightly shaded or unshaded in the partition above).

A Deontic-Aretaic Account of Supererogation and Offense and Mellema's Extensions

A *classical analysis* of supererogation and offense is easy to define with our mixed framework:

$$\mathbf{SU}^{\mathrm{a}}p \overset{\mathrm{def}}{=} \mathbf{OP}p \;\&\; \mathbf{PW}p \;\&\; \neg\mathbf{BW}\neg p$$

$$\mathbf{OF}^{\mathrm{a}}p \overset{\mathrm{def}}{=} \mathbf{OP}p \;\&\; \mathbf{BW}p \;\&\; \neg\mathbf{PW}\neg p.$$

On this account, the deontic and agent-evaluative notions are essential. Something is supererogatory (for Jane) if it is optional, praiseworthy, but not blameworthy to skip. Something is an offense if it is optional, blameworthy, but not praiseworthy to skip (Mellema, 1991) endorses a position much like this, and Chisholm offers something like this as an alternative to his choice of primitives (Chisholm, 1963b/1982).[26] To the extent that philosophers use the term 'supererogation', it seems they assume the right side of the first definition holds, and the second makes an offense an analog to supererogation so conceived. In DWE, we defined the notion of permissible suboptimality, but that can't play the role of an offense, since an action can be beyond the call and still be suboptimal, as when our hero saves Tiny Tara from the fire but does not go back in to save Tiny Tim (where either is beyond the call but saving both is best). So we seem to need the aretaic notions for an account of suberogation/offense, and I think we do for 'supererogation' as classically used by philosophers as well. (This classical style of analysis of **SU** and **OF** can be shown to fail in being too weak (the defining conditions are insufficient for the target concepts). See Hansson (2001) and McNamara (2011b) for different arguments. For example, morality's demands might be so great that it is praiseworthy to do the minimum, beyond the call to do more, and so optional to do the minimum, but not blameworthy to not do it. In such a case, we clearly do not have something supererogatory, since it involves doing the least one can do, yet it meets the conditions of the classical analysis.

In logical space, there are still more distinctions, even if for substantive reasons some might be rejected. (We need the conceptual space to raise the questions in the first place.) Mellema (1987), (1991) argues for acts of *quasi-supererogation* and *quasi-offense*, which we define as follows:

[26] Not all account of supererogation follow this path (e.g., Heyd, 1982) offers a nuanced alternative). (McNamara, 2021) provides a detailed analysis of Chisholm's positions in Chisholm (1963a, 1963b), and in Chisholm (1964, 1974).

$$\mathbf{QS}p \stackrel{\text{def}}{=} \mathbf{OP}p \,\&\, \mathbf{PW}p \,\&\, \mathbf{BW}\neg p$$

$$\mathbf{QO}p \stackrel{\text{def}}{=} \mathbf{OP}p \,\&\, \mathbf{BW}p \,\&\, \mathbf{PW}\neg p$$

So, something is quasi-supererogatory if it is optional, praiseworthy, but *is* blameworthy to skip, and something is a quasi-offense if it is optional, blameworthy, but *is* praiseworthy to skip.

Finally, we introduce one more mixed concept, *optional aretaic indifference*:

$$\mathbf{OI}p \stackrel{\text{def}}{=} \mathbf{OP}p \,\&\, \mathbf{AI}p.$$[27]

Normative Positions in This Expanded Framework and Two More Reductive Schemes

The five new operators defined above are already sitting ripe for the picking in our prior 21-fold partition. (See the middle row parentheticals.). So lingering behind the classical conception of supererogation is a potential for at least twenty-one mutually exclusive categories.

Now, some argue for the rejection of supererogation by endorsing:

(c) $\mathbf{PW}p \rightarrow \mathbf{OB}p$.[28]

(c) Entails that nothing is supererogatory, quasi-supererogatory or a quasi-offense, thus eliminating the first two and the last two of the lightly shaded cells in the middle row of our 21-fold partition. Only the three middle cells of the middle row would remain (only offenses and the fully indifferent categories in the middle. Given *No Conflicts* for **OB** (c) also entails our earlier (a) $\mathbf{PW}p \rightarrow \mathbf{PE}p$, so the last two shaded cells of the top row, and the first two of the bottom row are ruled out as well. Principle (c) is *highly* eliminative.

The following bridging principle is often endorsed (in arguing for the rejection of suberogation, but also in recent discussions of determinism[29]):

(d) $\mathbf{BW}p \rightarrow \mathbf{IM}p$

(d) Entails nothing is an offense, quasi-supererogatory, or a quasi-offense, thus eliminating all but the central and end cells of the middle row. Only the supererogatory and full indifference categories remain in the middle row. Given *No Conflicts* for **OB**, (d) also entails our earlier (b) $\mathbf{BW}p \rightarrow \mathbf{PE}\neg p$, so the first two shaded cells of the top row and the last two shaded cells of the bottom row also go. Principle (d) is also highly eliminative.

Summarizing, if *either* (c) *or* (d) hold, Mellema's quasi-notions are out, along with supererogation or suberogation, and two of the three shaded boxes in the top

---

[27] We could also sensibly introduce *weak-supererogation* and *weak-offense*: $\mathbf{WS}p \stackrel{\text{def}}{=} \mathbf{OP}p \,\&\, \mathbf{PW}p$ (i.e., $\mathbf{SU}p \vee \mathbf{QS}p$) and $\mathbf{WO}p \stackrel{\text{def}}{=} \mathbf{OP}p \,\&\, \mathbf{BW}p$ (i.e., $\mathbf{OF}p \vee \mathbf{QO}p$).

[28] C) is an old saw in objections to supererogation (Cf. (Pybus, 1982); Cf the first puzzle in (Munoz, 2021).

[29] E. g. (Widerker, 1991).

and bottom rows; in each case, there are eight eliminations, leaving only thirteen normative positions; if *both* (c) *and* (d) hold, there are twelve eliminations, leaving just the nine white unshaded positions in our 21-fold partition.

Weaving in the categories of DWE

What happens if we combine the twenty-one aretaic normative positions above with DWE's deontic positions? If we make the combination in question, we have seven aretaic positions and twelve deontic positions, we get eighty-four combined positions, provided we add no eliminating additional constraints like those explored above. The logical space for the normative positions is large, as seen in Fig. 17 further below. The seven possible aretaic positions range horizontally across the top of the main rectangle, and the twelve possible DWE deontic positions range vertically down the left side. This includes improved conditions for the classical analysis of supererogation

	PW & BN:	PW & BW¬:	PN & BW:	AI:	PN & BW¬:	PW¬ & BW:	PW¬ & BN:
OB:	PW & BN & OB	PW & BW¬ & OB	PN & BW & OB	AI & OB	PN & BW¬ & OB	PW¬ & BW & OB	PW¬ & BN & OB
MA & MI¬:	PW & BN & MA & MI¬	PW & BW¬ & MA & MI¬	PN & BW & MA & MI¬	AI &MA & MI¬	PN & BW¬ & MA & MI¬	PW¬ & BW & MA & MI¬	PW¬ & BN & MA & MI¬
MA¬ & MI¬:	PW & BN & MA¬ & MI¬	PW & BW¬ & MA¬ & MI¬	PN & BW & MA¬ & MI¬	AI & MA¬ & MI¬	PN & BW¬ & MA¬ & MI¬	PW¬ & BW & MA¬ & MI¬	PW¬ & BN & MA¬ & MI¬
¬MA & ¬MA¬ & LE¬:	PW & BN & ¬MA & ¬MA¬ & LE¬	PW & BW¬ & ¬MA & ¬MA¬ & LE¬	PN & BW & ¬MA & ¬MA¬ & LE¬	AI & ¬MA & ¬MA¬ & LE¬	PN & BW¬ & ¬MA & ¬MA¬ & LE¬	PW¬ & BW & ¬MA & ¬MA¬ & LE¬	PW¬ & BN & ¬MA & ¬MA¬ & LE¬
MA & MI:	PW & BN & MA & MI	PW & BW¬ & MA & MI	PN & BW & MA & MI	AI & MA & MI	PN & BW¬ & MA & MI	PW¬ & BW & MA & MI	PW¬ & BN & MA & MI
MA¬ & MI:	PW & BN & MA¬ & MI	PW & BW¬ & MA¬ & MI	PN & BW & MA¬ & MI	AI & MA¬ & MI	PN & BW¬ & MA¬ & MI	PW¬ & BW & MA¬ & MI	PW¬ & BN & MA¬ & MI
¬MA & ¬MA¬ & MI:	PW & BN & ¬MA & ¬MA¬ & MI	PW & BW¬ & ¬MA & ¬MA¬ & MI	PN & BW & ¬MA & ¬MA¬ & MI	AI & ¬MA & ¬MA¬ & MI	PN & BW¬ & ¬MA & ¬MA¬ & MI	PW¬ & BW & ¬MA & ¬MA¬ & MI	PW¬ & BN & ¬MA & ¬MA¬ & MI
MA & ¬MI & ¬MI¬:	PW & BN & MA & ¬MI & ¬MI¬	PW & BW¬ & MA & ¬MI & ¬MI¬	PN & BW & MA & ¬MI & ¬MI¬	AI & MA & ¬MI & ¬MI¬	PN & BW¬ & MA & ¬MI & ¬MI¬	PW¬ & BW & MA & ¬MI & ¬MI¬	PW¬ & BN & MA & ¬MI & ¬MI¬
MA¬ & ¬MI & ¬MI¬:	PW & BN & MA¬ & ¬MI & ¬MI¬	PW & BW¬ & MA¬ & ¬MI & ¬MI¬	PN & BW & MA¬ & ¬MI & ¬MI¬	AI & MA¬ & ¬MI & ¬MI¬	PN & BW¬ & MA¬ & ¬MI & ¬MI¬	PW¬ & BW & MA¬ & ¬MI & ¬MI¬	PW¬ & BN & MA¬ & ¬MI & ¬MI¬
¬MA & ¬MA¬ & ¬MI & ¬MI¬:	PW & BN & ¬MA & ¬MA¬ & ¬MI & ¬MI¬	PW & BW¬ & ¬MA & ¬MA¬ & ¬MI & ¬MI¬	PN & BW & ¬MA & ¬MA¬ & ¬MI & ¬MI¬	AI & ¬MA & ¬MA¬ & ¬MI & ¬MI¬	PN & BW¬ & ¬MA & ¬MA¬ & ¬MI & ¬MI¬	PW¬ & BW & ¬MA & ¬MA¬ & ¬MI & ¬MI¬	PW¬ & BN & ¬MA & ¬MA¬ & ¬MI & ¬MI¬
IN:	PW & BN & IN	PW & BW¬ & IN	PN & BW & IN	AI & IN	PN & BW¬ & IN	PW¬ & BW & IN	PW¬ & BN & IN
IM:	PW & BN & IM	PW & BW¬ & IM	PN & BW & IM	AI & IM	PN & BW¬ & IM	PW¬ & BW & IM	PW¬ & BN & IM

Fig. 17 An eighty-four-fold aretaic-deontic partition

and suberogation (offense) that I have argued for in McNamara (2011a, 2011b):

$$\mathrm{SU}'p : \mathbf{BC}p \,\&\, \mathbf{PW}p \,\&\, \neg\mathbf{BW}\neg p$$
$$\mathrm{OF}'p : \mathbf{PS}p \,\&\, \mathbf{BW}p \,\&\, \neg\mathbf{PW}\neg p.$$

So, something is now supererogatory if it is beyond the call (per DWE), praiseworthy, and not blameworthy to skip, and something is now an offense if it is permissibly suboptimal (per DWE), and blameworthy and not praiseworthy to skip. This, for example, resolves the problem raised above with the classical analysis of supererogation since it adds a further condition to the right side that rules out the problematical case where doing the minimum comes out as supererogatory. Supererogation here becomes a praiseworthy case of going beyond the minimum that is not blameworthy to skip. Similarly, something is now an offense if it is a blameworthy case of permissible suboptimality that is not praiseworthy to skip. I am proposing a distinction between going beyond the call (doing more than the minimum) and supererogation, with the latter involving an *agent*-evaluative element (in addition to doing more good than you have to), and similarly for suboptimality and suberogation/offense (McNamara, 2011a, 2011b). This provides a coherent conception of an offense/suberogation. Mere suboptimality, a notion that is more act-evaluative, surely is not enough given that one can go beyond the call and yet be doing less than the best. For suberogation to be coherent enough to argue about, it must presumably be both suboptimal and permissible, but more is needed, and it is hard to see what is more other than something aretaic.[30]

4 Digging a Bit Deeper and Wrapping Up

4.1 *Interlude: Revisiting the DWE Structures*

Here we briefly return to reflecting on the acceptable and unacceptable worlds and their ordering. Three questions:

Q1. What is the basis for the singular ranking of all the accessible/reachable worlds (i.e., the acceptable and unacceptable worlds?

Q2. In our ranking of acceptable and unacceptable worlds, what determines the cut off?

Q3. Should all worlds ranked as high as an acceptable world be themselves acceptable?

We will return to the last question further below. First, I want to illustrate how the first two questions might be answered. We have been assuming that we have an ordering of all the worlds that are *accessible/ reachable* for Jane given her abilities

[30] See McNamara (2011b), 226–228 on the language of 'offence' and the contention that suberogation is a symmetric analog to supererogation.

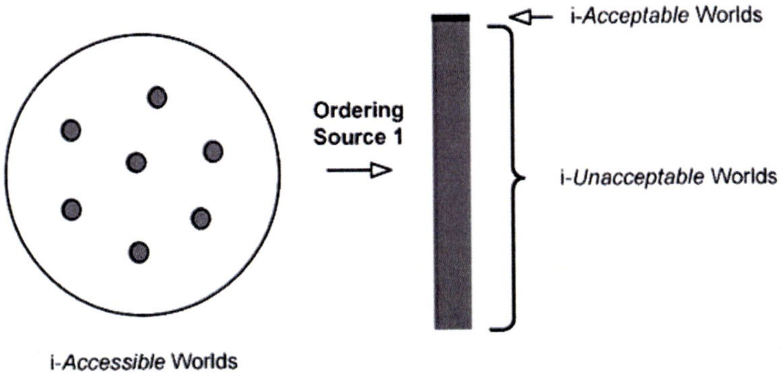

Fig. 18 How the acceptable worlds might emerge from the accessible ones

and situation in *i*, but they are also divided so that the *i*-acceptables ones are all above the *i*-unacceptable ones (given GOO). We might also ask do we rank the *i*-unacceptable worlds in the same way as the acceptable ones—using the same ranking consideration(s)? Let us work our way to a possible answer to Q1 and Q2. Set aside the i-unacceptable worlds for the moment, and let us suppose that we have a non-trivial antecedent ordering source yielding the *i-acceptable* worlds from the *i-accessible/reachable* ones (Fig. 18).

For the sake of concreteness, assume the ranking of the *accessible* worlds is by *justice* (e.g., minimizing justice violations relativized to Jane Doe in some way), so that the *acceptable* worlds among her accessible worlds are just those most just in this ranking. How might we now rank her acceptable worlds? Not by justice as they are all tied in that respect. We need a second ranking source. For the sake of concreteness: assume a ranking by *social welfare* (again relativized to Jane Doe).[31] OK, so we now have an example of how the acceptable worlds might be determined, and then ranked; but this just invites the question: *How then do we rank the missing i-unacceptables? By (a) a social welfare ranking or (b) a justice ranking?* Classic contrary to duty (CTD) style reasoning serves here: Let *X* be the proposition that characterizes the de facto *i*-acceptable worlds (say, that Jane does her justice best, or that she minimizes justice violations). We ask what *would be i*-acceptable *given that X* is foreclosed (the justice-best world will not be selected).[32] We naturally turn to the *next justice-best* worlds.

This means a simple social welfare ranking of the *i-un*acceptables cannot be right. For that might rank some worlds (e.g., organ-harvesting worlds) that are neither justice-best nor next-justice-best *above* the next-best justice-ranked worlds, and that

[31] An alternative conceptualization might be that the acceptable worlds are determined by the best attunement to the balance of *requiring* reasons, and the ranking in turn of those worlds is by the balance of *favoring* reasons; no doubt there are others. The strategy here does not depend on the exact interpretation, but on the role of the orderings.

[32] The particulars for contrary-to-duty reasoning will depend on what specific violations occur (See McNamara, 2022), but here we are asking what is next best given only that some violation will occur.

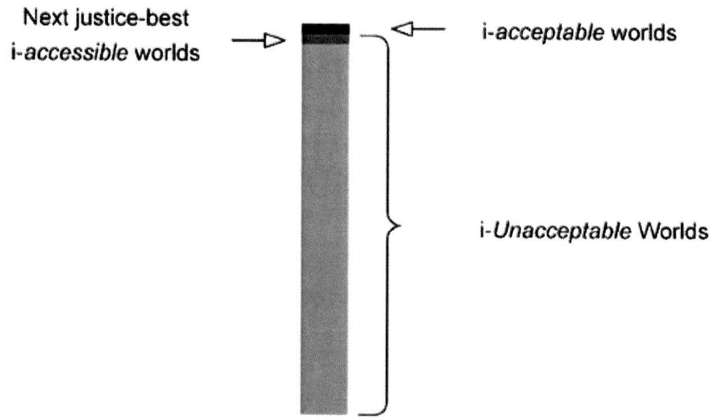

Fig. 19 Next-best justice worlds and ranking the unacceptable worlds

does not fit the suggested priority of the *first* ordering source over the *second*. So option (a) fails. That leaves (b): order them by justice ranking. But (b) cannot be right either. For this would ignore morally relevant distinctions in *social welfare* ranking among the unacceptable worlds that are *equi-ranked* as *next-justice-best* worlds. What to do? (Fig. 19).

This is where CTD-style reasoning blends with that called for by supererogation and kin. The picture that emerges is indicated in Fig. 20.

We rank the second-best justice-ranked worlds via social welfare and place those so-ranked 'just below' their earlier cousin—the best-justice-ranked worlds (the acceptable ones) ranked via social welfare; we then rank the third-best justice-ranked worlds via social welfare under the social welfare ranked second-best justice worlds, and so on, thus ordering all Jane's reachable worlds.

Let us regiment a bit. Imagine we have three ordering relations:

$^1\succsim_i$: 1st ordering per i (e.g., justice)
$^2\succsim_i$: 2nd ordering per i (e.g., social welfare)
$^3\succsim_i$: 3rd composite ordering per i (e.g., moral all-in-all ordering).

Take the first two as primitive here, and analyze the third as follows:

$$j\,^3\succsim_i k \overset{def}{=} j\,^1\succ_i k \text{ or both } j\,^1\approx_i k \;\&\; j\,^2\succsim_i k.$$

For concreteness, j is morally (at least) as good as k if and only if and because either j is more just than k or j is equi-just to k and *it* has (at least) as much social welfare value as k. For each ordering relation n (1, 2, 3), we assume these familiar definitions of outranking and equi-ranking relations:

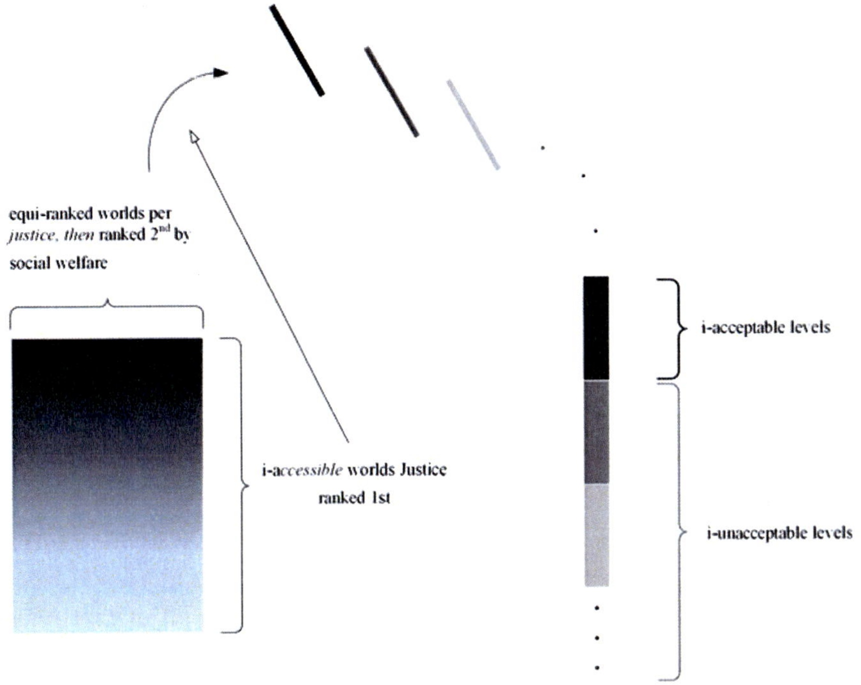

Fig. 20 Ranking of all Jane's accessible/reachable worlds via two orderings [33]

$$j\ {}^{n}\succ_i\ k \stackrel{\text{def}}{=} j\ {}^{n}\succsim_i k \,\&\, \neg\left(k\ {}^{n}\succsim_i j\right)$$
$$j\ {}^{n}\approx_i\ k \stackrel{\text{def}}{=} j\ {}^{n}\succsim_i k \,\&\, k\ {}^{n}\succsim_i j.$$

Assume ${}^{1}\succsim_i$ and ${}^{2}\succsim_i$ are reflexive and transitive relations on the *i*-accessible worlds. Then these follow:

(A) $j\ {}^{1}\succ_i\ k \rightarrow j\ {}^{3}\succ_i\ k$
(B) $j\ {}^{3}\succsim_i\ k \rightarrow j\ {}^{1}\succsim_i\ k$
(C) $j\ {}^{3}\succsim_i\ k \nrightarrow j\ {}^{2}\succsim_i\ k$
(D) $j\ {}^{2}\succ_i\ k \,\&\, j\ {}^{1}\approx_i\ k. \rightarrow j\ {}^{3}\succ_i\ k$
(E) ${}^{3}\succsim_i$is reflexive, and transitive: (i) $j\ {}^{3}\succsim_i j$ and (ii)$(j\ {}^{3}\succsim_i k \,\&\, k\ {}^{3}\succsim_i l) \rightarrow j\ {}^{3}\succsim_i l$

Via our sample interpretation of the first two relations, (A) says if *j* justice outranks *k*, then it does all-in-all too; (B) *j* is ranked at least as high as *k* all-in-all only if it is so as well when justice ranked; (C) *j*'s outranking *k* all-in-all does *not* imply it outranks *j* per social welfare; (D) if *j* outranks *k* per social welfare and they are equi-ranked per justice, then *j* outranks *k* all-in-all; (E) the resulting derivative all-in-all moral ranking is also reflexive and transitive.

[33] Darker hear is justice-better lighter, justice-worse.

We have now provided some ideas on how to answer Q1. In the process, we have also indicated how one might answer Q2: "In our singular ranking of acceptable and unacceptable worlds, what determines the cut off?" What determines the cut off is the lowest ranked worlds per the secondary ordering (e.g., welfare) of the highest ranked worlds per the primary ordering (e.g., justice). What is next down is the highest ranked per the secondary ordering (e.g., welfare) of the set of second-best worlds per the primary ordering (e.g., justice), and so forth, all the way down.

This also lends itself to an interpretation of *must* and *ought*, along with the other notions we analyzed. For the former pair, what Jane *must* do is what she does in the best accessible worlds per the primary ordering (say where she minimizes her justice violations), and what she *ought* to do is what is best all-in-all via the derived tertiary ordering, which means she ought to do the best per the secondary ordering (via welfare) of the best per the primary ordering (via justice) since given our work above on how to order them all: the best of the acceptables are the best all-in-all. With this deeper conception of the underpinnings, notice also that GOO is sustained: any world ranked as high *all-in-all* as an acceptable one must be acceptable too. For if *j* is an acceptable world, then it is one of the best per the primary ordering (say justice), but given how the third ordering is derived, any world ranked at least as high in the third all-in-all ordering must be one ranked among the best per the primary ordering, even if ranked lower per the social welfare ranking in isolation. The all-in-all ordering is derivative and preserves the idea that the secondary (welfare) ranking is subordinate to the primary (justice) ranking, and the third all-in-all ordering preserves this priority, so nothing ranked all-in-all as high as an acceptable world is unacceptable.

4.2 Some Brief Additional Reflections and Conclusion

Some Puzzles

It has long been asked "how can something that is beyond the call, since good, not be obligatory?" Framed this way, the question is naïve and muddled. For there is no reason pre-theoretically or on reflection for friends of supererogation to deny that graded supererogation is possible, that there can be mutually exclusive good alternatives with one better than the other, each beyond the call, as we noted in Sect. 2.3, and as I have noted elsewhere (e.g. McNamara, 1996b). This also indicates why we cannot assume, as is often done tacitly, that if A is beyond the call, then A should be done (morally speaking). The question really needs to be revised to: How can anything less than the best be permissible? This is an old saw, and here I note that answering 'it can't' would seem to imply that these three concepts collapse effectively (at least up to necessary equivalence): *What one ought to do* and *the best one can do* is *what one must do* and is *the least one can do.* The default is surely that *ought* and *must* and *the least one can do* are semantically distinct and can diverge, that we must model those distinctions, and doing so seems to clearly allow a logical space for graded permissible alternatives, and with that will fall out going beyond the call, permissible suboptimality, etc. And indeed, almost any of the single reflection points (in Sect. 2.3) used to motivate the DWE semantic structures can be used independently to do so yet they converge on the same (or very similar) structures so that there is a mutual corroboration suggesting that at least to a first approximation, we are on track.

What of Horton's all or nothing problem or Ulla Wessell's supererogatory holes (Horton, 2017; Wessels, 2002, 2015)? Here I must be brief. I will focus on Horton's example.[34] Suppose (as some have about this case) that rescuing *just one* from death at high cost when one can rescue two with no further effort or cost is deemed impermissible but rescuing none is deemed permissible as well as rescuing two. Then if the structural suggestions above are on track (framed again using our concrete interpretation), rescuing one will be ranked *lower* than rescuing none in the *overall moral ranking*, although obviously not so on a *purely social welfare ranking* of options. We might say that, in such a case, Jane rescuing one is a gross violation of justice as she treats the other as if of no value at all, but Jane's rescuing none does not involve any such violation of justice, though she falls short of what is morally ideal since she does not opt for the highest welfare option among the justice-best ones (the permissible ones).

That said, this take on the case is not without its challenges. Suppose Jane rescues just the one. Does it sound right to say: "Jane, *the least you could have done* was rescue none and just call rescue services" (with the previously noted pragmatic implication: 'and you didn't even do that!"). I have argued above and elsewhere that the 'could'/ 'can' here is that of permissibility and 'the least' is the least among such, so this ought to sound right if the preceding position on the case is correct. The alternative is to say it *is* permissible to rescue just one, and likewise for rescuing none (just calling rescue services say), as well as for rescuing both. We then, again in keeping with the resources of the DWE framework, say that you *ought* to have rescued both—that would have been best, and you *ought not to* have rescued just one—that ruled out the best, and likewise for Jane rescuing none, but we don't say she *must* either rescue just one or both, nor that she *must not* rescue just one. It then makes no sense to say to Jane when she rescues the one, "the least you could have done was rescue none and just call rescue services!"

However, here is a minor variation of a case I described in Q & A at the Formal Ethics 2014 conference in discussion with Ulla Wessels and Wodek Robinowicz on her position on supererogatory holes: I rescue two, but I have a skewed sense of justice and think that since many others die when not rescued if I rescue either of the two, there should be some loss to them (as a matter of justice), so I break an arm for each in the process of rescuing them. Do we say that the value of the rescue results in a state of affairs where I comported myself better all-in-all than not rescuing, so that it is permissible although again it *ought not* be done since the harm is de facto gratuitous and the rescue without the arm breaking is equally accessible and better, or do we instead say that it is better overall morally for me to do nothing—the

[34] I exposit and discuss Wessels theory of supererogation in McNamara (2021).

least I could have done was do nothing, and I did not even do that in saving the two cum arm breaks? Perhaps a better path here is to stick with the first option and invoke the agent-evaluative components and say that although the agent was overall praiseworthy for the rescue even as made (lives were saved and the rescuer's arm was lost and his overall intention was good), it was nonetheless *blameworthy to break the arms*. It does not follow from that that it was overall blameworthy to perform the rescue (as done).[35]

These brief remarks must suffice here. The cases that motivate 'supererogatory holes' and 'all or nothing' scenarios are indeed puzzling cases, but we should make sure we are considering the full gamut of our resources for evaluating agents' actions in exploring possible assessments.

Conclusion

Supererogation is a key mark of moral exemplars, so it warrants careful study for that reason alone. It is also a rich and subtle notion. Reflecting on supererogation calls on us to consider multiple concepts in its neighborhood, some of which are pregnant with implications (like 'the least one can do'), and it calls on us to develop models for these notions that allow their intuitive connections to shine through and be explained. The notions in the neighborhood are themselves often neglected notions, sometimes conflated with one another traditionally, at times thereby periling even the possibility of supererogation. Getting a coherent framework is hard and can be elusive, but the attempt can be expansive, widening the realm of conceptual possibilities in the moral domain and at once deepening our sense that it is a rich and coherent domain.

References

Archer, A. (2016). Are acts of supererogation always praiseworthy? *Theoria, 82*, 238–255.
Chisholm, R. M. (1963a). Supererogation and offence: A conceptual scheme for ethics. *Ratio, 5*, 1–14.
Chisholm, R. M. (1963b/1982). Supererogation and offence: A conceptual scheme for ethics. *Brentano and Meinong studies. (studien zur österreichinschen philosophie iii)*, 98–113. Amsterdam, Rodopi B.V. Originally published in Ratio, 5, 1–14, 1963b.
Chisholm, R. M. (1964). The ethics of requirement. *American Philosophical Quarterly, 1*, 147–153.
Chisholm, R. M. (1974). Practical reason and the logic of requirement. In S. Korner (Ed.), *Practical reason* (pp. 1–17). New Haven.
Hansson, S. O. (2001). *The structure of values and norms*. Cambridge University Press.
Heyd, D. (1982). *Supererogation: Its status in ethical theory*. Cambridge University Press.
Heyd, D. (2012). 'Supererogation.' *The Stanford Encyclopedia of Philosophy (Winter 2019 Edition)*. http://plato.stanford.edu/archives/win2012/entries/supererogation/
Horton, J. (2017). The all or nothing problem. *The Journal of Philosophy, 114*, 94–104.

[35] If it did, given the framework for aretaic appraisal we added to DWE, the rescue could not be all in all praiseworthy too, but surely we do not want to endorse the idea that a blameworthy component of a complex act entails that the complex act is overall blameworthy thereby and that it cannot be overall praiseworthy despite a blameworthy/faulted component.

Krogh, C., & Herrestad, H. (1996). Getting personal: Some notes on the relationship between personal and impersonal obligation. In M. A. Brown & J. Carmo (Eds.), *Deontic logic, agency and normative systems* (pp. 134–153). New York.

Mares, E. D., & McNamara, P. (1997). Supererogation in deontic logic: Metatheory for dwe and some close neighbours. *Studia Logica, 59*, 397–415.

McNamara, P. (1990). *The deontic quaddecagon.* Dissertation. University of Massachusetts.

McNamara, P. (1994). *Society for exact philosophy.* University of Texas-Austin.

McNamara, P. (1996a). Doing well enough: Toward a logic for commonsense morality. *Studia Logica, 57*, 167–192. Earlier version In A.J.I. Jones & M. Sergot (Eds.), *DEON '94: Second International Workshop on Deontic Logic in Computer Science (Workshop Proceedings)* (pp. 165–197), Oslo 6–8 January 1994, 1993. Oslo: Tano.

McNamara, P. (1996b). Making room for going beyond the call. *Mind, 105*, 415–450.

McNamara, P. (1996c). Must I do what I ought? (or will the least I can do do?). In M. A. Brown & J. Carmo (Eds.), *Deontic logic, agency and normative systems* (pp. 154–173). New York.

McNamara, P. (2000). Toward a framework for agency, inevitability, praise and blame. *Nordic Journal of Philosophical Logic, 5*, 135–160.

McNamara, P. (2004). Agential obligation as non-agential personal obligation plus agency. *Journal of Applied Logic, 2*, 117–152.

McNamara, P. (2011a). Praise, blame, obligation, and DWE: Toward a framework for the classical conception of supererogation and kin. *Journal of Applied Logic, 9*, 153–170.

McNamara, P. (2011b). Supererogation, inside and out: Toward an adequate scheme for common sense morality. In M. Timmons (Ed.), *Oxford studies in normative ethics* (Vol. i, pp. 202–235). Oxford.

McNamara, P. (2021). Logics for supererogation and allied concepts. In D. Gabbay & J. Horty, X. Parent, R. van der Meyden, L. van der Torre (Eds.), *Handbook of deontic logic and normative systems, volume 2*. London, College Publications.

McNamara, P. (2022). A natural conditionalization of the DWE framework. In P. McNamara, M. Brown, A. Jones (Eds.), *Agency, Normative Systems, Artifacts, and Beliefs: Essays in honor of Risto Hilpinen* (pp. 113–136). Switzerland, Springer Nature, Synthese Library 454.

McNamara, P., Van De Putte, F. (2022). *The Stanford Encyclopedia of Philosophy (Fall 2022 Edition).* E. N. Zalta. https://plato.stanford.edu/archives/fall2022/entries/logic-deontic/.

Mellema, G. (1987). Quasi-supererogation. *Philosophical Studies, 52*, 141–150.

Mellema, G. (1991). *Beyond the call of duty: Supererogation, obligation, and offence.* Albany, SUNY Pr.

Munoz, D. (2021). Three paradoxes of supererogation. *Nous, 55*, 699–716.

Pybus, E. M. (1982). Saints and heroes. *Philosophy, 57*, 193–200.

Urmson, J. O. (1958). Saints and heroes. In A. I. Melden (Ed.), *Essays in moral philosophy* (pp. 198–216). Seattle.

Widerker, D. (1991). Frankfurt on 'ought implies can' and alternative possibilities. *Analysis, 51*, 222–224.

Wessels, U. (2002). *Die gute samariterin. Zur struktur der supererogation.* Berlin, de Gruyter.

Wessels, U. (2015). Beyond the call of duty: The structure of a moral region. In C. Crowley (Ed.), *Supererogation* (pp. 87–104). Cambridge.

Supererogation and the Limits of Reasons

Nathaniel Baron-Schmitt and Daniel Muñoz

> ...the model of reasons for action is inadequate to the explanation of supererogation.
> –David Heyd, Supererogation, p. 170

Abstract We argue that supererogation cannot be understood just in terms of reasons for action. In addition to reasons, a theory of supererogation must include prerogatives, which can make an action permissible without counting in favor of doing it.

Keywords Supererogation · Reasons · Prerogatives · Commendatory reasons · Reasons first · Paradox of supererogation · All or nothing problem

1 Introduction

Supererogatory acts are good deeds beyond the call of duty, ranging from friendly favors to saintly sacrifices to risky rescues. As any reader of this handbook will have noticed, philosophers disagree about how exactly supererogation should be defined, and whether it is even a coherent idea.

To some extent, this is a verbal dispute. 'Supererogation' is not a term of ordinary language, like 'good' or 'wrong.' It is a 'quasi-technical term' (Heyd, 1982), whose

The authors were equal contributors to this project. Daniel would like to thank Kerah Gordon-Solmon and Theron Pummer (as always); from Nathaniel, thanks to Benjamin Kiesewetter, Thomas Schmidt, and other colleagues at the Human Abilities Center in Berlin. Both of us thank Joe Horton and David Heyd for very helpful comments.

N. Baron-Schmitt (✉)
The Free University of Berlin, Berlin, Germany
e-mail: n.baron-schmitt@fu-berlin.de

D. Muñoz
The University of North Carolina at Chapel Hill, Chapel Hill, USA
e-mail: daniel.munoz@unc.edu

D. Heyd (ed.), *Handbook of Supererogation*,
https://doi.org/10.1007/978-981-99-3633-5_10

meaning is somewhat up for grabs.[1] If you say that supererogation must spring from a noble motive, while someone else says it only needs to be a good thing to do, the two of you might not disagree on any matters of substance; you probably just prefer different definitions.

In our view, however, the big questions about supererogation are substantive, not verbal, and they can be asked in plain language. The fundamental question is the:

Classic Paradox of Supererogation

If A is a better thing to do than B, how could it be permissible to do B?

For example, if donating a kidney to a stranger is better than keeping it, how could donating be optional rather than obligatory? How could morality permit us to do the worse thing? These are perfectly good questions, and we can't dodge them by fiddling with definitions. We confront, as Dancy puts it, 'a *philosophical* boggle' (1993, 131, emphasis added). Nor can we define our way out of more recent boggles, like Horton's (2017) All or Nothing Problem.

How do philosophers try to solve the paradoxes of supererogation? In recent decades, most attempts have drawn on a single source: the theory of reasons for action.[2] And so we find a flurry of distinctions between kinds of reasons: agent-relative versus agent-neutral (Dancy, 1993), perfect versus imperfect (Portmore, 2011), requiring versus justifying versus favoring (Archer, 2016; Horgan & Timmons, 2010; Little & Macnamara, 2017, 2020; Tucker, Forthcoming), and on the list goes. By mixing and matching varieties, and by linking them to other concepts like permissibility and value, ethicists have tried to make sense of supererogation.

This chapter provides a tour of some of the main paradoxes of supererogation, as well as the main solutions provided by reasons-ologists. We end with a twist: *the paradoxes of supererogation cannot be solved with reasons alone.* Supererogation, we think, is a counterexample to the 'Reasons First' program, which tries to reduce ethics to the study of reasons.

None of our arguments will turn on the definition of 'supererogation.' That said, a definition will come in handy. Since we are here to talk about the classic paradox and its descendants, we define 'supererogatory acts' as those that give rise to the paradox: they are optional and better than a permissible alternative. Also, we are talking about morality here, so we mean that supererogation is *morally* better than a *morally* permissible alternative.

As for 'reason,' we can use it in the familiar way: to be a reason is to count in favor of some way of acting, making it more choiceworthy, or in other words, making it a

---

[1] By contrast, when Horgan and Timmons define 'supererogation,' they do so with an eye towards the 'common-sense usage of the term' (2010, 31).

[2] Some philosophers, especially old-school consequentialists, do see supererogation as intolerably paradoxical (see, e.g., Kagan, 1984, 1989). But this view has its costs. "In commonsense moral reasoning, we take it for granted that there are supererogatory acts, and it would be incredible if the very idea of supererogation turned out to be incoherent" (Dreier, 2004, 145).

better thing to do.[3] Reasons, so understood, may not be enough to solve the paradoxes of supererogation.

We begin with the classic paradox.

2 The Classic Paradox

Why does supererogation feel paradoxical? Or to put it another way, why would anyone worry that heroic sacrifices and kindly favors might turn out to be obligatory?

Let's start with a textbook case of an obligatory act. Consider:

Scarce Drug

You own a scarce drug, which you can use either to save your acquaintance Alex, who needs it all to survive, or five others, who need only a fifth each. (Foot, 1967).

Most people think you have to save the five, other things being equal.[4] Saving one life is good; saving five is better still; and so, people infer that you have to save the five.

Behind this easy inference looms a principle: we have to do the best thing we can. The 'best' thing, in the relevant sense, is the one that is most choiceworthy, the most favored by the balance of reasons. (The best thing to do needn't have the best *outcome*. Suppose you could kill Alex to save the five, or spare Alex and let the five die. Deontologists might say killing is worse, since it violates Alex's rights, even if saving the five leads to a better outcome with less death.)

So we get:

The Most Reason Principle

If you have most reason to do A, then A is obligatory.

A corollary is that, if you have more reason to do A than B, B is wrong—obligatory not to do.[5]

This principle can seem undoubtable, and so it should be no surprise that we find people asking how one could possibly doubt it:

How can one be permitted to refrain from action which is required by reason? (Raz, 1975, 165)

---

[3] We have in mind 'normative reasons,' as opposed to 'motivating reasons,' which are the grounds on which an agent acts. The two concepts can come apart. One might not be motivated by the normative reasons in favor of helping those far away (for example), and one might instead be motivated by mere whims or prejudices. For more on this distinction, see Dancy (2000).

[4] For exceptions, see Taurek (1977) and Anscombe (1967).

[5] When we refer to options like A and B, we have in mind options that are fully specific, rather than options that can be carried out in relevantly different ways. Think: 'saving Alex,' not 'either saving Alex or saving the five.' Also, we'll stick to choices from finitely many options. (Our 'corollary' wouldn't follow in infinite cases.)

> …how can an action that is morally best to perform fail to be what one is morally required to do? (Horgan & Timmons, 2010, 29)
>
> …how can supererogatory acts be so valuable and important, and yet not obligatory? (Heyd, 1982, 4; he calls this the "good–ought tie-up")

You get the idea.

All of this leads us to the classic paradox. The problem is that supererogatory acts, if they exist, must be better than some permissible alternative. But the alternative *can't* be permissible if supererogating would be better. Doing best is obligatory, given the Most Reason Principle, which is just a basic principle linking duties to reasons. So there can't be such a thing as a supererogatory act.

And yet, such acts seem not only conceivable but actual! A friend of ours once donated a kidney to a stranger, who was probably spared years of painful dialysis. This was a wonderful thing to do, far better than the alternative of not donating. But it seems extreme to say that the stranger was entitled to the kidney, or that our friend was merely doing her duty. Don't people have the freedom to decide what happens to their bodies, even when their decision isn't quite optimal?

More generally, people often seem to enjoy a permission to favor their own interests over the greater good of others. Contrast our earlier example with:

> *Alex's Scarce Drug*
>
> Alex owns a scarce drug, which she can use either to save herself or five others.

Now, saving the five does *not* seem like an obligation. It seems optional, like our friend's donation. (Though Alex *would* be obligated to sacrifice if the stakes were different. Presumably, she wouldn't be allowed to cure her own headache rather than save a million lives.)

The classic paradox, in effect, gives us two problems at once. First, we need to explain how the very idea of supererogation is coherent. This means, if we are reasons theorists, that we need to replace the Most Reason Principle with a more complex account of how reasons give rise to requirements. Second, our account needs to fit with core instances of supererogation and obligation, such as the drug cases above. Our theory might be coherent but problematic when applied to the relevant examples.

In our view, these problems are central to normative ethics, and it is a good thing that ethicists have tried to solve them using reasons.[6] The idea of a reason nicely brings out the tension in the idea of supererogation. Reasons traditionally play three roles: they tend to favor, justify, and require. Supererogation seems to upset this presumption. Perhaps the reasons to supererogate merely favor (without requiring), or the reasons against supererogation merely justify (without favoring or requiring)—unless we add *something* like that to our theory, the very idea of supererogation will be incoherent.

---

[6] Reasons took a while to catch on. (See Pybus, 1982, for an example of a rejection of supererogation that doesn't use the concept of a reason.)

3 Equal Weight

Is that right? Why can't we just say that supererogation involves a clash of reasons, understood in the traditional way?

Again, reasons are traditionally thought to play three roles at once:

1. Reasons favor actions, in the sense of tending to make them choiceworthy.
2. Reasons justify actions, in the sense of tending to make them permissible.
3. Reasons require actions, in the sense of tending to make them obligatory.

One simple way to think of this is with a trio of numbers. We assign to each option **X** a triple $<\mathbf{X_F}, \mathbf{X_J}, \mathbf{X_R}>$, where these are numbers representing how much the relevant reasons favor, justify, and require that option, respectively.

A word of caution about what these numbers mean. If we give **X** a score of <2, 2, 2>, we aren't saying that **X** is *doubly* justified, or *doubly* required. These numbers get their meaning from how they figure in comparisons, which then determine permissions and requirements.

> *Obligation*
>
> An option is obligatory if and only if its 'requirement' score beats any alternative's 'justification' score.
>
> *Permission*
>
> An option is permissible if and only if its 'justification' score is not beaten by any alternative's 'requirement' score.
>
> *Choiceworthiness*
>
> An option is more choiceworthy than an alternative if and only if it has a higher 'favoring' score than the alternative has.[7]

This three-way distinction gives us a deeper way to understand the classic paradox. Notice that the three principles above do *not* by themselves conflict *in the least* with the possibility of supererogation. For they don't entail anything like the Most Reason Principle—not until we add a further assumption. That assumption—the real crux of

---

[7] We can write this out a bit more formally, if you like, using 'O(**X**)' to mean '**X** is obligatory,' and 'P(**Y**)' to mean '**Y** is permissible.'

> *Obligation*
>
> O(**X**) *iff*, for any alternative **Y**, $\mathbf{X_R} > \mathbf{Y_J}$.
>
> *Permission*
>
> P(**X**) *iff*, for no alternative **Y**, $\mathbf{Y_R} > \mathbf{X_J}$.
>
> *Favoring*
>
> **X** is more choiceworthy than **Y** *iff* $\mathbf{X_F} > \mathbf{Y_F}$.

We assume that the three dimensions can be measured using a common unit, as we can use meters to measure a thing's length, width, and height. We might also want to assume that the ratios of numbers matter, and that the scales have a meaningful zero point.

the paradox—is the traditional equivalence between the three kinds of weight. We can call it:

Equal Weight

$\mathbf{X_F} = \mathbf{X_J} = \mathbf{X_R}$.

In other words, reasons favor, justify, and require in equal measure. So there cannot be reasons that purely favor, purely justify, or purely require.[8]

With this, we get the classic paradox in a fancier form, which distinguishes the kinds of reasons at play. Supererogation is optional and better than a permissible alternative. So, there must be more favoring reason to supererogate than to do the alternative. It follows, given Equal Weight, that there must be more *requiring* reason to do the supererogatory act than there is *justifying* reason to do the alternative. But then the alternative cannot be permissible after all, given our principle of Permission. Contradiction!

How can we make room for supererogation? Nowadays, the most popular move is to deny Equal Weight, opting for reasons that purely justify or purely favor.[9] But another option, more popular in the 1970s and 1980s, did not rely on any such revision to our idea of reasons. Let us take a closer look at this move, along with its problems.

4 The Shadow Paradox

As we have just seen, there is no way to get supererogation, as we have defined it, given the traditional view of reasons. But remember: what counts as 'supererogation' is up for grabs, so even if supererogation is impossible *as we define it*, it might be possible given a less stringent definition. Here we have to use our judgment to see whether the less stringent claim is still interesting.

Consider, for example, the idea that self-sacrifice involves a conflict between two kinds of moral reasons: reasons to help others and reasons to respect duties to oneself.[10] Given Equal Weight, this sort of view cannot say that self-sacrifice is

[8] Equal Weight is, strictly speaking, overkill. It implies that we can make comparisons of strength between favoring reasons and justifying/requiring reasons, using a common unit. (Like saying 'X's height in meters = Y's length in meters.') To get the Paradox, we don't need such comparisons. We just need to say that $\mathbf{X_F} > \mathbf{Y_F}$ entails $\mathbf{X_R} > \mathbf{Y_J}$. But this principle is a bit unfamiliar, and the technicalities here don't matter much, so in the text we use the stronger, simpler Equal Weight.

[9] For a different approach, see Bedke's (2011) 'Millian inversion,' which analyzes requirements in terms of 'reasons to require,' not reasons to do the thing required. See Snedegar (2016) for a powerful critique. (Snedegar also gives an important critique of Gert (2007); his arguments clearly complement ours, though he chooses different targets, and he focuses on accounts of 'ought' rather than solutions to our paradoxes.)

[10] This idea is popular in Kantian ethics. Patricia McGoldrick, for example, argues that heroic sacrifices are not obligatory because such an obligation "would come into conflict with our obvious duty to recognize our own intrinsic worth" (1984, 527), and she says that, in the ethics of self-sacrifice, the "heart of the matter is the Kantian argument that we have duties to ourselves as well as others" (ibid., 527).

supererogatory (in our sense). At best it can say that, in some cases, self-sacrifice is optional. Giving one's kidney can be optional if the reason to keep it is equal to the reason in favor of giving; the duty to help the recipient balances off against the donor's duty to self.

But this move does not get us very far, for two reasons.

First, the optionality is not robust enough. If the reasons to give *exactly* balance off against the reasons to keep, then—assuming that the weights of reasons are measured with single numbers, like physical weights—any increase on one side will tip the scales. If we were to make an optional sacrifice even barely less costly, that would make it obligatory. This is an absurd result; self-sacrifice should be optional across a range of costs and benefits, not just at one precise balance point (Hurka & Shubert, 2012, 8; Kagan, 1989, 378–79).

To address this problem, reasons-ologists have come up with several ideas as to how a tie between reasons might be stable over small additions to either side. For example, some say that the two clashing reasons are only *imprecisely* equal in weight and that small boosts don't break imprecise ties (Parfit, 2011, 137–41).[11]

But even if these moves can solve the classic paradox—and we have doubts[12]—they lead us straight into a second, deeper paradox. Let's grant that the reason against donating undercuts the duty to donate. Given Equal Weight, won't it also undercut the *betterness* of donating? On a view like Parfit's, there is nothing more choiceworthy about making the sacrifice. We save the optionality of supererogation at the cost of its superiority.

We call this the:

Shadow Paradox of Supererogation

> If there is a strong reason against typical supererogatory acts, why should these acts be more choiceworthy than the moral minimum?[13]

As with the classic paradox, the shadow paradox has a conceptual side as well as a normative one. Our theory has to be coherent while also being a good fit for the relevant cases, most notably self-sacrifice.

Any theory of supererogation, then, has to steer between these two paradoxes, the classic and its shadow. We need to find something that can make supererogation optional in the relevant cases without detracting from its value.

---

[11] Portmore has a more complicated view: the clashing reasons are *imperfect*, in the sense that they "do not support performing any specific alternative, but instead support performing any of the alternatives that would each constitute an equally effective means of achieving the same worthy end" (2011: 156). We focus on Parfit's view because it is simpler and more general.

[12] See Muñoz (ms) for a critique of the appeal to imprecise weights.

[13] Some views face a particularly grave version of the shadow paradox called the 'Wrongness Problem' (Muñoz & Baron-Schmitt, ms; Muñoz, 2021a, 615). If the reason against donating is weightier than the reason to donate, then, given Equal Weight, the sacrifice we wanted to call supererogatory is in fact *wrong*. Postow (2005, 246) discusses a version of the shadow paradox where the sacrifice is 'irrational'—opposed by decisive non-moral reasons. Dale Dorsey (2013, 2016) argues that supererogatory acts should be understood as rationally optional yet morally required.

5 Favoring Reasons

To solve the paradoxes, we have our work cut out for us. We need to prevent the reasons to supererogate from generating a duty (classic), while also ensuring that supererogation is favored over the moral minimum (shadow). There are two basic ways to do this. We could say that the reasons to supererogate favor more than they require, or that the reasons against supererogating justify more than they favor. Let's start with the first option.

According to the 'pure favoring view,' supererogation is possible because the reasons to supererogate count in favor without tending to require anything. Such reasons are 'purely favoring reasons.'[14] These reasons are almost tailor-made to solve our pair of paradoxes. Because pure favorers do not even tend to ground requirements, acting on them is optional by default; they do not need to be offset by contrary reasons, which would then threaten the value of supererogating.

The pure favoring view, we think, is conceptually coherent. But does it fit the cases? It seems like a good fit for low-stakes kindnesses, gifts, and personal favors, which are the focus of Horgan and Timmons (2010). They argue that pure favoring fits the phenomenology of kindnesses, such as in their case of Olivia:

> Olivia... meets a recently widowed woman, Mary, a neighbor who lives a few doors down... [who] lost her husband to cancer... [and who] is an avid baseball fan... But without anyone to go with, [Mary] doesn't go [to baseball games] anymore. The next day, it occurs to Olivia that it would be a nice gesture to offer to go to a Cardinals game with Mary, although [Olivia] herself has no particular interest in the game. But she thinks: "Here is a chance to do something nice for someone... Why not?" She calls Mary, who is delighted by the invitation, and they end up going to a game. (2010, 47)

Horgan and Timmons suggest that "[in] contrast to cases of obligation, Olivia does not experience the reasons she has to [invite Mary to the game] as requiring her to do so, although, of course, the reasons in question are experienced as favoring [inviting her]. Such reasons, then, are experienced differently than are the reasons involved in experiences of obligation" (2010, 48). And if the experience is accurate, we can conclude that the reasons to perform such kindnesses are purely favoring.[15]

What about costly beneficence? Horgan and Timmons do not try to extend the pure favoring view to risky rescues and saintly sacrifices. Dreier (2004), however, suggests that these can be morally optional if beneficence is purely favoring. He does not think all moral reasons purely favor; reasons of justice, in his view, do ground requirements. But beneficence, in particular, is purely favored, and that is why we don't have to make costly sacrifices.

The pure favoring view is coherent, and it seems plausible in certain cases. But not all cases (Snedegar, 2016, 165). What if beneficence is cheap? Suppose you could

---

[14] Other terms for 'favoring' include 'merit-conferring' (Horgan & Timmons, 2010) and 'commendatory' (Little & Macnamara, 2017). See also Dancy (2004a) on 'enticing reasons' and Heyd (1982, 171–72) on the 'commendatory' versus 'prescriptive' senses of 'ought.'

[15] The argument shows that Olivia's reason is, in a sense, non-requiring. But must it be *intrinsically* non-requiring? What if it only fails to require because Olivia is justified in spending her time as she likes? We'll come back to this idea later when we discuss justifying reasons and prerogatives.

rescue a child at the cost of getting your suit muddy (Singer, 1972). Or suppose it is 100 children, at no cost at all. Surely *costless* rescues are obligatory.[16] Beneficence, on reflection, is only optional when the costs are fairly high relative to the benefits. This suggests that costly sacrifices are optional precisely *because they are costly*, not because the reasons to help others are inherently non-requiring. It is not that beneficence is inherently optional, but rather that one can sometimes justifiably refuse to pay big costs.

We take this 'costless rescue problem' to be a decisive objection to the pure favoring view. Pure favoring cannot be the solution to the paradoxes of supererogation, because many cases of supererogation do not involve pure favoring. But some writers have tried to amend the view to get around the problem. They posit *primarily* favoring reasons, which favor more than they require, while still requiring to some degree. Little and MacNamara (2017) take this approach (see also Archer, 2016, 459): they say that the reasons to save lives have *some* requiring weight, but not as much as they have favoring weight: $\mathbf{X_F} > \mathbf{X_R} > 0$. So, rescues are optional when costly but obligatory when cheap.

The 'primarily favoring view,' as we'll call it, is the best view we've seen yet. It can handle the classic and shadow paradoxes and a whole range of cases, including costless rescues. But we don't think the view ultimately works. It struggles with a more complicated paradox, Joe Horton's All or Nothing Problem, to which we will now turn.

6 All or Nothing: The Need for Justifying Reasons

Horton gives a case with three options:

Two Kids

> Suppose that two children are about to be crushed by a collapsing building. You have three options: do nothing, save one child by allowing your arms to be crushed, or save both children by allowing your arms to be crushed. (2017, 94)

Saving neither ('**0**') seems to be permissible, as does saving both ('**2**'); but saving only one ('**1**') is wrong. You can't justify saving only one when you could save a further child at no extra cost. Yet, despite its wrongness, **0** is surely not more choiceworthy than **1**. If you save a life by sacrificing your arms, that does not seem worse than doing nothing and saving no one. In our terms: $\mathbf{0_F} \leq \mathbf{1_F}$.[17] This is the All or Nothing Problem.

---

[16] Here we follow Archer (2016, 460) and Portmore (2008, 381), who make the same point against Dreier's (2004) version of the pure favoring view. Our objection to the primarily favoring reasons view, in the next section, is original.

[17] Horton himself would not put the point this way. His preferred term is 'ought rather,' and the claim he denies is that one ought to save zero rather than only one. While we are on the subject, one understated part of Horton's view is his eschewal of Reasons First. He thinks we have 'justifications' not to pay costs, which may or may not be 'reasonable' to appeal to in defense of one's actions

In our view, solving the problem means finding a way to make these moral 'seemings' fit together. But the primarily favoring view by itself cannot do this. If we suppose that justifying weight and favoring weight are equal, it follows from $\mathbf{0_F} \leq \mathbf{1_F}$ that $\mathbf{0_J} \leq \mathbf{1_J}$. Since **0** is not favored over **1**, **0** is not more strongly justified. But since **0** is permissible, whereas **1** is not, the justifying reason for **0** must be greater than the justifying reason for **1**. $\mathbf{2_R} > \mathbf{1_J}$ and $\mathbf{2_R} \leq \mathbf{0_J}$, so $\mathbf{0_J} > \mathbf{1_J}$. Contradiction!

What went wrong? The issue here is not that favoring is allowed to outstrip requiring; it is that justifying is not allowed to outstrip favoring. The primarily favoring view, like the pure favoring view, leaves intact the assumption that $\mathbf{X_F} = \mathbf{X_J}$. It is this assumption, we think, that we need to reject if we want to solve the All or Nothing Problem.[18]

Let's suppose that there can be 'purely justifying reasons,' so that an option's justifying weight can outstrip its favoring weight.[19] Then we can say that at least some of the self-centered reasons to do nothing justify more than they favor (though not so for the ordinary moral reasons for saving one, which, let's assume, all justify and favor to the same extent). In other words, $\mathbf{0_J} > \mathbf{0_F}$. This allows us to say that $\mathbf{0_J} > \mathbf{1_F} = \mathbf{1_J} > \mathbf{0_F}$. This is an intuitive idea, once you get past the symbols. It follows from the natural thought we touched on earlier: costly sacrifices are optional because of their costs. Although **1** is at least as choiceworthy as **0**, **1** is more costly, making **0** easier to justify.

With purely justifying reasons, we can tell a perfectly coherent story about the All or Nothing Problem. The weights of reasons could be as follows:

	Save 0	Save 1	Save 2
Favoring	0	5	10
Requiring	0	5	10
Justifying	15	5	10

Of course, this is too simplistic. Most of us think there is *something* favoring the choice to do nothing; we might fix this by making the zeros into, say, twos. (See the

(2017, 98), but he never relies on the idea of reasons for action, much less does he try to reduce everything to reasons. This is just one further way in which Horton's paper is original.

18 See Rulli (2020) and Muñoz (2021b), which go into more depth about the idea that a wrong option could be no worse than a permissible alternative.

19 The recent literature on 'justifying reasons' starts with Joshua Gert (2007), who distinguishes two distinct 'strength values' a reason may have: justifying and requiring. But the idea of pure 'justifying strength' is not so new. Before Gert, 'prerogatives' (Kamm, 1996, Chap. 8; Scheffler, 1994), or 'agent-relative permissions' (Parfit, 1978), played the same role as justifying reasons, but without the name 'reason.' Others have developed similar concepts; Snedegar's (2021) version of a justifying reason is one that counts 'for' an option but not 'against' alternatives; Greenspan (2005) distinguishes 'positive' from 'negative' reasons; Portmore (2011) applies the justifying/requiring distinction to moral reasons in particular; and so on. The story, in short, is that pure justifiers were developed before Reasons First, and they were rediscovered later and only then conceived as a part of the theory of reasons, rather than something beyond its limits.

discussion below of 'primarily justifying reasons.') Some might be skeptical of the idea that *any* numbers could accurately represent the weights of reasons; in that case, we would need a more complicated formalism.

Still, even if the details need to be finessed, the above table tells us something important. Notice that the favoring weight for all three options is equal to the requiring weight. We can give a coherent story about the All or Nothing Problem without distinguishing favoring weight from requiring weight. Solving the problem doesn't require primarily favoring reasons (or purely favoring reasons). We just need one departure from Equal Weight: a reason to do nothing that justifies more than it favors.

This is a striking result. With purely justifying reasons, we can solve Horton's problem, as well as the classic and shadow paradoxes and the costless rescue problem.[20] We do not need to distinguish requiring weight from favoring weight, or to allow that requiring (or favoring) weight can exceed justifying weight.[21] (In fact, we may want to deny that this last inequality is possible since it allows for moral dilemmas. If $\mathbf{X_R} > \mathbf{Y_J}$ and $\mathbf{Y_R} > \mathbf{X_J}$, then one must do each of two incompatible things.)

7 Are Justifying Reasons Reasons?

Purely justifying reasons allow us to solve the paradoxes of supererogation. Supererogation is better, yet optional, because the reason against merely *justifies* refraining; it doesn't undercut the betterness of going ahead, much less make it wrong to do so. This, in our view, is the most promising approach for the 'reasons first' account of supererogation. But there is something funny, we think, about the very idea of purely justifying reasons: they don't *favor*. Isn't favoring, understood as making an option more choiceworthy, supposed to be an essential mark of reasons?

Many ethicists think so:

> *A* is better than *B* iff there is more reason to choose *A* than to choose *B*, given the choice between *A* and *B*. (Snedegar, 2017, 93)

---

[20] Just in case it's not clear how justifying reasons help with the other paradoxes: costless rescue is obligatory, whereas costly rescue is optional, because one has a purely justifying reason not to pay the costs. (The reason to rescue can have equal justifying, favoring, and requiring weight.)

[21] One further paradox does complicate things: Kamm's Intransitivity Paradox (Archer, 2016; Dorsey, 2013; Kamm, 1985, 1996; Portmore, 2017). Elsewhere, one of us argues that we can only solve the paradox if justification is *comparative*, in the following sense: how far I can justify doing **X** rather than **Y** is not intrinsic to **X**, but depends on the relative costs of **X** and **Y**; so, I might have a powerful justification to do nothing rather than give my arms to save a life, but no such justification to do nothing rather than costlessly save somebody's hand (Muñoz, 2021b). The idea that justification is comparative is, we think, fairly intuitive. The idea that it is nontransitive goes back to Parfit's (1982) reply to Kavka (1982). Parfit's insight has been overlooked, though his example has had a comeback lately as a precursor to (wait for it) the All or Nothing Problem.

> I will take the idea of a reason as primitive. Any attempt to explain what it is to be a reason for something seems to me to lead back to the same idea: a consideration that counts in favor of it (Scanlon, 1998, 17).

> ...reasons...count in favor of possible *properties* of agents—things that agents can do, in a very broad sense of 'do' (Schroeder, 2021a, 2021b, 34).

> [I]t's commonly accepted that a normative reason for action is *a consideration that counts in favor of the action* (Markovits, 2014, 2).

Now, to be sure, not everyone wants to define reasons as favorers. Some think 'counting in favor' is problematically ambiguous (Hieronymi, 2005); some think reasons can be understood simply in terms of what we ought to do (see, e.g., Hurka, 2014, 32, though see also Dancy, 2004a, 2004b, Chap. 2); others find a subtle difference between favoring and being a reason, which applies in certain many-option cases (see Schroeder, 2021a, 2021b; Snedegar, 2013, 42).

Still, we think most readers will agree that (at least part of) the business of reasons is to count in favor of actions. But then how could there be purely justifying reasons? The whole point of such reasons is that they do *not* favor (or require). They only justify.[22]

There is a possible way around this problem. Perhaps there are no *purely* justifying reasons. But there might be *primarily justifying reasons*—reasons that 'justify more than they require' (and favor), in the words of Chris Tucker.[23] The idea here is that the reasons against supererogation count as genuine reasons because they do genuinely favor, but only a bit, so it is still better overall to act against them. And so we might think that the cost of being the hero, in Two Kids, is a primarily justifying reason not to save anyone. The costs favor somewhat ($\mathbf{0_F} > 0$), and they justify even more than they favor ($\mathbf{0_J} > \mathbf{0_F}$).

But we do not think this move can save Reasons First.[24]

---

[22] It won't help to say these reasons "count in favor" of an action's being permissible, or merely that these reasons "compete" with other reasons to determine what is permissible (see Schroeder, 2021b, Sect. 3). Favoring means counting in favor of *doing* an action, making it more choiceworthy.

[23] Tucker (Forthcoming) defends the existence of such reasons, which he calls "justifying heavy requiring reasons."

[24] There are other relevant views that do not appeal to justifying reasons. Dancy thinks supererogation is opposed by an agent-relative reason we can "discount" at will. This is like our view, except we prefer waivable rights to discountable reasons, because such reasons would not be "stable" over the choice of whether to act on them (see Muñoz 2020, 700, 2021a, 616). Another view is Raz's (1975, 167). He thinks supererogation is made possible by "exclusionary permissions," which come from reasons that "entitle disregarding" ordinary reasons to act. The balance of (first-level) reasons favors supererogating, but one may disregard the balance (thanks to second-level exclusionary reasons). Raz's view struggles with the shadow paradox: if one has a (decisive) reason to disregard the reasons to give one's kidney, then giving the kidney isn't choiceworthy, and may be irrational. (Raz might also have to reject the principle, mentioned above, that reasons must be "stable" over our decisions.) For another important proposal, which we hope to discuss in future work, see Schmidt (Forthcoming a, Forthcoming b); see also Portmore (2011) on dual-ranking act consequentialism.

For one thing, even if costs favor, not all justifiers work like costs. Consider body rights. You have the right to keep your spare kidney, if you choose, because it is *yours*. But this in no way counts against donating. As Jeremy Waldron puts it:

> …the fact that P has a right to do A does not of itself give rise to any reason in favor of A which is capable of competing with and being balanced against the reason for not doing A. (Waldron, 1981, 28)

The fact that your organs are rightfully *yours*, then, would appear to be a pure justification for not donating.[25] (See Muñoz (2021a) for more on this idea.) In our view, the same can be said about body rights in Two Kids; the fact that it's *your arms* at stake is a pure justification for not sacrificing them.[26]

Second, even if costs are 'primarily justifying reasons,' such reasons are an awkward fit for Reasons First. For they seem to be *more* than just reasons. Costs, in this view, are supposed to justify actions more than they favor. So how can we reduce their justifying to their favoring, that is, to their being reasons? Their excess justifying weight is a further independent feature that they have, in addition to their being reasons.

We do agree with Tucker about some things. In Two Kids, there is some favoring weight behind saving no one, and still more justifying weight. But from this, we don't want to infer that *one and the same thing* is doing the favoring and the justifying, or that the justifying *reduces to* the favoring. The fact doing the favoring could be the cost of heroism; the source of the excess justification could be a pure justifier, like the right not to harm oneself; and at any rate, the justifying seems like an independent factor, not reducible to the favoring.

What are we to make of the appeal to 'justifying reasons,' if they aren't just reasons? We could give up on them, and anything like them, on the grounds that Reasons are First. But then we would leave some of the paradoxes unsolved. Solving the paradoxes, in our view, is more important than keeping Reasons First.

And we're close to a solution! We've seen that so-called justifying reasons can solve paradoxes because *they justify more than they favor*. It is only this feature of them, not their status as reasons, that we need for our theory.

---

25 Waldron himself would not say that body rights 'justify.' For Waldron, to justify an action is "to show that the standard to which in the circumstances it conformed or the worthiness of the goal that it was intended to advance" (1981, 28). Clearly, 'it's mine' does not show that keeping a spare organ is a *worthy* goal. But it does help show that keeping the kidney is *justifiable*, and this is all we mean by saying that it 'justifies.' (We might say 'it's mine' reduces the need for justification, rather than providing justification. After all, we do say things like 'it's mine, so I don't *owe* you a reason.')

26 Heyd, in another pioneering passage, links supererogation to *autonomy* (1982, 172–78), which seems like a plausible source for pure justifiers. Bodily autonomy, for example, does not *favor* keeping one's kidney, or tell against donation; it just creates a presumption that choices about your body are *yours to make*. (See especially p. 173, where Heyd discusses actions that go against the overall good; for Heyd's thoughts on Raz and reasons, see pp. 167–172, and our epigraph.)

8 If Not Reasons, then What?

We have argued that the best view of supererogation is one where justification can outstrip the favoring power of reasons. If giving one's spare kidney is supererogatory, for example, then one has a sufficient justification to keep it, even though one has more reason to give than to keep. On such a view, Reasons are not really First. Alongside reasons, we need pure justifiers—which are traditionally known as prerogatives (Kamm, 1996; Muñoz, 2021a; Scheffler, 1994), or 'agent-relative' permissions (Parfit, 1978, 287; Slote, 1984).[27]

But what is a prerogative, if not a kind of reason? Hurka and Shubert (2012) argue that we should take the concept of a prerogative, like that of a reason, to be primitive. Prerogatives cannot be derived from any other element of ethics, though they can be described with analogies and metaphors. Hurka and Shubert give a helpful physics analogy: whereas reasons are like independent forces acting on an object, prerogatives are like friction (2012, 7–8). Friction cannot move an object on its own; it can only mitigate other forces. Prerogatives, analogously, cannot favor or require an action on their own; they can only prevent reasons from grounding an obligation.

Hurka and Shubert argue that prerogatives are worth having even if they have to be left primitive. We agree. But in our view, there is still hope for deriving prerogatives from something deeper—not reasons, but *rights*. In our view, prerogatives can be derived from waivable rights against oneself (see Muñoz, 2021a for the details).[28] The key is that whether such rights are in play depends on what you decide; your kidney is off limits if, and only if, you decide not to donate.[29]

To solve the paradoxes of supererogation, we need not only reasons, which count in favor of actions, but also prerogatives, which purely justify. The classic and shadow paradoxes can be solved using purely or primarily favoring reasons, but these lead to other problems: the costless rescue problem and Horton's All or Nothing Problem. Justifying 'reasons' solve the paradoxes, but they aren't really reasons. Instead, we need something other than reasons to justify: either primitive prerogatives, rights against oneself, or some other justifiers waiting to be discovered.

---

[27] Hurka and Shubert (2012) prefer the term 'prima facie permissions,' meaning 'things that tend to make permissible,' as an homage to Ross's (1930) 'prima facie duties,' which are things that tend to make an option a duty. (Although we do not follow Hurka and Shubert in taking prerogatives to be primitive, we have learned much from their paper.)

[28] See also Kagan (1989, 206–216); Muñoz and Baron-Schmitt (ms). For more on waiving duties to/rights against oneself, see Kanygina (2022), Muñoz and Baron-Schmitt (Forthcoming), Muñoz (2020), Schaab (2021), and Schofield (2021).

[29] We think another possible source of prerogatives is the happiness of merely possible people (supposing their happiness only matters *if you create them*); here we take inspiration from Spencer (2021) on attractive permissions. Another promising account of prerogatives ('permitting reasons') is due to Theron Pummer (2023, 25), who suggests that they are a kind of defeater of requiring reasons. One of us (Daniel) will be discussing this view in more detail in a book review.

References

Anscombe, G. E. M. (1967). Who is wronged? Philippa foot on double effect: One point. *The Oxford Review, 5*, 16–17.
Archer, A. (2016). Moral obligation, self-interest and the transitivity problem. *Utilitas, 28*(4), 441–464.
Bedke, M. (2011). Passing the deontic buck. In R. Shafer-Landau (Ed.), *Oxford studies in metaethics* (Vol. 6, pp. 128–152). Oxford University Press.
Dancy, J. (1993). *Moral reasons*. Blackwell.
Dancy, J. (2000). *Practical reality*. Oxford University Press.
Dancy, J. (2004a). Enticing reasons. In R. Jay Wallace, P. Pettit, S. Scheffler & M. Smith (Eds.), *Reason and value: Themes from the moral philosophy of Joseph Raz* (pp. 91–118). Clarendon Press.
Dancy, J. (2004b). *Ethics without principles*. Oxford University Press.
Dorsey, D. (2013). The supererogatory, and how to accommodate it. *Utilitas, 25*(3), 355–382.
Dorsey, D. (2016). *The limits of moral authority*. Oxford University Press.
Dreier, J. (2004). Why ethical satisficing makes sense and rational satisficing doesn't. In M. Byron (Ed.), *Satisficing and maximizing: Moral theorists on practical reason* (pp. 131–54). Cambridge University Press.
Foot, P. (1967). The problem of abortion and the doctrine of the double effect. *Oxford Review, 5*, 5–15.
Gert, J. (2007). Normative strength and the balance of reasons. *Philosophical Review, 116*(4), 533–562.
Greenspan, P. (2005). Asymmetrical practical reasons. In J. C. Marek & M. E. Reicher (Eds.), *Experience and Analysis: Proceedings of the 27th International Wittgenstein Symposium* (pp. 387–394). ÖBV and HPT.
Heyd, D. (1982). *Supererogation: Its status in ethical theory*. Cambridge University Press.
Hieronymi, P. (2005). The wrong kind of reason. *Journal of Philosophy, 102*(9), 437–457.
Horgan, T., & Timmons, M. (2010). Untying a knot from the inside out: Reflections on the 'paradox' of supererogation. *Social Philosophy and Policy, 27*(2), 29–63.
Horton, J. (2017). The all or nothing problem. *Journal of Philosophy, 114*(2), 94–104.
Hurka, T. (2014). *British ethical theorists from Sidgwick to Ewing*. Oxford University Press.
Hurka, T., & Shubert, E. (2012). Permissions to do less than best: A moving band. *Oxford studies in normative ethics* (Vol. 2, pp. 1–27). Oxford University Press.
Kagan, S. (1984). Does consequentialism demand too much? Recent work on the limits of obligation. *Philosophy and Public Affairs, 13*(3), 239–254.
Kagan, S. (1989). *The limits of morality*. Oxford University Press.
Kamm, F. M. (1985). Supererogation and obligation. *Journal of Philosophy, 82*(3), 118–138.
Kamm, F. M. (1996). *Morality, mortality: Volume II: Rights, duties, and status*. Oxford University Press.
Kanygina, Y. (2022). Duties to oneself and their alleged incoherence. *Australasian Journal of Philosophy, 100*(3), 565–579.
Kavka, G. S. (1982). The paradox of future individuals. *Philosophy and Public Affairs, 11*(2), 93–112.
Little, M. O., & Macnamara, C. (2017). For better or worse: Commendatory reasons and latitude. In C. Mark (Ed.), *Oxford studies in normative ethics* (Vol. 17, pp. 138–160). Oxford University Press.
Little, M. O.,& Macnamara, C. (2020). Non-requiring reasons. In Chang & Sylvan (Eds.), *The Routledge handbook of practical reasons*. Routledge.
Markovits, J. (2014). *Moral reason*. Oxford University Press.
McGoldrick, P. M. (1984). Saints and heroes: A plea for the supererogatory: Discussion. *Philosophy, 59*(230), 523–528.

Muñoz, D. (2020). The paradox of duties to oneself. *Australasian Journal of Philosophy, 98*(4), 691–702.

Muñoz, D. (2021a). From rights to prerogatives. *Philosophy and Phenomenological Research, 102*(3), 608–623.

Muñoz, D. (2021b). Three paradoxes of supererogation. *Noûs, 55*(3), 699–716.

Muñoz, D. (ms). Values as vectors.

Muñoz, D., & Baron-Schmitt, N. (Forthcoming). Wronging Oneself. *Journal of Philosophy.*

Muñoz, D., & Baron-Schmitt, N. (ms). Why isn't supererogation wrong?

Parfit, D. (1978). Innumerate ethics. *Philosophy and Public Affairs, 7*(4), 285–301.

Parfit, D. (1982). Future generations: Further problems. *Philosophy and Public Affairs, 11*(2), 113–172.

Parfit, D. (2011). *On what matters* (Vol. 1). Oxford University Press.

Portmore, D. W. (2008). Are moral reasons morally overriding? *Ethical Theory and Moral Practice, 11*(4), 369–388.

Portmore, D. W. (2011). *Commonsense consequentialism: Wherein morality meets rationality.* Oxford University Press.

Portmore, D. W. (2017). Transitivity, moral latitude, and supererogation. *Utilitas, 29*(3), 286–298.

Postow, B. C. (2005). Supererogation again. *Journal of Value Inquiry, 39*(2), 245–253.

Pummer, T. (2023). *The rules of rescue.* Oxford University Press.

Pybus, E. M. (1982). Saints and heroes. *Philosophy, 57*(220), 193–199.

Raz, J. (1975). Permissions and supererogation. *American Philosophical Quarterly, 12*(2), 161–168.

Ross, W. D. (1930). *The right and the good.* Oxford University Press.

Rulli, T. (2020). Conditional obligations. *Social Theory and Practice, 46*(2), 365–390.

Scanlon, T. (1998). *What we owe to each other.* Harvard University Press.

Schaab, J. (2021). On the supposed incoherence of obligations to oneself. *Australasian Journal of Philosophy, 99*(1), 175–189.

Scheffler, S. (1994). *The rejection of consequentialism: A philosophical investigation of the considerations underlying rival moral conceptions.* Oxford University Press.

Schmidt, T. (Forthcoming a). How reasons determine moral requirements. In R. Shafer-Landau (Ed.), *Oxford studies in metaethics* (Vol. 18). Oxford University Press.

Schmidt, T. (Forthcoming b). The balancing view of ought. *Ethics.*

Schroeder, M. (2021a). *Reasons first.* Oxford University Press.

Schroeder, M. (2021b). The fundamental reason for reasons fundamentalism. *Philosophical Studies, 178*(10), 3107–3127.

Schofield, P. (2021). *Duty to self: Moral, political, and legal self-relation.* Oxford University Press.

Singer, P. (1972). Famine, affluence, and morality. *Philosophy and Public Affairs, 1*(3), 229–243.

Slote, M. (1984). Morality and self-other asymmetry. *Journal of Philosophy, 81*(4), 179–192.

Snedegar, J. (2013). Reason claims and contrastivism about reasons. *Philosophical Studies, 166*(2), 231–242.

Snedegar, J. (2016). Reasons, oughts, and requirements. In R. Shafer-Landau (Ed.), *Oxford studies in metaethics* (Vol. 11, pp. 155–181). Oxford University Press.

Snedegar, J. (2017). *Contrastive reasons.* Oxford University Press.

Snedegar, J. (2021, September 19). Reasons, competition, and latitude. In R. Shafer-Landau (Ed.), *Oxford studies in metaethics* (Vol. 16, pp. 134–56). Oxford University Press. https://doi.org/10.1093/oso/9780192897466.003.0006

Spencer, J. (2021). The procreative asymmetry and the impossibility of elusive permission. *Philosophical Studies, 178*(11), 3819–3842.

Taurek, J. (1977). Should the numbers count? *Philosophy and Public Affairs, 6*(4), 293–316.

Tucker, C. (Forthcoming). Parity, moral options, and the weights of reasons. *Noûs.*

Waldron, J. (1981). A right to do wrong. *Ethics, 92*(1), 21–39.

The Evaluative Condition for Supererogation

Claire Benn

Abstract Supererogatory actions must go beyond duty not only by being optional, but also by being good to do. Understanding the evaluative condition that supererogatory actions must meet is vital in order to understand the very concept of supererogation. I argue for two key features of the goodness of supererogatory actions: firstly, that they are comparative, and secondly, that they are relative. Specifically, I argue that an action meets the evaluative condition of supererogation if and only if it is (i) better than some permissible alternative and (ii) better than (or at least as good as) all permissible alternatives that are (a) as costly or (b) less costly; where the sense in which it is better is *relative* to a particular beneficiary. Seeing why this is so reveals the complexity of our notion of the supererogatory and captures our intuitions on core and complex cases. Furthermore, it makes room for supererogation's mirror: the suberogatory.

Keywords Supererogation · Goodness · Cost · Beneficence · Permissibility · Suberogation

1 Introduction

Supererogatory actions—those that go above and beyond the call of duty—are good to do and yet optional (that is, neither morally required nor morally forbidden).[1]

[1] I use the term 'optionality', following Paul McNamara's terminology ("Making Room for Going Beyond the Call").

C. Benn (✉)
Australian National University, Canberra, Australia
e-mail: cmabenn@gmail.com

D. Heyd (ed.), *Handbook of Supererogation*,
https://doi.org/10.1007/978-981-99-3633-5_11

Both the evaluative and deontic conditions are necessary.[2] Some actions are good and yet not supererogatory, like keeping a promise to help a friend.[3] Others, despite being optional, fail to be supererogatory because they are not truly good. Consider Feinberg's "greedy adventurer who sets off on an arduous journey into the heart of the jungle, determined to brave all dangers in order to find a buried treasure" and his "dedicated crackpot who nearly freezes to death trying to convert the indifferent Eskimos to Caribbean Voodooism."[4] These actions are optional: they are neither required nor forbidden. However, they are not supererogatory, because they fail to aim at something morally good.

Much has been written about the deontic condition for supererogatory actions: why and how supererogatory actions are optional.[5] We know less about the evaluative condition, which is just as key and just as complex. I address this gap. Note that I will not be giving a substantive account of goodness in general. Instead, I argue for two structural points about the goodness of supererogatory actions, both of which are compatible with different accounts of what makes an action good. Specifically, I will show that the goodness of supererogatory actions must be judged (1) in comparison to the goodness of alternative, less- (or as-) costly actions the agent could have performed and (2) relative to the interests of a particular beneficiary.

My account reveals the complexity of our notion of the supererogatory and its relationship to other normative considerations: the actions that could have been performed instead, the costliness of actions, and the identity of a beneficiary. Furthermore, my account captures our intuitions on core and complex cases of supererogation and makes room for supererogation's mirror: the suberogatory.

In Sect. 2, I defend my view that supererogatory actions should be understood comparatively—as 'better than permissible alternatives'—by rejecting two alternatives. In Sect. 3, I argue that supererogatory actions are not only better than a permissible alternative, but that they must be better than all less-costly permissible alternatives. I then consider further complex cases where the cost or benefits are the same, propose two modifications, and address two potential objections. In Sect. 4, I defend why we should understand the evaluative condition relative to specific beneficiaries. In Sect. 5, I demonstrate that my account, unlike others, can make space for the concept of suberogation in a way that meets important desiderata.

---

[2] I assume for simplicity that these two conditions are sufficient as well as necessary. Note, however, that there have been discussions about whether or not supererogatory acts must also be praiseworthy (Archer, "Are Acts of Supererogation Always Praiseworthy?"; McNamara, "Supererogation, inside and out: Toward an Adequate Scheme for Common-Sense Morality"; Mellema, "Supererogation and the Fulfillment of Duty"; Cowley, "The Agents, Acts and Attitudes of Supererogation.") and/ or intentional (Heyd, *Supererogation*; Benn, "Intentions, Motives and Supererogation"; Archer, "Supererogation and Intentions of the Agent"; Jackson, "The Nature of Supererogation").

[3] I set aside the claim that, in some specific situations, it can be supererogatory to keep a promise (Harman, "Morally Permissible Moral Mistakes").

[4] Feinberg, "Supererogation and Rules," 281.

[5] Benn, "Supererogation, Optionality and Cost"; Stocker, "Acts, Perfect Duties, and Imperfect Duties"; Mellema, "Supererogation and the Fulfillment of Duty"; Horgan and Timmons, "Untying a Knot from the Inside Out"; Chisholm, "The Ethics of Requirement".

2 A Comparative Account

I propose that supererogatory actions are better than *some* permissible alternatives, and also better than *all* less-costly permissible alternatives. This is a *comparative* account of supererogation. To defend it, I first show that its non-comparative counterpart is wrong and then defend my particular comparative account.

2.1 *Rejecting a Non-comparative Account of Goodness*

For simplicity, let us adopt a welfarist axiology and suppose that human welfare—happiness—makes actions good. Additionally, let us focus solely on the amount of happiness experienced by persons *other than the agent.*[6] In order to lay the groundwork for a more promising alternative, I start with an obviously implausible view: that an action is supererogatory if it is optional, and it realizes a state of affairs where there is, on balance, more happiness than unhappiness.

This non-comparative view has to be wrong. First, it is far too dependent on the background levels of happiness and unhappiness already being experienced. Imagine a world where everyone is so happy that there is nothing I can do to decrease the overall amount of happiness such that I would bring about a state of affairs where there is more unhappiness than happiness overall. In this world, every optional action would be supererogatory because the result of any action would always be net happiness. But clearly, there are times when, despite there remaining more happiness than unhappiness in the world, our actions should not be thought of as supererogatory because we *reduce* the amount of happiness. For example, suppose I choose to say something sarcastic and hurtful to a friend who has such a sunny disposition that, although they are *less* happy than they would have been, they are not hurt enough to be unhappy. Nevertheless, my action should not be considered good simply because it failed to make my friend more unhappy than happy, given that they would have been much happier if I had, permissibly, held my tongue.

Or imagine the opposite world, where everyone is so deeply miserable that nothing I can do can decrease the overall amount of unhappiness such that I can bring about a state of affairs where there is more happiness than unhappiness overall. Every optional action, no matter how good, would fail to count as supererogatory on this non-comparative account.

Again, clearly, there are times when making someone *less* miserable when we need not have exerted the effort to do so can rightly be thought of as supererogatory, even though it doesn't bring about an on-balance happy world. For example, suppose that instead of spending my Sunday afternoons doing what I like, I spend time with my elderly, depressed neighbor. Surely this is supererogatory if my presence and

[6] This altruistic condition is endorsed by numerous theorists. For just two, see New, "Saints, Heroes and Utilitarians"; Heyd, *Supererogation.*

company help reduce her depression a little, despite there being nothing I can do to make her actively happy.

A wholly non-comparative approach to supererogation, then, is clearly mistaken. We need to understand the goodness of supererogatory actions comparatively. However, as I will demonstrate, there are better and worse ways to do that too.

2.2 *Rejecting a Before-and-After Comparison*

Instead of focusing on the total value of the state of affairs that an action realizes, we might consider the world before and after we act, and focuses on the difference our action makes to the overall levels of happiness. This account deals well with the cases of the exceptionally happy and unhappy worlds discussed in Sect. 2.1. However, it overlooks cases where we make someone less happy than they were, and yet have still done something supererogatory. For example, suppose I need to tell you some bad news. I could let you know via email, which would be a deeply impersonal and upsetting way to receive the news, or I could tell you face-to-face. While receiving the news face-to-face would still make you unhappy, it would not make you as unhappy as finding out via email. Telling you face-to-face is a better action—and plausibly supererogatory—despite the fact that giving you this news will make you unhappier than you were. It is plausibly supererogatory precisely because I could permissibly have told you the news in a way that would make you *even* unhappier.

Thus, it can be supererogatory to make someone less happy than they were before in those cases where it would have been permissible to make them *even less happy than that.*

2.3 *Good, Better, Best*

We, therefore, need an account of the evaluative condition of supererogation that focuses on *better* rather than simply good.

Before turning to answer the question 'better than what?', I want to note that this view does not entail that a supererogatory action must be the *best* of all available, permissible actions.[7] The idea that supererogatory actions are limited to the *best* action we could have performed rules out many intuitive cases. Most would agree that it is perfectly supererogatory to donate blood once a year, even though it would be *best* to donate as frequently as allowed.

In addition to avoiding the intuitive problems outlined above, distinguishing supererogatory actions from the *best* actions we could perform accords with the justification offered for why we are permitted to perform suboptimal actions at all (a

[7] Those who focus on doing what is best (such as McGinn, "Must I Be Morally Perfect?") normally believe doing so is *required* and thus reject the notion of supererogation completely.

necessary condition of supererogation), which applies not only to the best action we could perform, but also to many of the other actions available to us. Take, for example, the argument that room must be made in our moral theory for supererogation on pain of otherwise being implausibly over-demanding.[8] If this provides a reason for why we are not required to do what is morally best, it plausibly also releases us from being required to do what is morally second or third best, given that this is likely to also be extremely demanding in many cases. The same is true for other arguments for supererogation: the argument from autonomy[9]; the argument that requirements to act optimally are counter-productive[10]; the argument that supererogatory actions arise as 'spandrels' (a side-effect of other normative commitments we have, such as rights)[11]; and Urmson's arguments that acting well must be within the reach of ordinary folk.[12] All of these justifications for supererogation apply to actions that are not necessarily morally best.

3 The Point of Comparison

3.1 *Better than What?*

As I have argued above, supererogatory actions need to be better than some—as yet unspecified—permissible, alternative action. The question is which alternative action or actions should serve as the point of comparison by which to judge whether an action is good enough to qualify as supererogatory?

I begin by only considering cases in which the costs and benefits differ (I return to cases where they are the same in Sect. 3.2). I will defend the following necessary condition against some alternatives (I offer some small modifications and additions later):

> **Better-than-Less-Costly-Alternatives**: an action meets the evaluative condition for supererogation if and only if it is (i) better than some permissible alternative and (ii) better than all less-costly permissible alternatives.

You might think that (i) is all we need.[13] If so, that would imply:

---

[8] Scheffler, *Human Morality*; Murphy, *Moral Demands in Nonideal Theory*; Kagan, *The Limits of Morality*. This objection, as I have argued elsewhere, is in fact better understood as the claim that theories without the supererogatory are overly *confining* (Benn, "Over-Demandingness Objections and Supererogation").

[9] Raz, "Permissions and Supererogation"; Heyd, *Supererogation*; Clark, "The Meritorious and the Mandatory".

[10] Benn, "The Enemy of the Good: Supererogation and Requiring Perfection".

[11] Benn, "Supererogatory Spandrels".

[12] Urmson, "Saints and Heroes".

[13] One reason for this is the belief that it is unnecessary to introduce the notion of cost in order to determine whether an action is good. Some like Horgan and Timmons have rejected that costs play

> **Better-than-the-Worst-Permissible**: an action meets the evaluative condition of supererogation if and only if it is better than the action that, of all the permissible actions available, produces the least good.[14]

To see that this view is insufficient as an account of the evaluative condition of supererogation (and thus, that part (ii) of my account is needed), consider this case: I am about to return home from a holiday in the US, sadly never to return. I have in my possession a $10 note. There is another $10 note somewhere in my hotel room, though it would take a lot of effort to find it. Furthermore, I have no use for dollars in my home country. While they have no value to me, they would benefit the hotel porter if I give them to him before I leave. Now suppose that I have the following four options:

(1) Give the porter the $10 bill I have handy: this involves little in the way of time or effort;
(2) Change my $10 bill into two of $5 at the front desk; give one to the porter and throw the other away: this involves a small cost in terms of my time and effort;
(3) Change my $10 bill into coins at the laundromat down the street, give $2 to the porter and throw the rest away: this is slightly more costly to me in terms of my time and effort; and

any role in determining whether or not an act is supererogatory (Horgan and Timmons, "Untying a Knot from the Inside Out."). However, elsewhere, I have demonstrated the importance of cost for determining the optionality of supererogatory (that is, which *permissible* actions are optional) (Benn, "Supererogation, Optionality and Cost."). None have explored whether cost is important for determining the way in which a supererogatory act is morally good. Note that while some might reject the relevance of cost in its entirety in determining normative matters, most proponents of supererogation accept at the very least that considerations of cost to the agent can determine the extent and limit of our duty (Stanlick, "The Nature and Value of Supererogatory Actions"; Straumanis, "Duties to Oneself: An Ethical Basis for Self-Liberation?"; McGoldrick, "Saints and Heroes: A Plea for the Supererogatory"; Jackson, "The Nature of Supererogation"; Rawls, *A Theory of Justice*; Jacobs, "Obligation, Supererogation and Self-Sacrifice"; Portmore, "Position-Relative Consequentialism, Agent-Centered Options, and Supererogation."). Others, like Barry Curtis, have argued that the disparity between costs and benefits also determines the limits of *supererogation*, as actions that are extremely costly and provide relatively insignificant benefits are, he claims, 'foolish' rather than supererogatory (Curtis, "The Supererogatory, the Foolish and the Morally Required."). These include risking your life to keep a trivial promise, getting into debt for an extravagant present and quitting your job over a minor social injustice.

[14] This is one interpretation of McNamara's view that supererogation is an actions whose "performance is superior to any performance you might have put in while doing the minimum" (McNamara, "Making Room for Going Beyond the Call," 429).

(4) Look for the other $10 bill and give both it and the $10 bill that I have handy to the porter: this involves substantially more time and effort.[15]

Suppose for now that these are the only permissible options; just keeping the $10 is impermissible. (Later, I consider the case where it is permissible for me to do nothing).

The Better-than-the-Worst-Permissible view claims that the point of comparison for meeting the evaluative condition is the action that involves the least amount of good of all the permissible actions available to us. This is action (3), where I give $2. Thus, actions (1), (2), and (4) will meet the evaluative condition because they bring about more good than this action, namely $5, $10, and $20.

On my account—the Better-than-Less-Costly-Alternatives account—an action meets the evaluative condition if and only if it is (i) better than some permissible alternative and (ii) better than (or at least as good as) all less-costly permissible alternatives. Both this view and the Better-than-the-Worst-Permissible view agree that option (1) meets the condition: this action is better than some permissible alternative—namely, option (3)—and there is no option that is less costly and better than it because there is no permissible option less costly than it.[16] They also agree that option (4) meets the evaluative condition: it is better than some permissible alternative—namely, options (1), (2), and (3)—and because it is in fact the best action, there is no less-costly permissible action that is *better* than it. There is also agreement that option (3) fails to meet the evaluative condition: it is the action that does the least good and so fails to meet the condition of being better than some permissible alternative.

The disagreement lies with option (2). My claim is that, contrary to the Better-than-the-Worst-Permissible view, option (2)—to break the $10 note into two $5 notes and give one to the porter and throw the other away—should not be classified as meeting the evaluative condition, and therefore, should not be classified as supererogatory.[17]

---

[15] The cost in terms of time and effort are judged to be small, medium and large *comparative to the alternatives*. I do not suppose that, generally speaking, looking through a hotel room for a lost $10 note involves a large amount of effort. Also, for the sake of this example, let us assume that all four options are permissible, leaving aside any legal or moral prohibitions on the destruction of currency.

[16] This is not to say that (1) in this example is supererogatory. A supererogatory action has to be, not only good, but also *optional*. On my account of optionality for example then the least-costly act constitutes the 'bare minimum' and, as such, would not be optional (Benn, "Supererogation, Optionality and Cost," 2404.). Option (1) in this example, therefore constitutes the bare minimum and while meeting the evaluative condition on both views under discussion would not be optional and therefore not supererogatory.

[17] Parfit has argued that, if you decide to enter a burning building to save the life of one child and could for the same cost save two, you are morally required to save two, even if you are not in fact morally required to enter the burning building and save even one child. I do not go so far as to say that you are under a moral *duty* to give $10 rather than $5 (i.e. that the option of giving $5 is in fact *im*permissible), instead offering a milder claim that it is simply not supererogatory to offer $5 when you could more easily offer $10. However, Parfit's example have caused some to suggest (Heyd, private correspondence) that it might be possible for an agent to both do something supererogatory (save one child) and by the very same action something that violates a duty (failing to save the

Doing option (2) brings about *more* good than we could otherwise have done, but this does not mean it should be considered to be *good*. Choosing not to be as bad as we could be does not make us good when it involves—as option (2) does—going to a greater effort to bring about less good. As the Better-than-the-Worst-Permissible view misclassifies option (2) as meeting the evaluative condition, this view should be rejected. I maintain, therefore, that the Better-than-Less-Costly-Alternatives view is preferable to the Better-than-the-Worst-Permissible view as an account of the evaluative condition for supererogatory actions.

It might be thought that the problem with option (2) could be dealt with by a weaker condition than the one I give in the Better-than-Less-Costly-Alternatives view. Perhaps we only need to add that it must be better than the *least-costly* permissible alternative. This view is as follows:

> **Better-than-the-Least-Costly-Alternatives**: an action meets the evaluative condition of supererogation if and only if it is better than the *least-costly* permissible action available.

This view would also deny that (2) is supererogatory. (1) is the least-costly action and (2) is not better than it. What, then, can be said for my Better-than-Less-Costly-Alternatives view over this Better-than-*Least*-Costly-Alternatives view?

Suppose now that I could, instead, just throw away my $10 bill. Call this option (0). This is now your least-costly alternative, and the one that will do the least good. If (0) is permissible, then (1) to (4) all satisfy the Better-than-the-Least-Costly view: they are better than some permissible alternative—namely, option (0)—and also better than the least-costly permissible alternative—again, option (0). However, this is implausible. It classifies (2) as meeting the evaluative condition, which was the reason that the Better-than-the-Worst-Permissible view was rejected. Additionally, option (3), where we go to even greater trouble to bring about even less good, would also be classed as good on this view. Going to great lengths to do less good than you could have done cannot be considered supererogatory. My view—Better-than-Less-Costly-Alternatives—gets the answer right. If (0) is permissible, then (1) is supererogatory, as is (4), but (2) and (3) are not.

This example establishes the intuitive basis for part (ii) of my account: that an action must be better than all less-costly permissible alternatives. But, you might wonder, if we have part (ii), do we need part (i) (that an action must be better than some permissible alternative)? We do, because the *least*-costly permissible action always meets the second part of the definition—of being better than all less-costly permissible alternatives—because there are by definition no less-costly permissible

other child). The explanation for this however is based on the centrality of reactive attitudes (it is appropriate for the mother of the first child to consider the agent's action heroic and worthy of praise and gratitude but at the same time appropriate for the mother of the second child to harshly blame the agent for not saving her child's life). While I agree with the aptness of these reactions, it therefore emphasizes to me that reactive attitudes are not the appropriate measure by which we should determine deontic categories of permissibility and impermissibility. There are strong theoretic reasons, I believe, to refuse to accept that supererogatory actions can be duty-violating. For more on the deontic condition of supererogation, see my Benn, "Supererogation, Optionality and Cost".

alternatives. Thus, the first part of the definition is needed because it shouldn't be trivially true that the least-costly permissible action meets the evaluative condition.

That the least-costly permissible action should not *always* meet the evaluative condition does not mean that it can *never* meet the evaluative condition. It is often the case that the least-costly permissible action is also the worst permissible action, but that need not always be the case. When the least-costly permissible action is also *better* than some permissible alternative, then it can meet the evaluative condition. This seems intuitively right and is captured by my account.[18]

3.2 Equally-Costly Actions and Equally-Beneficial Actions

In the above discussion, the only cases considered are ones where the costs and benefits differ. Let's turn to cases where they are the same.

First, costs. The Better-than-Less-Costly-Alternatives view states that for an action to meet the evaluative condition, it need only be better than all actions that are *less* costly than—and of course better than some—permissible alternatives. But must it also be better than all *equally*-costly permissible alternatives? I argue that it must. To see why, consider the classic case in which you are faced with a burning building with people trapped inside. Here are your options. You could do nothing (or at least, nothing more than call the emergency services). You could go inside, risking your life, to save Beth. You could go inside, *at no extra risk*, to save Ben as well as Beth. All three options are permissible. I propose that it is not supererogatory to choose to save Beth when you could, for the same cost, save Ben and Beth.[19] Thus, I restrict the Better-than-Less-Costly-Alternatives view by requiring that to meet the evaluative condition of supererogation, an action must be better than all less-costly and all *equally-costly* permissible alternatives.[20]

---

[18] Note that even if it is accepted that the least-costly act is good on some occasions, this does not entail that it is supererogatory. Recall that on my account of the optionality of supererogatory acts, the least-costly permissible act constitutes the bare minimum and is therefore not optional (Benn.). On this view, therefore, the least-costly permissible act is never supererogatory.

[19] Horton goes further and claims that not only is it not good and not supererogatory to save one when you could for the same cost save two, but that it would be *wrong* to do so (Horton, "The All or Nothing Problem." See also Pummer, "Whether and Where to Give"; McMahan, "Doing Good and Doing the Best."). I do not suppose here that it is wrong or forbidden. I simply claim that it fails to meet the evaluative condition for supererogation. This is important because, as I will discuss in Sect. 5, the possibility of being entitled to perform worse actions, even at no less cost, opens the way for suberogation.

[20] Some might worry that this restricted version of the evaluative condition is a low bar for supererogation. Suppose that (A) saving Ben or (B) saving Ben and Beth were the *only* permissible options. On my account, it is true that saving Ben alone fails to meet the evaluative condition, but even in this modified example (where it is impermissible to do nothing) saving Ben *and* Beth still meets the evaluative condition: it is better than a permissible alternative, namely (A), and it is better than all equally-costly permissible alternatives (and there are no less-costly alternatives). This might make the evaluative condition too easy to fulfill. However, I am happy to bite this particular bullet,

What about benefits? Here, I propose that we expand the Better-than-Less-Costly-Alternatives account to include those actions that are merely as good as all less-costly (and all equally-costly) permissible alternatives. Whether you agree will depend on whether you accept that irrational actions can be supererogatory. This expansion allows an action to be supererogatory even if the good it brings about could have been brought about at less cost to the agent. For example, suppose that you could (A) help a friend on the first weekend of the month when you have no prior commitments, (B) help on the second weekend when you would miss a much-anticipated reunion with friends, or (C) do neither. It makes no difference to your friend whether you help them on the first or second weekend. On the Better-than-Less-Costly-Alternatives account, (A) would not meet the evaluative condition because it is not better than a less-costly alternative (namely B), although it is *as* good. In some sense, (A) is irrational: why would you choose to bring about the same good for another person at a greater cost to yourself?[21] However, while this cost seems problematic from a rational point of view, it is hard to explain why it is problematic from a moral point of view (we are, after all, often allowed latitude over self-imposed costs).[22] My intuition pulls me toward allowing that such moderately irrational actions could be supererogatory. This is more compelling if we also allow for limits on self-imposed costs through, for example, self-regarding duties based on cost (it is permissible to unnecessarily miss the weekend with friends, but impermissible to unnecessarily miss a chemotherapy appointment) or the interaction between cost and benefit (it is permissible to unnecessarily risk your life to save the life of another, but impermissible to unnecessarily risk your life in order to get someone milk for their tea).[23]

I have so far accounted for the comparative nature of the goodness of supererogatory actions. If we accept the two modifications proposed in this section, an action is good if and only if it is (i) better than some permissible alternative and (ii) better than or as good as permissible alternatives that are (a) as costly or (b) less costly.

---

especially if we note that just because (B) meets the evaluative condition, this does not entail that it meets the *deontic* condition for supererogation. On my account of optionality, for example, as both (A) and (B) are the least-costly actions, they constitute the 'bare minimum' and thus (B) would not count as supererogatory, even if it does meet the evaluative condition (Benn, "Supererogation, Optionality and Cost."). This raises the bar for supererogation as the action must not only be better (or as good as) all less-costly or as-costly alternatives, it must also be more *costly* than some permissible alternative. I thank Seth Lazar for prompting me to consider this case, a similar one appearing in his Lazar, "Deontological Decision Theory and Agent-Centered Options." (Note that Lazar's discussion of this case asks how uncertainty should be factored into determining permissibility. I leave aside all considerations of uncertainty).

21 One reason is that the beneficiaries are different and you would go to greater lengths to help this person over that person, even if the benefit you bring about would be the same. I discuss the case of different beneficiaries in Sect. 4.

22 In this, I agree with Lazar that "if the moral difference between two acts depends only on *my* well-being, then it should be up to me (perhaps within limits) which of those acts to choose" (Lazar, "Accommodating Options," 250).

23 See for example Curtis's discussion of 'foolish' acts: where the cost to the agent far outweighs the benefits and therefore, he argues, should not be considered supererogatory (Curtis, "The Supererogatory, the Foolish and the Morally Required").

To put it another way: action A meets the evaluative condition of supererogation if and only if it is better than some permissible alternative, and for every permissible alternative P, either A is at least as good as P or P costs more than A.

3.3 *An Objection*

Here is an objection that might be raised against my account: that performing an action that is *vastly* better (and perhaps much more costly) than a permissible alternative (say, doing nothing) is not good enough to be supererogatory when it is possible to do marginally better for marginally less. That my account does not appear to take into consideration the comparative degrees of betterness might seem counterintuitive: surely it is supererogatory to (A) sacrifice my arm and a finger to rescue someone, even if I could (B) just sacrifice an arm and benefit them marginally *more*, if it was (C) permissible for me to have done neither?[24]

However, this seemingly counterintuitive position can be accounted for. If (A) and (C) were the only options, (A) would indeed meet the evaluative condition on my account. It is easy to think of the case *as if* (A) and (C) were the only options (that is, as if (B) were not an option at all) because the marked difference in costs and benefits makes the comparison between (A) and (C) very salient. Furthermore, because bringing about less benefit at a greater cost is intuitively irrational, it is easy to suppose in these highly contrived cases that the agent has a justification for their choice (that some cost or some benefit exists that has not yet been accounted for). Finally, the intuition that (A) could be supererogatory may reflect a difference in beneficiaries. I will explore in Sect. 4 how, if the beneficiaries are different in each option, then bringing about less benefit at a greater cost can sometimes be classed as meeting the evaluative condition on my account.

So let us turn now to the *relativity* of goodness: that is, we must determine if the evaluative condition should be sensitive to differences in beneficiaries.

4 A Beneficiary-Relative Account of Goodness

I now address whether, to be considered supererogatory, an action must be 'better' regardless of the beneficiaries involved. I argue that it must only be better relative to a specific beneficiary.

Note that for those who do not share my intuition regarding the following cases, this part of my definition of the goodness of supererogatory actions is detachable from the previous discussion. However, as I go on to articulate, there is a serious cost to be borne if the relativity of the goodness of supererogatory actions is rejected;

---

[24] This is close to the 'production cost' view outlined in Lazar, "Accommodating Options".

namely, an extremely narrow definition that cannot accommodate many classic cases of supererogation.

4.1 Cases Where the Beneficiaries Differ

On the view defended in Sect. 3.2, an action meets the evaluative condition for supererogation if and only if it is (i) better than some permissible alternative and (ii) better than or as good as permissible alternatives that are (a) as costly or (b) less costly. This makes sense of the examples discussed above, where the beneficiary of the action in each of the options is the same.

However, a problem arises when examples are given where the beneficiary changes depending on which course of action is chosen. To see why this is an issue, imagine that you have a choice between (1) doing nothing, (2) giving £5 to Oxfam and saving the life of a stranger or (3) spending £20 on a book for your friend. Intuitively, both action (2) and (3) are supererogatory. As discussed, most accounts accept that supererogatory actions do not have to be the best actions that we can perform. However, the account proposed thus far, where the 'better' option is insensitive to the difference in beneficiaries, cannot accommodate the intuition that, in the above case, action (3) is supererogatory. This is because not only does action (2) bring about more good, but it also does so *at a lesser cost to the agent* (under the assumption that money is the only relevant cost in question): we do a lot more good—namely, save a life—for a quarter of the cost of giving the book to a friend.

Should we, therefore, accept that (3) is not supererogatory?[25] If we do, the implications for supererogation are radical. If we could always save a life at very little cost, say by donating to a charity, almost no action normally considered supererogatory would turn out to be so. Helping in soup kitchens, helping friends, giving gifts, and so on would no longer be considered supererogatory, as we could perform other actions that are less costly but do more good, namely saving a life. This is deeply problematic if we want to preserve our usual understanding of the supererogatory.[26]

Of course, it is better to give the money to charity than to give a gift to your friend when the former saves a life. This position is perfectly acceptable, as some supererogatory actions are better than other supererogatory actions. The problem is to capture the idea that giving the gift to your friend *is supererogatory at all.*

---

[25] Of course, the difficulty will be how to individuate beneficiaries. Take giving to charity: is it only supererogatory to give to the charity that does the most good for the amount we give? If we consider the beneficiaries in this case to be same—that is, 'recipients of my charity donation'—then my account would say that it would not be supererogatory to give to a less efficient charity. However, if we consider the beneficiaries of different charities to be distinct, then my account declares that giving to even an inefficient charity is good enough to be considered supererogatory. I do not propose to settle the question here.

[26] Heyd explores how it is a core part of the concept of supererogation that we are able to apply discretion over the beneficiary of our actions in his "Ethical Universalism, Justice, and Favouritism".

4.2 Building in the Relativity of Goodness

The solution is to opt for an understanding of 'better' that is relative to a particular beneficiary. We could accommodate the example in Sect. 4.1 by stating that giving the present to your friend is supererogatory because it is non-comparatively good, or because it is better on a before-and-after comparison, or because it is better than the worst or least-costly option. However, we have seen the shortcomings of these approaches. After all, suppose the options were (1*) doing nothing, (2*) spending £5 to save your friend's life, and (3*) spending £20 to give that same friend a book as a present.[27] Additionally, suppose that it is not possible to perform more than one of these actions. I would argue that it is *not* supererogatory to perform action (3*). To give someone a comparatively expensive gift is not good when you could have saved their life at a quarter of the cost. This follows from the discussion in Sect. 3.

The only difference between these two versions is that the beneficiary, in this case, is the same across the options, whereas, in the original, they were different. To account for this, we need to understand goodness as being better *for* particular persons—i.e., to take into account the differences between beneficiaries. If a reference to the identity of the beneficiary is not built into the account of goodness, the claim that it is supererogatory to give presents to our friends or help others in non-vital ways would have to be rejected, given that we could instead save a life at very little cost. This is too far removed from our understanding of supererogatory actions. I, therefore, propose that the following relativity should be accommodated on our understanding of supererogation: an action meets the evaluative condition of supererogation if and only if it is (i) better than some permissible alternative and (ii) better than (or at least as good as) all permissible alternatives that are (a) as costly or (b) less costly, *where 'better' is to be understood as relative to specific beneficiaries.*

Take the original example where you have a choice between (1) doing nothing, (2) giving £5 to Oxfam and saving the life of a stranger, and (3) spending £20 on a book for your friend. Action (2) is better for the stranger than an alternative action and it is also better for that person than all as costly and all less-costly permissible alternatives. It, therefore, meets the evaluative condition. Action (3) is better for your friend than a permissible alternative (namely, action (1)). Also, there is no as-costly or less-costly permissible alternative that is better than it. I can, therefore, accommodate the intuition that action (3) is good, *despite* the fact that we could instead have saved a life for less.

---

[27] Given that it is a friend's life at stake this time, it may be suggested that we are required to spend £5 to save their life. The amount can be changed to a figure where it is thought to be permissible not to spend the money to save their life (and the cost of the gift larger than this amount); alternatively the example could be changed so that the beneficiary is a stranger—importantly the same stranger in both option (2*) and (3*).

5 Suberogation

As evidenced in Sects. 1–4, my account of the evaluative condition for supererogatory actions avoids the problems of other views in accounting for intuitive cases of supererogation. It also has a second virtue: it can help us account for *sub*erogation.

What are suberogatory actions? As Julia Driver claims, "Whereas supererogatory acts are good to do, but not required, suberogatory acts are bad to do, but not forbidden."[28] Many examples can come under this definition. Mary Forrester includes "all those small sins which are not forbidden in one's society or for which sanctions of any kind would bring about more harm than good. Small discourtesies which are or ought to be overlooked fall into this category."[29] Roderick Chisholm and Ernest Sosa, who use the terms 'offence' and 'permissive ill-doing' to describe suberogatory actions,[30] give examples of such actions as "those minor acts of discourtesy which most of us feel we have a right to perform (e.g., taking too long in the restaurant when others are known to be waiting)."[31] They also give examples of more extreme, 'villainous' behavior, as in the case where, despite the fact their business is already flourishing "an industrialist may introduce a line of advertising which is calculated to drive a minor competitor out of business, and he may thus 'win without actually cheating'."[32]

No one, to my knowledge, accepts the category of supererogation while rejecting the possibility of suberogatory action. I, therefore, take those committed to the existence of supererogatory actions to also be committed to the existence of suberogatory actions. Here are some desiderata for an account of suberogation:

A. Some actions are suberogatory: they are permissible but bad.
B. It is not *always* the case that, when faced with a range of permissible actions, each is either supererogatory or suberogatory. We are not always faced with a 'morally charged situation', to use Driver's terminology.[33]
C. We can have multiple suberogatory options.

The Better-than-Less-Costly-Alternatives view fulfills these desiderata, but the Better-than-the-Worst-Permissible and Better-than-the-Least-Costly-Alternative views cannot.

---

[28] Driver, "The Suberogatory," 286.

[29] Forrester, "Some Remarks on Obligation, Permission, and Supererogation," 225. She also includes, to my mind highly controversially, "such morally offensive practices as slavery and human sacrifice, when these are condoned in one's society" (Forrester, 225).

[30] Mellema also uses the term "offence", defining suberogatory actions in terms of praise and blame; as actions that are blameworthy to do and not praiseworthy not to do (Mellema, "Quasi-Supererogation," 145).

[31] Chisholm and Sosa, "Intrinsic Preferability and the Problem of Supererogation," 326.

[32] Chisholm and Sosa, 326.

[33] Driver, "The Suberogatory." Even Driver, who defends the possibility of morally charged situations—where we face the option of either doing something supererogatory or something suberogatory—allows that there are occasions where the omission of a supererogatory act is not suberogatory, but rather "perfectly fine" (Driver, 290).

Let us begin with Better-than-the-Worst-Permissible view. On this view, all permissible actions that are better than the worst we could permissibly have done meet the evaluative condition for supererogation. This leaves only the worst permissible action as a candidate for being suberogatory. The Better-than-the-Least-Costly view has the same outcome: all permissible actions better than the least-costly action meet the evaluative condition for supererogation and thus only the least-costly permissible action is a candidate for being suberogatory.

Both views run into the same problem. If they deny that the worst permissible action or the least-costly action is suberogatory, they leave no room for the suberogatory, contradicting (A) above. If they say that these actions are suberogatory, then given that there will always be a worst permissible or least-costly option, these actions will always be suberogatory. All other actions will meet the evaluative condition and thus we will always be faced with a morally charged situation where all our choices are between supererogatory and suberogatory actions—violating (B). Moreover, even if they find some middle way such that *sometimes* the worst permissible or least-costly action is suberogatory, they will not leave room for multiple suberogatory options, violating (C).

There is one possibility that avoids this: multiple permissible options are possible on even maximizing act consequentialism, as two actions could be tied for 'best'. Thus, two permissible actions could be equally worst or least costly and we would have a choice between two suberogatory actions. However, such a possibility prompts me to formulate the above desideratum (C) more precisely, to reflect that, just as we prefer a theory that allows for multiple *super*erogatory actions with a range of values, so too should we prefer a theory that allows for multiple *sub*erogatory actions with a range of values. That two actions are tied for 'worst' fails to meet this modified desideratum. Two actions tied for 'least costly' could potentially entail a range of values. However, while this would help satisfy (C), it will still violate (B): for the reasons above, we will always be in morally charged situations.

The Better-than-Less Costly-Alternatives view, by contrast, allows for a variety of bad actions: the worst permissible option, but also those that are both more (or as) costly and worse than some permissible alternative.[34] Take the holiday money example first introduced in Sect. 3.1. Options (2) and (3) are paradigm cases of the suberogatory: they are within your rights (it is your money after all) but it is petty and mean-spirited to go to extra effort to bring about less good than an alternative permissible action (namely option (1) of just giving the $10 to the porter). Thus, the Better-than-Less-Costly-Alternatives view makes room for the possibility of suberogation, fulfilling desiderata (A). It also allows for cases where there is no suberogatory

[34] Some like Lazar agree with me about the relevance of comparative costs (Lazar, "Accommodating Options."), as well as the importance of agent-centered options both to favor our own interests (that is, where we can take into account the cost to ourselves) and also to sacrifice them (that is, we are allowed to do what leaves us worse off, within reason). Nevertheless, this view, unlike mine, fails to make room for suberogation. It simply rules out as permissible actions where there are better moral alternatives that are less costly, as costly or only marginally more costly. By allowing on my view that such actions could well be permissible (as they could be 'within your rights'), I can meet the desiderata above.

option, as perhaps there are no more costly, less good permissible actions, thereby fulfilling desiderata (B). Finally, the Better-than-Less-Costly-Alternatives view also allows for a range of suberogatory actions without relying on them being tied for being equally costly and equally bad, as there can sometimes be multiple more (or as) costly, less good permissible actions. Both (2) and (3) fit this category but differ in terms of just how bad they are, thereby fulfilling desiderata (C).[35]

Thus, my account deals not only with problem cases, but also makes room for an important category of moral action—suberogation. Alternative accounts cannot do either of these things adequately.

6 Conclusion

I have shown that the goodness of supererogation depends on a complex interaction of benefits and costs. Thus, I have argued for the following:

> **The Evaluative Condition of Supererogation**: An action meets the evaluative condition of supererogation if and only if it is (i) better than some permissible alternative and (ii) better than (or at least as good as) all permissible alternatives that are (a) as costly or (b) less costly, where the sense in which it is better is *relative* to a particular beneficiary.

It is not enough that supererogatory actions go beyond duty insofar as they are permissible and not required: they must also go beyond duty in terms of their goodness. While this is a core part of almost every account of supererogation, it rarely gets explored.

Costs have often been seen as playing an important role in determining the limits of the call of duty. A deep dive into the contours of the evaluative condition reveals cost is also relevant for understanding how some actions go above and beyond that call. My account explicates the intricate relationship between the supererogatory and other ethical considerations, such as the actions that could have been performed instead, the costliness of actions, and the identity of a beneficiary. It captures our intuition about both core and complex cases of both the supererogatory and the suberogatory. This not only provides much-needed theoretical symmetry, but also illuminates, as Driver says, those "judgements lying in the dark corners between right and wrong."[36]

[35] I focus here on more costly, less good options being suberogatory, leaving aside whether or not the *worst* permissible action should never, sometimes, or always be considered suberogatory. It is beyond the scope of this paper to give a precise formulation of suberogation. However, it should be noted that no specific position needs to be taken on the question of the worst permissible alternative in order to meet the three desiderata laid out.

[36] Driver, "The Suberogatory," 295.

Bibliography

Archer, A. (2016). Are acts of supererogation always praiseworthy? *Theoria, 82*(3), 238–255. https://doi.org/10.1111/theo.12085

Archer, A. (2013). Supererogation and Intentions of the Agent. *Philosophia, 41*(2), 447–462. https://doi.org/10.1007/s11406-013-9422-9

Benn, C. (2019). Intentions, motives and supererogation. *The Journal of Value Inquiry, 53*(1), 107–123.

Benn, C. (2016). Over-demandingness objections and supererogation. In M. van Ackeren & M. Kühler (Ed.), *The limits of moral obligation* (pp. 68–83). Routledge Studies in Ethics and Moral Theory. New York and London: Routledge.

Benn, C. (2017a). Supererogation, optionality and cost. *Philosophical Studies, 175*(10), 2399–2417. https://doi.org/10.1007/s11098-017-0965-7

Benn, C. (2017b). Supererogatory spandrels. *Etica & Politica / Ethics & Politics, 19*(1), 269–290.

Benn, C. (2018). The enemy of the good: Supererogation and requiring perfection. *Utilitas, 30*(3), 333–354.

Chisholm, R. M. (1964). The ethics of requirement. *American Philosophical Quarterly, 1*(2), 147–153.

Chisholm, R. M., & Ernest, S. (1966). Intrinsic preferability and the problem of supererogation. *Synthese, 16*, 321–331. http://www.springerlink.com/index/n315h744735h5174.pdf

Clark, M. (1978). The meritorious and the mandatory. *Proceedings of the Aristotelian Society, 79*, 23–33. http://www.jstor.org/stable/https://doi.org/10.2307/4544931

Cowley, C. (2015). The agents, acts and attitudes of supererogation. *Royal Institute of Philosophy Supplementary, 77*, 1–24.

Curtis, B. (1981). The supererogatory, the foolish and the morally required. *Journal of Value Inquiry, 15*, 311–318.

Driver, J. (1992). The suberogatory. *Australasian Journal of Philosophy, 70*(3), 286–295.

Feinberg, J. (1961). Supererogation and rules. *Ethics, 71*(4), 276–88. http://www.jstor.org/stable/https://doi.org/10.2307/2379643

Forrester, M. (1975) Some remarks on obligation, permission, and supererogation. *Ethics, 85*(3), 219–226. http://www.jstor.org/stable/https://doi.org/10.2307/2380049

Harman, E. (2016). Morally permissible moral mistakes. *Ethics, 126*(2), 366–393. https://doi.org/10.1086/683539

Heyd, D. (1978) Ethical universalism, justice, and favouritism. *Australasian Journal of Philosophy, 56*(1), 25–31. http://www.tandfonline.com/doi/pdf/https://doi.org/10.1080/00048407812341021

Heyd, D. (1982). *Supererogation: Its status in ethical theory*. Cambridge Studies in Philosophy. Cambridge: Cambridge University Press.

Horgan, T., & Mark, T. (June 16, 2010). Untying a knot from the inside out: Reflections on the 'paradox' of supererogation. *Social Philosophy and Policy Foundation, 27*(02), 29–63. https://doi.org/10.1017/S026505250999015X

Horton, J. (2017). The all or nothing problem. *The Journal of Philosophy, 114*(2), 94–104.

Jackson, M. W. (1986) The nature of supererogation. *The Journal of Value Inquiry, 20*, 289–296. http://link.springer.com/article/https://doi.org/10.1007/BF00146118

Jacobs, R. A. (1987). Obligation, supererogation and self-sacrifice. *Philosophy, 62*(239), 96–101. http://www.jstor.org/stable/3751025

Kagan, S. (1989). *The limits of morality*. Oxford University Press.

Lazar, S. (2019). Accommodating options. *Pacific Philosophical Quarterly, 100*(1), 233–255. https://doi.org/10.1111/papq.12252

Lazar, S. (2017). Deontological decision theory and agent-centered options. *Ethics, 127*(3), 579–609. https://doi.org/10.1086/690069

McGinn, C. (1992). Must I Be Morally Perfect? *Analysis, 52*(1), 32–34.

McGoldrick, P. M. (1984). Saints and heroes: A plea for the supererogatory. *Philosophy, 59*(230), 523–528. http://www.jstor.org/stable/3750369

McMahan, J. (2018). Doing good and doing the best. In *The ethics of giving: Philosophers' perspectives on philanthropy*. New York: Oxford University Press. https://doi.org/10.1093/oso/9780190648879.001.0001

McNamara, P. (1996). Making room for going beyond the call. *Mind, 105*(419), 415–450. http://mind.oxfordjournals.org/content/105/419/415.short

McNamara, P. (2011). Supererogation, inside and out: Toward an adequate scheme for common-sense morality. In M. Timmons (Ed.), *Oxford studies in normative ethics* (vol. 1, pp. 45–66). Oxford: Oxford. https://doi.org/10.1093/acprof:oso/9780199693269.001.0001

Mellema, G. (1987). Quasi-supererogation. *Philosophical Studies, 52*, 141–150. http://www.springerlink.com/index/NQ90065355311077.pdf

Mellema, G. (1991). Supererogation and the fulfillment of duty. *The Journal of Value Inquiry, 25*, 167–175. http://www.springerlink.com/index/N475156P93R4V4N2.pdf

Murphy, L. (2000). *Moral demands in nonideal theory*. Oxford University Press.

New, C. (1974). Saints, heroes and utilitarians. *Philosophy, 49*(188), 179–189. http://journals.cambridge.org/production/action/cjoGetFulltext?fulltextid=3513600

Portmore, D. W. (2003). Position-relative consequentialism, agent-centered options, and supererogation. *Ethics, 113*(2), 303–332. http://www.jstor.org/stable/https://doi.org/10.1086/342859

Pummer, T. (2016). Whether and where to give. *Philosophy and Public Affairs, 44*(1), 77–95. https://doi.org/10.1111/papa.12065

Rawls, J. (1999). *A theory of justice*. Rev. Ed. Cambridge, Mass.: The Belknap Press of Harvard University Press.

Raz, J. (1975). Permissions and supererogation. *American Philosophical Quarterly, 12*(2), 161–168. http://www.jstor.org/stable/https://doi.org/10.2307/20009570

Scheffler, S. (1992). *Human morality*. New York: Oxford University Press. https://doi.org/10.1093/0195085647.001.0001

Stanlick, N. A. (1999). The nature and value of supererogatory actions. *Journal of Social Philosophy, 30*(1), 209–222. https://doi.org/10.1111/0047-2786.t01-1-00013

Stocker, M. (1967). Acts, perfect duties, and imperfect duties. *The Review of Metaphysics, 20*(3), 507–517. http://www.jstor.org/stable/https://doi.org/10.2307/20124315

Straumanis, J. (1984). Duties to oneself: An ethical basis for self-liberation? *Journal of Social Philosophy, 15*, 1–13.

Urmson, J. O. (1969). Saints and heroes. In J. Feinberg (Ed.), *Moral concepts* (pp. 60–73). Oxford University Press.

Supererogation and Forgiveness

Christopher Cowley

Abstract Forgiveness is widely considered a paradigm of supererogation: it seems to be morally permissible without being obligatory, and it seems to be almost always admirable and praiseworthy. I want to show that the phenomenon is a bit more complicated, and that many instances are hard to describe as supererogatory. First, I will distinguish forgiveness from some other responses to the transgression (ignoring, excusing, letting go). Second, I will examine the philosophical debate over the question of whether or not the victim should wait for the transgressor to fulfil some kind of condition (e.g., repentance, apology, compensation) before forgiving, and how this might affect the supererogatory status. Third, I look at more serious cases of transgression and ask what it might mean for something or someone to be unforgivable.

Keywords Supererogation · Forgiveness · Excuses · Understanding · Unforgivable

Forgiveness is widely considered a paradigm example of supererogation. It seems to be elective, neither obligatory nor prohibited, and genuinely up to the victim; and it seems to be almost always a positive thing, certainly if one is to judge from the tone of the hundreds of titles on Amazon. So as elective and positive, it will be praiseworthy if performed, but not blameworthy if omitted. (That's enough of a definition of 'supererogation' to be getting on with.) The only difficult thing would seem to be overcoming the psychological resistance of one's understandable pride and pain. This seems to be basically similar to the way that supererogatory heroism has to overcome the psychological resistance of one's understandable caution and fear.

C. Cowley (✉)
University College Dublin, Dublin, Republic of Ireland
e-mail: Christopher.cowley@ucd.ie

D. Heyd (ed.), *Handbook of Supererogation*,
https://doi.org/10.1007/978-981-99-3633-5_12

The philosophical literature on forgiveness is by now quite substantial, and I cannot hope to do it justice here.[1] Suffice to mention Hughes and Warmke's entry 'forgiveness' in the *Stanford Encyclopedia* and the recent Pettigrove and Enright (eds) *The Routledge Handbook of the Philosophy and Psychology of Forgiveness* (2023).[2] My aims in this chapter are more modest. I want to ask whether forgiveness, or certain kinds of forgiveness, or forgiveness in certain kinds of situation, might or actually might not be supererogatory, what that means, and what follows on from it. I have two broadly skeptical conclusions: first, forgiveness is more complicated and often more mysterious than it first appears; second, many instances of forgiveness are less obviously supererogatory than they first appear, even when they are still touching and admirable.

At some point I need to define forgiveness, of course; but it will prove helpful to start negatively, by looking at several responses that are similar to but distinct from forgiveness, and which themselves might be supererogatory: ignoring, excusing, and letting go. This approach allows us to take a victim's-eye view of the 'practical space' of deliberation among the possible responses. Along the way we can introduce some of the main philosophical issues relevant to forgiveness and supererogation. To complicate things, the victim's response might be a hybrid of responses, and this might change the supererogatory status.

Later, I want to consider one central distinction in the philosophical literature: between *conditionalist* and *unconditionalist* accounts of forgiveness, framed around the question for the victim: should she wait for the transgressor to fulfil certain conditions (e.g., apologizing) before she forgives him? The two accounts raise different issues about supererogation. My third aim will be to look at *egregious* transgressions, and how we are to make sense of victims and third parties who forgive the transgressors. Here I will argue that such forgiveness is characteristically personal in a way that defies understanding as elective and praiseworthy, and therefore that defies understanding as supererogatory too.

---

[1] There is also a substantial theological (especially Christian) literature on forgiveness. One important text would be Anthony Bash's *Forgiveness: a Theology* (2015), which also contains a good deal of philosophical argument. The Templeton organization has sponsored a large project on Forgiveness (https://forgiveness.templeton.org/), which is very Christian in tone, but also claims to find scientific support on how healthy it is to forgive. In what follows I will mention some *popular* Christian ideas, but will not discuss the theology in any depth.

[2] Unlike Pettigrove and Enright, I will not be dealing with any of the psychological literature on forgiveness, and will only be considering 'folk' psychological ideas. For completeness I should mention that Routledge offers another recent volume, The *Handbook of Forgiveness* (2nd ed. 2020), edited by Worthington and Wade.

1 What is Forgiveness? What is It *not*?

To organize the discussion, I'll start with a basic schema and bring in complications as we go along:

- A (male) Transgressor encounters a (female) Victim, and commits a Transgression against her. (The genders are merely to allow efficient use of pronouns.)
- According to the schema, the Victim is morally justified in her anger against the Transgressor.[3]

The question for the Victim will eventually be whether she can and should forgive. Note the double-object structure of the verb: the Victim would forgive *the Transgressor* for *the Transgression*, in normal cases both objects are necessary (we will later ask whether the Victim can have different responses to Transgressor and Transgression). However, logically and chronologically prior to the question of whether to forgive, the Victim will have the option of at least five other responses. More positively, she may choose (i) to ignore the Transgression, (ii) to excuse the Transgressor, (iii) to 'let go' of the whole encounter. I group these responses together since they resemble forgiveness, they may sometimes be mixed with each other or with forgiveness, and they may turn (partly) into forgiveness over time. Alternatively, the Victim may choose something more negative: (iv) to retaliate against the Transgressor or (v) to withdraw and nurse her justified anger against him.

1.1 *Ignoring*

In our busy lives it is inevitable that putative transgressions will often result from ordinary friction and imperfect knowledge; the crowded bus jostling or the ambiguous social media post or the lack of recognition from the stressed boss. And most of the time, most of us can overcome our momentary anger and ignore the encounter, we move on, and eventually we probably forget it. (If the transgression is more serious, the victim might still insist on ignoring, in a spirit of 'it's beneath me to respond', although probably without forgetting.) This is not forgiveness, however, despite the mechanical use of expressions such as 'forgive me'.[4] If the Transgressor

[3] Two comments. First, some object to the word 'victim', which seems to imply a passivity that would hinder growth after the transgression; the better word is often 'survivor'. I share the concern, but since my discussion focuses very much on the moment of choice among possible responses, 'victim' seems more appropriate. Second, the relevant emotion (reactive attitude) is often described as 'resentment', but this has always struck me as too modest, too prim, too contrived to capture the rawness and the retributive force of the victim's response in many scenarios. Other emotions might be relevant in some situations, e.g., contempt, disappointment, sadness, loathing, hatred. For brevity I will use 'anger' throughout.

[4] The Templeton Forgiveness project (https://forgiveness.templeton.org/) defines forgiveness on its homepage as "to stop feeling anger toward or blaming someone who has done something wrong"—but by itself, that defines no more than ignoring, forgetting or letting go of the transgression.

is a stranger, this might make it easier to move on, since we have things to do and places to go; if they are a colleague or a family member, we can give them the benefit of the doubt since we have to continue working or living with them. None of this is problematic; beyond a basic social duty to 'get along' there will be clear admirable and non-obligatory instances of ignoring that may be supererogatory.[5]

Reconciliation is often taken as a key component of forgiveness. But sometimes I can forgive without seeking reconciliation (especially if the Transgressor was a stranger to me), and sometimes I can seek reconciliation without forgiveness, as when the relationship is too useful to break off, or when I am 'stuck' in a family or work relationship with high exit costs.

This issue of psychological resistance has to be a major element in any discussion of forgiveness—and of ignoring, excusing and letting go. The Victim might *try* to ignore the Transgression, might *want* to ignore it, might judge her anger to be petty and vain, but still she can't ignore it. Indeed, the force of the transgression might linger, as she obsessively revisits the sequence of events over and over again. Or maybe the Victim succeeds in ignoring it for a while, if necessary by avoiding the Transgressor, until she bumps into a striking reminder and the anger flares up again, much to the Victim's frustration. Because of this longer term psychological resistance, one should really not speak of ignoring or forgiving as a once-off autonomous decision but as a partly unpredictable process that can only be attempted and renewed. Importantly, such efforts might indeed last a lifetime, and might become a central project in the Victim's self-concept.

1.2 Excusing (and Justifying)

Sometimes the Transgressor might offer, or the Victim might make efforts to discover or imagine, some relevant excuse. An excuse is something that reduces or even eliminates the Transgressor's moral responsibility for the Transgression. The Transgressor might have been (partly) ignorant of something, or might have been (partly) unfree to refrain from transgressing. If the Victim accepts such a (partial) excuse, such acceptance will not constitute forgiveness; it will involve an appropriate recognition of the moral reality. If the excuse is only partial, however, then the Victim can excuse that proportion of the wrong, and will face a new decision of whether to generously ignore or even forgive the remainder.[6]

---

[5] In passing, we have the injunction to 'forgive and forget'. Most philosophers have dismissed this as incoherent. First of all, one cannot choose to forget: one has to *ignore* and hope that the memory fades. But second, to forgive involves facing the transgression, and the wrongness of the transgression, squarely, and this would seem to hinder forgetting, at least in the short term. As a result, the injunction should read either "ignore and hopefully forget" or perhaps "forgive, then ignore, then hopefully forget".

[6] For a detailed analysis of the difference between excusing, justifying and forgiving, see Allais (2008) pp. 33–41.

There is a popular expression 'to understand all is to forgive all', and most philosophers seem to reject this as incoherent. In simple cases, when the Victim seeks to *understand* the Transgressor, then she is looking for relevant excuses. Once she understands 'all', then she has found sufficient excuses (noun) to excuse (verb) him—rather than forgive him. So the popular expression should perhaps read "to understand all is to excuse all." As such, it seems more plausible to say that the Victim's dilemma of whether to forgive only begins when she is unable to understand the Transgressor, that is, unable to find a relevant excuse: the Victim says to herself "How *could* he do this to me? It's incomprehensible, inexcusable."[7]

However, the notion of understanding another person is often more complicated. The Victim's efforts to understand a putatively accidental Transgression might reveal excuses, and she can excuse; on the other hand, they might reveal that the Transgression was more intentional or reckless—and less excusable—and with such an understanding she would then have to decide whether to ignore or forgive. But even when the Victim comes to understand the Transgression as deliberate, there will be a further question of whether she can understand the degree of *malice*. Perhaps the Transgressor was seeking revenge for a real or imagined prior transgression on him, combined with stress or over-sensitivity in the Transgressor's wider life—and this might be an unthinking lashing-out and therefore excusable. But the precise degree, focus and causal origins of the malice (real or imaginatively inferred) might remain stubbornly incomprehensible even after sympathetic and epistemically responsible investigation.

But insofar as the Victim thinks in terms of unconfirmed excuses, this is still a kind of ignoring, or an ignoring-excusing hybrid; at any rate, it is still not quite forgiveness, although it can be supererogatory. Paradoxically, someone who regularly deploys this sympathetic ignoring-excusing strategy might be described as a 'forgiving' sort of person (i.e., as possessing the virtue).[8] However, as we will see below, this might not

[7] Westlund (2019 p. 186) thinks there *is* room for taking "to understand all is to forgive all" at face value. This is because 'understanding' is not only about possible excuses, but also about distinctive cognitive and emotional failures to which I too am vulnerable, and so 'forgiveness' here amounts to a kind of non-judgementalism:

> [One kind of forgiveness] marks a kind of forbearance, or nonjudgmentalism, toward others, which may be rooted in a sense of the fragility of our grasp on the full moral import of anyone's actions, including our own. Forgiveness, in this sense, expresses a principled refusal to place oneself morally above others, as in a position to condemn or dismiss others in light of their moral (and narrative) failings.

See also Pettigrove (2007) for a similar argument.

[8] Or she might be called *merciful*. Mercy is normally the prerogative of judges and presidents. But a personal merciful attitude will be relevant for transgressors who are offensive without transgressing against the merciful person. Bennett (2004) offers the example of a 'genteel racist grandfather', who uncritically inherited his racism during his childhood in 1950s Britain. He is not a dangerous thug, and he has not harmed the grandson, so forgiveness is not at issue. Instead, the grandson mercifully refrains from fully condemning his racism. Bennett's point is that such merciful restraint is actually a kind of insult, for it implies that the grandfather is no longer able to accept moral criticism and to reflect on his objectionable attitudes.

be as positive as it sounds, since it might be a strategy for avoiding conflict or turning a blind eye to disturbing injustice, at which point it would lose any supererogatory status.

The Victim might talk to the Transgressor or investigate the putative Transgression and find that it was *justified*, e.g., by a greater good. The classic example would be of a Transgressor violently pushing the Victim out of the way of an on-coming car: the momentary outrage at the violence would quickly turn to gratitude. A more difficult and often tragic class of cases would involve a justification that does not stand to benefit the Victim herself. So if a boss is forced to downsize in order to save the company, he is 'politically' (under the 'big picture') justified in firing one of the more recent hires, and the recently hired victim can understand this to a degree, but she is angry because she had been working demonstrably harder than some of the older colleagues who are to be spared. Assuming that the Victim is not happy to sacrifice herself for her boss or for the company, would the victim be in a position to *forgive* the boss? I don't think so, because the decision to fire the victim was not personal, and in many ways the boss's hands were tied. The Victim can only ignore the decision as best she can.[9] Would this be supererogatory? I'm not sure, to be honest. It might be admirable to watch the dismissed employee put a brave face on it, mutter 'stuff happens' and launch her job search with elan.

1.3 Letting Go

I would say that most of the 'forgiveness' books on Amazon are not actually about forgiveness but about *letting go*. They are not about forgiving because the Victim is primarily motivated by her desire for relief from her own suffering. Letting go is a kind of therapy—and of course a perfectly legitimate kind. The Victim's focus is on the suffering, and in that sense her efforts are unilateral. As such, letting go cannot be supererogatory, since it is effectively a kind of enlightened self-interest, without self-sacrifice. In contrast, and to anticipate a later argument, forgiveness has to be bilateral and has to somehow involve the Transgressor: the ideal involves the Victim forgiving the Transgressor *to his face*.[10] This is perhaps controversial, since it exposes the Victim to the risk of a second indignity of having her forgiveness rejected, and such failure of recognition might undermine its status *as* forgiveness. (Compare: if I admire Jones and leave a gift on her doorstep, she has to recognise it

---

[9] An extreme example of a 'political' transgression comes from the philosophical literature about moral dilemmas: the mythological Agamemnon, who chose (was 'obliged') to sacrifice his daughter Iphigenia to placate the Gods and save his troops from impending starvation. The philosophical discussion normally focuses on Agamemnon. But there is a real question of what Iphigenia thinks (or allows herself to think) about her father as he lifts the knife. Does she hate him? Pity him? Would she be able to forgive him for doing what he had to do *qua* commander, even though he betrayed her *qua* father?

[10] Such bilaterality is also essential to justice. In court, a defendant must be accused to his face, and must be allowed the chance to respond to the accusation.

as a gift for it to *be* a gift: if she thinks it's just rubbish and throws it away, it is no longer a gift.) However, it is important for the attempted forgiveness to be anchored to the reality of the Transgressor and the Transgression, to prevent the Victim's anger from distorting her perception. Consider the case of a person who enjoys *wallowing* in victimhood, in taking offence from this slight and that insult, and yet who, once in the presence of her favorite audience, magnanimously 'forgives' all.

The face-to-face requirement is additionally controversial, since it might be described as effectively giving control over the forgiveness to the Transgressor. In the best case, however, it might be the beginning of a mutually revealing conversation and reconciliation between Transgressor and Victim, but of course it might not; in which case, all the Victim will be left with is the option of letting go, away from the Transgressor.

Letting go is also distinct from ignoring, in two ways. First, under letting go, the anger characteristically involves a keener awareness of both sides of the Transgression: the Victim's thought is 'I can't believe *he* did that *to me*', and she has to let both transgression and transgressor go. Second, with letting go there is greater need for the Victim's deliberation and commitment (and re-commitment) over time, whereas ignoring is structurally simpler. While ignoring aims at forgetting, letting go tends to aim at motivational inertness. But there is a fuzzy boundary between letting go and ignoring, and there is room for the result to be supererogatory.

As a result of the face-to-face requirement, one cannot, strictly speaking, forgive the dead, one can only let go of them and their deeds. However, I think one can aspire to a quasi-forgiveness within the context of a continuing 'conversation' with the dead. I have in mind a widow who regularly visits the grave of her dead husband, in order to tell *him*—not tell *his body* or tell *his grave*—about how her week has gone. Importantly, the widow need not necessarily be superstitious or religious here; she would not deny the biological events happening to her husband's body underground, for example. But she will see herself as carrying on the same conversation as they had when he was alive, and as part of *that* conversation I think she could meaningfully forgive her husband, where such a forgiveness would require an imagined response from him too.

1.4 Retaliating and Protesting

The first three positive responses formed a group that were similar to forgiveness, and might indeed be mixed with forgiveness. (I'm aware I still haven't defined forgiveness positively, but bear with me.) To complete this account of the Victim's 'practical space' I now turn to the two other, more negative responses, which are clearly opposed to forgiveness.

If the Transgression genuinely angers the Victim, then perhaps retaliation might be the *first* option, spontaneously, rather than ignoring or excusing! And the Victim might well be justified, at least if the retaliation is proportionate to the Transgression. She then might have to 'talk herself down' from the urge, either in realistic acceptance of a power imbalance, or in pursuit of a higher pacifist ideal ("I'm not going to let myself be provoked.") Retaliation has the sound of vigilante justice, with all the risks of disproportionality and cycles of violence; but within the category 'retaliation' I am also including the possibility of protest. The Victim might simply declare to the Transgressor "you have treated me unfairly." This might well lead to an ignorant Transgressor apologising, and that might be the end of it. Or it might lead the Transgressor to reject or ignore the Victim, and the Victim will be back at the question of what to do about the Transgression. Importantly, the Victim's protest might lead to a new discussion about whether the putative Transgression was or was not unfair.[11] Even if the Victim chooses not to retaliate, awareness of the *option* of justified retaliation may well be morally healthy when we're talking about gross injustice, and about a transgressor likely to interpret forgiveness as weakness or condonation.

1.5 Nursing One's Justified Anger

The Victim might try and fail to ignore the Transgression. She might decide, for whatever reasons, to refrain from retaliation or protest. She might be unable to discover or impute excuses. And so she is left with her justified anger. (I am still calling the anger 'justified' by a real moral transgression, and not a matter of, say, mere injured vanity.) Sometimes long-term anger is as corrosive as the self-help books assume; sometimes sulking or wallowing merely makes the insult worse; sometimes strong emotions can blind one to one's better interests and important relationships. In all such cases, trying to ignore or let go will of course be advisable. But we should not assume that justified anger is always destructive. To lose the justified anger is sometimes to risk losing the motivation to change the world.

I am not just making the merely psychological claim that justified anger can be productively channelled. More than this, righteous anger is associated with a fundamental sense of self-respect. Even when the Victim has calculated that retaliation and protest are not in her self-interest, the continuing anger is part of an honest recognition of the moral reality of the world she lives in: one needs landmarks of good and evil to preserve one's moral orientation. I am taking the phrase 'moral reality' from the Holocaust survivor Jean Améry (1980), who believed that the magnitude of the offences against him demanded enduring resentment as the only appropriate response, precisely in order to avoid the crimes being 'lost to history' through an undiscriminating blanket letting-go, even if driven by the prima facie admirable need for reconciliation and co-operation in rebuilding Europe. A well-intentioned therapist who advised Améry to overcome or repudiate his anger for the sake of his own psychological health would have missed the point. Of course the Holocaust presents

[11] Here it has to be noted how I loaded the dice in my basic schema by stipulating that a Transgression had indeed been committed; for the putative Transgressor might reject the suggestion, and have good reason for believing the putative victim to be over-sensitive.

exceptional problems for victims' responses in virtue of the sheer scale of its evil, and I will return to it in the final section.

2 Forgiveness

The best place to start in any discussion of forgiveness is with examples, such as those in the 'stories' section of the online *Forgiveness Project*[12] or in Beverley Flanagan's (1994) book. Philosophers too often forget the sheer range of contexts in which people struggle to forgive, and the many different shapes that forgiveness can take. So let that be my first advice to the reader before continuing.

By laying out the different responses I have described the 'practical space' the Victim finds herself in; whether or not she considers each option, she has reached a position where she wonders whether to forgive. How is she supposed to think about forgiveness, exactly? It's surprisingly hard to specify in the abstract. For a start, we can reiterate the point that it is not a discrete action of just 'forswearing resentment' (Joseph Butler's seminal idea) or of repudiating justified anger,[13] but rather the launch of a gradual process, subject to unpredictable psychological resistance, and requiring self-monitoring and perhaps regular re-commitment. I am taking forgiveness as a scalar notion: I can forgive somewhat, then a bit more, then more again. (I can even go back on my forgiveness, when it becomes apparent later that, e.g., an apology was not fully sincere.) Some transgressions can be carried for years without the victim ever being really sure how much she has forgiven, and she may remain vulnerable to sudden painful reminders.

During this process, forgiveness requires some deft mental gymnastics. With ignoring, the process is guided by a 'single consciousness': I turn my attention away from both the Transgression and from my anger and I (try to) get on with my life amid other distractions. In contrast, forgiveness involves a 'double consciousness'; I remain acutely conscious of the Transgression *as* inexcusably and incomprehensibly wrongful, while at the same time repudiating my anger, and restraining its influence on my relationship with the Transgressor. The wrongfulness has to remain clear, even while (I hope) its painful and insulting hold on me loosens over time.

An obvious way of resolving this paradox is by invoking the Augustinian maxim to "hate the sin, and love the sinner." When the Transgressor apologizes and I forgive him, he repudiates the Transgression and invites me to join him in condemning it as a separate, loathsome object in the past. I've never been sure about this maxim, for the simple reason that the Transgression *came out of* the Transgressor, and the Transgression remains *his act*, even if it's now in the past. If the harm was intentionally directed at me, then the intention links the actor to the act. Even if it was unintentional

[12] https://www.theforgivenessproject.com/stories/.

[13] Benbaji and Heyd (2001) suggest the notion of 'suspending' the resentment (rather than attempting to silence it), taking the verb from Coleridge's description of theater-goers as 'suspending disbelief'.

(it was a case of recklessness or negligence), then it was still his act—whose else would it be? I might excuse him for it, of course, but insofar as it is inexcusable and he is a fitting object of forgiveness, then it is because of the act, and it cannot be conveniently separated away. And in the case where the Transgressor does not apologize, of course, then the separation would be even more difficult.

Consider an extreme case such as murder. Here the act transforms the agent into a murderer, and he will remain a murderer for the rest of his life, *even if* he has repented, *even if* he has served a fair prison sentence, *even if* he is forgiven by his victim's family. There is no such thing as an ex-murderer. In that sense he continues to carry the act with him, wherever he goes. And he also carries the potential to do it again, once the circumstances are right. The rest of us, without the experience of murdering, have grounds for believing that we would never do such a thing (even if we might not be able to predict our own response to, e.g., intense provocation).

There are two questions about the meaning of any attempted act of forgiveness. Earlier I argued that forgiveness cannot be unilateral but must be articulated to the Transgressor himself. There is a risk here. The Victim might be genuinely hurt by the Transgression, but when she goes to forgive the Transgressor to his face, he just laughs at her, deepens the wound, and degrades what was meant to be a noble gesture. Would this transform the Victim's action, against the Victim's intention, into a unilateral letting-go? It's true that this would seem to leave a lot of power in the hands of the still malicious Transgressor.[14]

Second, in my basic schema I did not say anything about the antecedent relationship between Transgressor and Victim. When there is such a relationship (be it one of mutual affection or mutual advantage, or merely being stuck in the same workplace or family), then forgiveness might be motivated by a desire for reconciliation. But the forgiveness, if successful, does not guarantee reconciliation; reconciliation will be a matter for a separate decision, more or less successful. (Perhaps in some cases the forgiveness can lead to the creation of a new relationship where none existed before the Transgression; both parties are moved by the experience to discover commonalities.) Of course, when there is an antecedent relationship and the Victim *does* want to repair it, it should be remembered that the Transgression will inevitably get in the way of full repair, since both parties will—and should—remember precisely that it was *this* Transgressor who did *that* against *this* Victim. As such the relationship has been changed irreversibly, even if the Transgressor's contrite apology and reparation might be an occasion for rapprochement.

For completeness, I should briefly mention a subsidiary debate about who has the 'standing' to forgive. If a person has standing it means that she is entitled to forgive, given her particular status in relation to the Transgression. At first glance, it would seem to be only the direct Victim who has such a standing. However, insofar as a third party *identifies* with the Victim, such a third party can have at least a secondary

[14] Indeed, this mirrors the risk from the other side: when asking to be forgiven the Transgressor also takes the risk of being rejected. However, the two risks are not symmetrical because of the different underlying attitudes: the Transgressor has to expect the possibility of rejection since he has no right to forgiveness; whereas the Victim, insofar as she has understood the facts correctly, can rightfully hope that her forgiveness will be welcomed.

standing. A paradigm of such secondary standing would be the parent of the Victim; it would make sense to say that any harm caused to the child will also be caused indirectly to the parent. Murphy (2009) also suggests that there might be those with *tertiary* standing: those who share a group membership with the Victim, even if they might not know her personally. For my purposes, I am happy to allow these multiple victims.[15]

3 Should the Victim Forgive?

I opened with the view that forgiveness seemed to be supererogatory because of its electivity and its positive valence. But things are not so simple. One long-standing philosophical debate that I want to focus on surrounds the *conditionality* of forgiveness: that is, the question of whether the Victim should wait for the Transgressor to fulfil certain conditions before forgiving him. A *conditionalist* will insist that true forgiveness, or the best kind of forgiveness, will wait on the fulfilment of the relevant condition. Under the simplest version, this would make forgiveness either obligatory or prohibited, and either way this would deny it supererogatory status.

In contrast, an *unconditionalist* account entails that the Victim may—or should—forgive without waiting for the Transgressor's fulfilment of any condition, and insofar as this would be admirable and elective, it would be supererogatory. The debate between the conditionalist and the unconditionalist presupposes a disjunction: 'true' forgiveness must be one or the other.[16]

3.1 The Conditionalist Accounts

There are actually two conditionalist accounts: action-focused and status-focused. The most famous action-conditionalists are Jeffrie Murphy (in Murphy & Hampton, 1988) and Charles Griswold (2007).[17] Under a more sophisticated version, the action-conditionalist will emphasize that 'earning' forgiveness is not the same as 'having

[15] The distinction between primary and secondary standing raises an interesting question in cases of disagreement between the two victim parties. Imagine the victim of an assault, who then forgives her assailant. The Victim's father is horrified by the assault, and refuses to forgive him. The daughter might say that she is the victim, that her decision should have priority, and that the father should respect that. The father might respond by saying: "what you decide is your business, but I cannot forgive him for what he did to you, and that's my business." For a discussion of broader issues of standing see Pettigrove (2009).

[16] I too (Cowley, 2010) accepted the disjunction, and argued for the unconditionalist account, but now I am a bit more pluralist. In what follows I will be accepting that both the conditionalist and unconditionalist accounts can be appropriate in different situations.

[17] Griswold (2007) offers a whole book on the subject, and includes a discussion of *political* forgiveness, which I will not be discussing at all.

a right to' forgiveness; there would be some normative pressure rather than the imposition of an obligation. There would thus remain *some* room for electivity even after the fulfilment of any relevant conditions. Such a constricted electivity would still allow supererogation, although the admirability of any resulting forgiveness would be reduced by its fittingness as a response to the normative pressure.

What sort of conditions? We already mentioned apology. There are various other candidates, and there will be a separate debate about which of them should be necessary or primary. Some philosophers focus on *repentance*, some on the apology, some on the offer of reparations, some on the explicit repudiation of the transgression, some on the corroborated desire for reconciliation, and some on the commitment to non-recidivism. Repentance is an internal state of mind, and it is often assumed that it will be reliably expressed in some or all of the external behaviors. I will not engage in this debate, and for the remainder I will speak only of the Transgressor repenting or not.[18]

We can understand the intuition behind earning forgiveness and thereby placing normative pressure on the Victim; a genuinely repentant Transgressor might voice his frustration as: "I've apologised! What more do you want me to do?" We can also imagine a third party importuning the Victim to forgive by vouching for the sincerity of the Transgressor's repentance. The worry here leads to the main criticism of the conditionalist account. In waiting for the Transgressor to do the moral work of repenting, the Victim's forgiveness, while elective, is not much of a moral achievement. The strongest version of this criticism would say that the genuinely repentant Transgressor does not technically *need* forgiveness, since he has understood everything of moral relevance and his repentance is appropriate; it is *his* job to get on with other appropriate responses and with character reform. However, he still needs forgiveness in two senses. First, in the same way that the forgiveness needs the Transgressor's recognition to work as forgiveness, so too the apology needs recognition in order to work as a full apology: the Transgressor has to apologize to the Victim face to face, and has to be ready for the real possibility of rejection. Second, although there is not much moral work in recognizing the Transgressor's repentance, there *is* some moral work in hoping that the Transgressor will actually improve (and possibly in supporting his reform efforts). For both these reasons, forgiveness would still be supererogatory, but perhaps not as admirable as the heroic or saintly acts in most discussions of supererogation.

[18] For completeness, here are some other relevant debates, which I do not have room to discuss properly: (i) about whether the apology is 'worth' only as much as the real or symbolic reparations that follow it (after all, 'talk is cheap'); (ii) about whether it is possible to apologize without repudiating the transgression (e.g., because the Transgressor still considers it to have been politically necessary); (iii) about whether the Transgressor sufficiently 'owned' the Transgression to apologize for it deeply; (iv) about whether the Transgressor does enough, following the Transgression, to promote reconciliation and/or avoid recidivism in the long term; (v) about the status of 'rehabilitationist' forgiveness, where the Victim believes that the forgiveness will itself motivate the Transgressor to apologize, (or at least to refrain from committing the Transgression again).

The other important intuition behind the conditionalist account takes the opposite situation. The Transgressor refuses to repent, and has therefore *not* earned forgiveness. At the same time, perhaps the Victim has not understood the insulting message, or has not realized just how harmful the Transgression was (or could have been); or she does not understand what the Transgression reveals about the Transgressor's character, and about the likelihood of future Transgressions; or she has low self-esteem, is too fearful of conflict, and too eager to please. If such a Victim forgives such a Transgressor, we would want to say that she has done wrong, despite the positive valence of all the self-help manuals with 'forgiveness' in the title. And as wrong, of course, it cannot be supererogatory. The stronger version of this intuition would describe unearned forgiveness as communicating two unfortunate messages to the Transgressor: first, a message of condonation of (or at least an indifference to) the wrong; second, a message that the Victim feels she does not deserve any better.

Following on from this second intuition, the conditionalist draws attention to the problem of repeated Transgressions, where the Victim forgives the Transgressor over and over again. Here it is the repetition that makes the forgiveness blameworthy, since the second Transgression already (i) falsified the Transgressor's implied commitment not to do it again, and (ii) thereby (at least partly) undermined the earlier repentance, however genuine at the time. One thinks of those awful cases of chronic domestic abuse, where the wife 'forgives' her husband again and again, motivated by fear or pragmatism or by a misplaced obligation of self-sacrifice: she might well have internalized various oppressive norms that expect more from women and especially from mothers. Surely we, as third parties, would want to advise the wife to get out of the relationship rather than forgive again.

So far I have been talking about an action-focused conditionalist account, i.e., which describes what the Transgressor must *do* to earn the forgiveness. In contrast, a status-based conditionalist account focuses on who the Transgressor *is*: either (i) he has been humiliated enough or has suffered enough or (ii) in virtue of his antecedent relationship to the Victim. (Perhaps the length of time since the Transgression will also be relevant.) So the Victim might try to forgive the Transgressor 'because he's family' or 'for old times' sake' (Murphy & Hampton, 1988 pp. 28–29).[19] The Victim would normally refuse to forgive this type of Transgression (assuming the Transgressor has not repented, or not repented enough), but the long-term family relationship or friendship is important enough for the Victim to overlook any implied disrespect.

Motivation will be relevant here. Consider a Victim stuck with the Transgressor in the same family. If her motivation is merely to preserve a useful contact, or to avoid awkwardness, then this will be too self-oriented to count as forgiveness and would

[19] Hughes and Warmke (2017) distinguish between 'wrongdoer-dependent' and 'victim-dependent' reasons to forgive. This is a separate distinction from the one I am making here. The wrongdoer-dependent reasons concern the wrongdoer's repentance. In contrast, there are two types of victim-dependent reasons—legitimate and illegitimate ones—and these refer to the victim's motivation. If the victim forgives out of fear or submission (as in the abused wife counter-example), then such reasons are illegitimate, as the unconditionalist claims; if the victim forgives out of a love that preserves self-respect, then such a reason is legitimate.

be better described as ignoring, and would not be supererogatory. If her motivation is focused on the preservation of the meaningful relationship with the other (even if only a remnant, even if only in memory), and therefore partly for the other's sake, then we could call it forgiveness, we could admire such a Victim, and it would be supererogatory. Between these two versions, however, there might be more complicated situations where the Victim *needs* the Transgressor precisely because their relationship is important to the Victim's sense of identity: such a need would reduce the electivity, perhaps to the point of making the forgiveness non-supererogatory.

3.2 The Unconditionalist Account

The alternative to the conditionalist account is the unconditionalist account. This begins with a popular understanding of what Christianity requires. In the Lord's Prayer, Christians ask God to "forgive us our trespasses, as we forgive those who trespass against us." This will be supported by the thought that it is a virtue to be 'forgiving' and a vice to be 'unforgiving', regardless of the Transgressor's behavior.[20] The most famous unconditionalist accounts are those of Holmgren (1994) and Garrard and McNaughton (2003); a very different unconditionalist account comes from Derrida (2001). In his discussion of forgiveness and supererogation, Gamlund (2010) describes unconditional forgiveness as 'typically' supererogatory.

Following the original basic schema, the Victim suffers the incomprehensible, unexcused Transgression, but then launches the forgiveness process without waiting for the Transgressor to fulfil any conditions. As we saw, the main criticism of the conditionalist account was that any forgiveness did not amount to much of an admirable moral achievement, once the Transgressor had repented. In contrast, the unconditionalist would see the Victim as doing the hard moral work precisely because there is no normative expectation on her to do so. Calhoun puts it like this:

> forgiveness is conceptually connected with supererogatory acts of generosity and charity … because [it] is an elective response to culpable wrongdoing … It is something we ask or hope for, rather than demand, for ourselves, and grant, rather than owe, to others. Forgiveness is a gift, not the paying of a debt or the remission of a debt whose collection would prove too costly. Calhoun (1992, p. 81)

As Derrida (2001) famously put it, conditional forgiveness entails no more than a 'transaction', guided by social norms. Such transactions certainly have a useful social function, but they are a far cry from the examples of the more substantial forgiveness cited in the Forgiveness Project. In almost all of those cases, there is no

[20] Heyd (1982, p. 115) argues that true forgiveness has to be "done voluntarily for the sake of someone else's good, and is thus meritorious." The worry is that if unconditional forgiveness is not guided by the Transgressor's repentance, it needs to be optimistically re-oriented toward its possibility by the Victim's benevolent motivation. Certainly some forgiveness is driven by benevolent motivation, but I disagree with Heyd in thinking that all forgiveness must be so. And this may well affect its status as supererogatory.

repentance, no apology, and the Victim is struck not only by the incomprehensible and inexcusable wrong they have suffered, but also compounded by the second wrong of the Transgressor's indifference or contempt. The Victim would be entitled to nurse her anger, to retaliate, and of course to ignore it—and yet she forgives. Despite the conditionalist's fears, such unearned forgiveness could be a show of strength, not just psychological but moral. Forgiving that double-wrong, without any transactional prompt or pressure, at risk of ridicule: that's the substantial moral achievement, and that's where meaningful supererogation lies.[21]

Once again, I don't think it's quite so simple. Consider the classic example of supererogatory heroism: running into the proverbial burning house to save the baby.[22] Here the electivity and the positive valence is clear to all. It makes sense for complete strangers to congratulate and celebrate the hero. She has performed an unequivocal good that nothing can take away. It makes sense for the baby, once she grows old enough to understand, to be deeply grateful. It even makes sense for the hero to feel rather chuffed with herself in front of the cheering crowd.

Here's the important point: I'm not sure this public response applies in the cases from the Forgiveness Project, because the forgiving responses seem so starkly *personal*. When I read the stories, sometimes I am appalled, sometimes I am horrified, often I am moved; but often I am mystified, and more often than not I honestly don't know what to make of the forgiveness. Was it right? Was it wrong? I don't know. Would I have forgiven the Transgressor in the Victim's shoes? I don't know. Sometimes yes, sometimes no.

But I can confidently say one thing: it would not be at all appropriate for me to go up to such a Victim and say "well done!"—in the way I would with the hero emerging with the baby from the burning house. And I don't think the Victim expects any such praise or even admiration from third parties, because even she might not be at all sure whether she did the right thing or wrong thing in forgiving. Just as she is not interested in what I would or would not have done. All we are left with are the stories; *their* stories. For me, that is what renders the supererogation question *idle* in a way that it is not for the hero rescuing the baby.

This 'personal' interpretation might be valid, but it runs a new risk explained by Gormley (2014). Not only is there the aforementioned risk of the Transgressor misinterpreting the unearned forgiveness as condonation and/or a lack of self-respect, but there is also the risk of the unearned forgiveness appearing to be *morally arbitrary*.

---

[21] Garrard and McNaughton (2003) argued for unconditional forgiveness motivated by 'solidarity'. However, by the time of their 2011 article (p. 105), they had come to be worried about 'defectively facile' unconditional forgiveness, where the Victim failed to grasp the moral significance of the transgression. So even without the need for the Transgressor's repentance, there is a minimal requirement of serious engagement with the reality of events. We can also consider Fricker's (2018) example of Torvald forgiving his wife Nora in Ibsen's *The Doll's House*, where he is oblivious to the oppressive power structures that have blinded him to Nora's true situation, with the result that his 'forgiveness' can only be condescending.

[22] I'm not speaking about a trained firefighter on duty, since such a person has an obligation to run into the burning house, and therefore—arguably—cannot be a hero. When praised, the firefighter ought to say 'I was just doing my job'. (It's true that a firefighter can be a hero by exceeding the norms of the job, and perhaps by becoming a firefighter rather than a stockbroker in the first place).

Gormley imagines himself as a third party, asking the Victim why she forgave. If the Victim cannot refer to the Transgressor's repentance or rehabilitation, if she cannot refer to the relationship between them ('because he's family', etc.), then there might not be anything she can say. This makes it sound arbitrary or whimsical. And even if she says 'because he's family' or even 'because I love him', that is hardly a *satisfactory* reason, since it depends on too many contingencies, as well as on the Victim's personal interpretation of the concepts of family and love. (It's also compatible with unhealthy forgiveness motivated by fear or moral resignation.).

Instead of this being a criticism, I actually think Gormley's point supports my conception of the Victim's *moral isolation* during the forgiveness process, to the point where no third-party evaluation, or question, or judgement is relevant. Again this is unlike the very public situation with the hero. And again I think this means that many of the acts of forgiveness in the Forgiveness Project cannot be easily classified as admirable, and therefore cannot be easily classified as supererogatory. It is a Kantian prejudice to think that the Victim's lack of a comprehensive reasoned account for why she forgave automatically demotes her decision into the realm of the morally arbitrary; it resists moral arbitrariness precisely because it is a serious and difficult decision that she wrestles with and which she will have to live with; so we should trust the Victim on this account. But her forgiveness can still be mysterious.[23]

4 Forgiveness and Choice

In the first two sections I looked at various responses to the Transgression, and then I looked at forgiveness. In the third section I looked at conditionality. In all three sections I assumed that it was possible to *choose* to forgive—or at least to choose to try to forgive, or to launch the forgiveness process, subject to unpredictable psychological limitations. Forgiveness being a matter of choice would seem essential to its being supererogatory. Many instances of forgiveness are clearly matters of choice; but in this section I want to look at situations without choice: those Victims who claim they 'had to' forgive, or 'could not' forgive (i.e., the Transgression was 'unforgivable').

Already the popular Christian conception of forgiveness would seem to make it a matter of obligation, not of choice. If so, that would make it more difficult for third parties to praise the act of forgiveness, and the Christian Victim would not be prepared to accept our words *as* praise. Although we might admire the Victim's action when compared to non-Christians refusing to forgive similar Transgressions, such comparisons will again be irrelevant to the Christian Victim herself. This has always been a problem in considering saintliness as supererogatory; for the saint will not see any of her acts as a sacrifice, but instead as dutiful service to her God. As Alfred Archer puts it, "these self-reports from moral exemplars present a challenge to

[23] For more discussion of the 'arbitrariness' charge see Garrard and McNaughton (2011).

the traditional view of supererogation as involving agential sacrifice" (Archer, 2015 p. 107).

Even if the Victim was not serving God, but some important identity-conferring commitment such as her family, then her forgiveness might be experienced as an obligation, and she might well use the words: "he's family, so of course I had to forgive him." And by rejecting choice and electivity, this would seem to deny supererogatory status.

And then there is the category of the 'unforgivable'.[24] Right away, following Hallich (2017 p. 40 *ff*), we have distinguish different meanings. (i) Some acts will be unforgivable in the *logical* sense because the act, on second look, turned out not to be a transgression, or the victim turns out not to have suffered. (ii) Some transgressions will be unforgivable in the *psychological* sense, even though the Victim wants very much to forgive. (iii) The *morally* unforgivable is the most relevant for our purposes. Perhaps one obvious reading of moral unforgivability is no more than the conditionalist conclusion about the unrepentant Transgressor: he *ought not to be forgiven.* If the Victim nevertheless forgives, she will be blameworthy for this. Hallich orients his whole discussion around this version, while missing two key points. First, if the Victim declares the Transgressor 'unforgivable' but then proceeds to forgive him, then he was not in fact unforgivable. Second, Hallich relies on a straightforward objectivist singularity to assume there is a right answer about every situation, without acknowledging the difficulties of trying to discover such a putative answer to the satisfaction of all parties involved, and without acknowledging the very different perspectives of Transgressor, Victim and Observer.

Some will reject the notion of moral unforgivability entirely. Presumably most Christians would, for example. Govier, in a secular vein, states (1999 p. 70): "From any moral point of view that incorporates respect for persons, there are no absolutely unforgivable persons." Murphy (2009 p. 570) agrees, and recommends that the putatively unforgivable Transgressor be approached in moral humility, in a spirit of "there but for the grace of God, go I." However, there are important limits to sympathetic imagination of oneself in another's shoes. Part of my self-conception is precisely that I would never carry out an unjustified killing, for example. Of course a lot of people in 1940s Germany thought that about themselves, and went off to commit atrocities (Murphy cites Christopher Browning's 1992 classic *Ordinary Men*). But the possibility that I too might have been transformed *that much* is simply too abstract to induce non-judgemental moral humility, and certainly too abstract to prevent me from

[24] Derrida famously declared that 'one can only forgive the unforgivable' (2001 p. 7). Many philosophers have dismissed this as meaningless and provocative nonsense. I think, however, there is an interesting insight here about the temporal progression of the Victim's experience, as I explained in Cowley (2010). After ruling out the conditionalist account on the grounds of its status as a transaction, Derrida argues that any conception of the Transgression as *in-principle forgivable* lets reasons and conditionality back in. The Victim's question about whether to forgive only begins when the transgression appears very much unforgivable. The Victim *then* moves onto the second stage, and this is logically a new situation, unconstrained from the first stage judgement. Ultimately the decision to forgive or not forgive is ineffable and impenetrable, both to outsiders and to the Victim.

confidently labeling such horrendous deed 'unforgivable'. (Moreover, if I had been so transformed, I cannot imagine the resulting person being *me* in any recognizable sense.).

However, there is a second reading of moral unforgivability, and that relates solely to the nature of the Transgression. Given that it is inexcusable and incomprehensible, the Victim might consider it *egregious* (my term) because it betrays one of her deepest identity-conferring commitments or relationships, and she will declare it unforgivable even if the Transgressor fulfils all the repentance conditions. Of course there is room for disagreement here between the Victim, the Transgressor, and any third-party observer, about the wrongness of the Transgression and its particular disvalue as a betrayal of the Victim's identity-conferring commitments. In the best case there will be mutually-enlightening dialogue. In the worst case, it will be the Transgressor, insofar as he cares, to come to terms with his new status as unforgiven.

Take the classic example of the Transgressor's sexual infidelity against the Victim, his spouse. Here we really could say that some people would find infidelity unforgivable; others would find it conditionally forgivable (with repentance); and yet others would find it a merely venial sin and easily forgivable. So much will depend on the individual person, on their wider values (both inherited and chosen), on the place of the marriage in their life, the particular agreements, explicit and implicit, within the marriage so far, and perhaps on the available options (financial and social) for a new start outside the marriage. Following on from the Augustinian maxim, Govier (1999 p. 65) suggests that it is possible to forgive the Transgressor while preserving the unforgivability of the Transgression: "Forgiveness is something we extend or do not extend towards *persons*, and it fundamentally affects the relationship between persons. And yet it is *deeds* that are said to be unforgivable." As I argued before, I am not sure this is coherent, and in this situation it is even less so, because insofar as the Victim forgives the Transgressor then she necessarily has to make some kind of peace with his status as the author of the Transgression.

Either way, there is a strong sense that the Victim should have the last word because it is her life; perhaps a sympathetic third-party friend might play down the Transgressor's infidelity as 'not a big deal', given that 'boys will be boys'; and yet the Victim declares it unforgivable and files for divorce. Has the Victim done something wrong? Or understood something incorrectly? Hard to say, especially given the opacity of marriage, both to outsiders but also, paradoxically, to the participants. Certainly the friend thinks the Victim is being hard-hearted or unsympathetic ('unforgiving' as a vice), but the friend hardly has an epistemically privileged vantage point, and the friend now has to accept and respect the Victim's decision and support her as she deals with the aftermath. If the Victim later changes her mind and forgives, is this necessarily a better outcome? Has she come to see the situation more clearly? Maybe she continued to describe herself under the opposite sense of necessity, as in "I took my marital vow seriously, and so I have to forgive him." Does she deserve praise for it? Again, hard to say. We have one person just trying to make the best sense of her life. So again, I think, the question of any possible supererogatory status falls idle.[25]

4.1 *The Sunflower*

There is a third reading of morally unforgivable, regardless of the Transgressor's repentance, and that concerns a universally egregious Transgression. The paradigm would be the Holocaust, where the crimes are unforgivable because of both the scale of the harm, and the evil of the intentions behind them. Again, whether or not the transgressor repents would normally be irrelevant to the status of unforgivability; and one would have good reason to doubt the sincerity or even the possibility of any clear-eyed repentance by someone capable of overcoming all moral restraint in committing the atrocity in the first place.

A central example in the forgiveness literature is *The Sunflower* (1998) by Simon Wiesenthal. In 1943 in Nazi-occupied Lemberg (in present-day Ukraine), a dying SS officer asks Wiesenthal, then a Jewish slave laborer, to forgive the officer for his participation in atrocities against other Jews, now dead. Wiesenthal says nothing, leaves the room, and the officer dies the next day. The bulk of the book comprises 53 responses from a range of public figures, including theologians, political scientists, jurists, Holocaust survivors, and even former Nazi Armaments Minister Albert Speer.

I confess to having never understood the central philosophical *problem* of the book (including, in my edition, the garish cover with the sensationalist sub-title). First there is the issue of standing. Surely the young Wiesenthal did not have the standing to forgive on behalf of the officer's dead victims, and any attempted 'forgiveness' on his part could only constitute a deep betrayal of the victims. Second, there is the question of whether Wiesenthal had *effectively* forgiven by refusing to refuse, as it were.[26] But leaving those two points aside, given the universal egregiousness of the crimes in question, given the status of the officer and of Wiesenthal with respect to those crimes, and given the timing of the request while the crimes were on-going, I would say that *of course* the crimes are unforgivable.[27] On the other hand Dith Pran

---

[25] Fricker (2018 p. 176) describes how forgiveness can even be used as a way to prevent healthy discussion about the state of the relationship, and impose a one-sided moral interpretation of the events. She imagines the Transgressor complaining: "I admit that what I did was pretty thoughtless, but to say it was a 'betrayal of our friendship' is melodramatic... But now I don't even have the chance to discuss it, because apparently I am 'already forgiven'".

[26] Wiesenthal's friend Bolek, who suffered as much as he did in the camps, says "through his confession [...] his conscience was liberated and he died in peace because you had listened to him" (Wiesenthal, 1998 p. 83). Bolek thought the officer deserved forgiveness, even if Wiesenthal remained ambivalent.

[27] One of the respondents, Cynthia Ozick, even described the officer's repentance as an *aggravating* factor, for it revealed that he had a conscience all along, and had actively suppressed it while committing the crimes. However, I would add that there is a sense that the officer still did not understand the enormity of the crimes, for if he had, he would have understood full well that they were unforgivable, and he would not have dared to ask for forgiveness.

(Wiesenthal 1998 p. 230), a survivor of the Cambodian terror, *would* be inclined to forgive the guards who carried out the terror, although he would not forgive the leadership. This is a remarkable claim, but for me frankly incomprehensible. And as incomprehensible, I find myself unable to praise it. And being unable to praise it, I would say it cannot be supererogatory: it is just a brute fact in the world.

There is a lot more one could say about *The Sunflower*, but that would take us too far into the particularities of the Holocaust, which would merit a much more detailed treatment. Suffice to say that the book sold well, and so its readers have evidently found some way of grasping the central question, and understanding the positive responses better than I.[28]

References

Allais, L. (2008). Wiping the slate clean: The heart of forgiveness. *Philosophy & Public Affairs, 36*(1), 33–68.

Améry, J. (1980). *At the mind's limits: Contemplations by a survivor on Auschwitz and its realities* (Sidney & S. P. Rosenfeld, Trans.). Bloomington: Indiana University Press.

Archer, A. (2015). Saints, heroes and moral necessity. *Royal Institute of Philosophy Supplement, 77*, 105–124.

Bash, A. (2015). *Forgiveness: A theology*. Cascade Books.

Benbaji, H., & Heyd, D. (2001). The charitable perspective: Forgiveness and toleration as supererogatory. *Canadian Journal of Philosophy, 31*(4), 567–586.

Bennett, C. (2004). The limits of mercy. *Ratio, 17*(1), 1–11.

Browning, C. (1992) *Ordinary men: Reserve police battalion 101 and the final solution in Poland.* Harper Collins.

Calhoun, C. (1992). Changing one's heart. *Ethics, 103*(1), 76–96.

Cowley, C. (2010). Why genuine forgiveness must be elective and unconditional. *Ethical Perspectives, 17*(4).

Derrida, J. (2001). *On cosmopolitanism and forgiveness*. Routledge.

Flanagan, B. (1994). *Forgiving the unforgivable*. Wiley.

Fricker, M. (2018). Ambivalence about forgiveness. *Philosophy, 84*, 161–185.

'Forgiveness', the Templeton Project. https://forgiveness.templeton.org/

Gamlund, G. (2010). Supererogatory forgiveness. *Inquiry: An Interdisciplinary Journal of Philosophy, 53*(6), 540–564.

Garrard, E., & McNaughton, D. (2003). In defence of unconditional forgiveness. *Proceedings of the Aristotelian Society, 104*, 39–60.

Garrard, E., & McNaughton, D. (2011). Conditional Unconditional Forgiveness. In C. Fricke (Ed.), *The ethics of forgiveness* (pp. 97–106) New York: Routledge.

Gormley, S. (2014). The impossible demand of forgiveness. *International Journal of Philosophical Studies, 22*(1), 27–48.

Govier, T. (1999). Forgiveness and the unforgivable. *American Philosophical Quarterly, 36*(1), 59–75.

Griswold, C. (2007). *Forgiveness: A philosophical exploration*. Cambridge University Press.

Hallich, O. (2017) '"I Cannot Forgive You." The unforgivable as an example of a practical necessity'. In H. Bauer, J. Varga, & S. Miethe (Eds.), *Dimensions of practical necessity*. Palgrave MacMillan.

Heyd, D. (1982). *Supererogation*. Cambridge University Press.

[28] My thanks to David Heyd for his comments on the first draft.

Holmgren, M. (1994). Forgiveness and the intrinsic value of persons. *American Philosophical Quarterly, 30*, 331–340.
Hughes, P., & Warmke, B. (2017). 'Forgiveness'. *The Stanford Encyclopedia of Philosophy*. https://plato.stanford.edu/entries/forgiveness/
Murphy, J., & Hampton, J. (1988). *Forgiveness and mercy*. Cambridge University Press.
Murphy, J. (2009). The case of Dostoyevsky's general: Some ruminations on forgiving the unforgivable. *The Monist, 92*(4), 556–582.
Pettigrove, G. (2007). Understanding, excusing, forgiving. *Philosophy and Phenomenological Research, 74*(1), 156–175.
Pettigrove, G. (2009). The standing to forgive. *The Monist, 92*(4), 583–603.
Pettigrove, G., & Enright, R. (2023). *Routledge Handbook of the Philosophy and Psychology of Forgiveness*. Routledge.
The Forgiveness Project. https://www.theforgivenessproject.com/
Westlund, A. (2019). Who do we think we are? *Philosophy and Literature., 43*(1), 173–191.
Wiesenthal, S. (1998). *The sunflower: On the possibilities and limits of forgiveness*. Schocken.
Worthington, E., & Wade, N. (Ed.). (2020). *Handbook of Forgiveness* (2nd ed.). Routledge.

Promise-Making and Supererogation

David Heyd

Abstract Studies of the practice of promising have concentrated on the reasons for keeping promises. This article focuses on promise-*making* and argues that the making of promises is typically supererogatory. It then addresses the question whether we can promise to perform supererogatory acts. Although once given, the promisor is under an obligation to perform the promised act, there is no paradox in describing the act as supererogatory. The proposed analysis is based on the distinction between the content of the promised act and the deontic status of its performance. In contrast to Jason Kawall and Claire Benn, I propose that there is no difficulty in the idea of promising to supererogate.

Keywords Supererogation · Promises · Individuation of actions · Kawall · Benn

1 Supererogatory Promises

There is a vast literature on the practice of promises and there is a similarly bulky study of the idea of supererogation. But surprisingly the intersection of the two concepts in the Philosopher's Index yields only very few items. The standard lists of examples of supererogatory behavior do not include promising. There are the saintly and heroic acts, typical cases of giving and charity, the less exciting forms of considerateness, the controversial case of forgiveness and pardon, the interesting case of volunteering, and even some forms of supererogatory omissions like toleration. But making a promise does not come to mind when we think about actions beyond the call of duty. The reason for this gap in the list of supererogatory acts is that almost all the studies of the practice of promises focus on the duty of keeping promises—its source, its scope, its force and its relevance to contracts. There is next to no inquiry about *promise-making*: why do we make promises? Is it ever our duty to make a promise? Could it ever be prohibited? And if promising itself is of value, and if it is not obligatory in at least many cases, isn't it a typically supererogatory act?

D. Heyd (✉)
The Hebrew University of Jerusalem, Jerusalem, Israel
e-mail: david.heyd@mail.huji.ac.il

D. Heyd (ed.), *Handbook of Supererogation*,
https://doi.org/10.1007/978-981-99-3633-5_13

In this article I wish to argue that despite its absence from the traditional inventory of supererogatory acts promise-making may turn out to be one of the clearest, non-controversial and very common cases of supererogatory action. Promise-making is universal and not culture dependent. Whether natural or conventional, we can hardly imagine human interaction and social life without this practice. Yet the universality and the social necessity of the practice do not make the act of promising obligatory. Promise-making, as we shall see, is in most cases free in the sense that it is not morally required and in the sense that one does not usually feel guilty for not having made a promise to ϕ even when one feels guilty for not ϕ-ying. I may feel guilty for not having invited you for dinner but have no regret about not having promised you to invite you for dinner.

It seems that one of the reasons for the concentration on promise-keeping rather than on promise-making is that promise-making has often been associated with contracts, that is, with deals which are mutual and which serve the interests of both parties. Hence Rawls, for example, identifies the wrongness of breaking one's promise in its *unfairness* (Rawls, 1971, 346), and Scanlon argues that such a violation is a form of *manipulation* or deception (Scanlon, 1998, 298–9). If promise-making is considered to be part of a contractual interaction between people then it is naturally analyzed in *pragmatic* rather than moral terms. For contracts are signed between two parties for self-interested reasons and making them has in itself no moral, let alone supererogatory, value. But mutual agreements, which may be described as involving a reciprocal promise of the parties, are not the common contexts of promise-making. We usually give promises to people who are not expected to do "their share" or to reciprocate. We promise with the sincere intention to do good to a friend as an unconditional favor. And in those everyday cases of unilateral promising the wrongness of failing to keep them is not a matter of unfair free riding or manipulation because we often lose more than we gain from breaking a promise; we could have avoided making the promise in the first place. Failing to keep a promise is caused more by forgetfulness or weakness of the will than by the intention to take a free ride or exploit the promisee.

If promise-making is (at least in most cases) supererogatory, the question arises whether one can make a promise to perform a supererogatory act. Jason Kawall raised this question a few years ago and answered it in the negative. His argument was that such a promise would involve a paradox. If I promise you to do a supererogatory act in your favor then when I actually do the act it is obligatory since I am under the obligation to keep my promise; however, in doing the promised act I cannot be said to have kept my promise to you since the promise was to perform a supererogatory act and now it turns out that the act is obligatory (Kawall, 2005, 390). Although Kawall is the first to connect promises and supererogation, he too is more concerned with keeping promises than with making them or, to be more precise, in examining the possibility of making a promise to supererogate in the light of the impossibility of keeping it.[1]

[1] I responded to Kawall's article in a short comment (Heyd, 2005). Then a second round between us ensued in 2006—all on the pages of *Philosophia*. Some of the arguments made in the second part

The term 'supererogatory promises' is ambiguous and can refer to two different issues: the deontic status of the act of making a promise, on the one hand, or the content of the promised act, on the other hand. The plan of this article is to discuss in its first part the former issue, namely, the deontic status of promise-making and then, in the second part, to analyze the second issue, namely, the content of the promised act, and show how Kawall's paradox can be solved and to claim that there is nothing contradictory in promising to do a supererogatory act.[2]

2 The Deontic Status of Promise-Making

Since we are interested here in promise-making rather than in promise-keeping we should first ask what are the motivations or reasons for making a promise. Both Hobbes and Hume focus on contracts and agreements and accordingly analyze the motivation to make a promise in utilitarian terms of self-interest. For Hobbes, the expectation that someone give a promise is part of a contract in which that person is supposed to do his or her share in the agreement sometime later, which requires trusting the other party. Hobbes calls such a mutual contract "a pact, or a covenant" (Hobbes, 1996, 94). Promise-making is thus a constitutive part of signing contracts or covenants rather than a *further* separate act of the contractors. Hobbes adds that promises are not merely declarations of intentions or resolutions but commitments with direct normative implications. "Tomorrow I will give" is merely an expression of an intention and does not change the normative relations in an interpersonal context, but "I have given, or do give to be delivered tomorrow" is a promise that puts one under an obligation (94–5). In modern terms, Hobbes adheres to the authorization model (Owens, 2006; Raz, 1977) according to which a promise consists of the transfer of normative power from the promisor to the promisee, the power or the right over the promisor's action. And this transfer of right occurs *now*, though the promised act will be carried out only later.

Although the central context of promise-making is for Hobbes contractual, and patently utilitarian, he leaves room for other motivations to promising. A promise may be made as a free *gift* ('grace'). Here the transfer of normative power is not mutual as in the case of contracts but a unilateral commitment motivated by "the hope to gain friendship" or in order to make a reputation for oneself of being magnanimous (94). These cases open the door to a supererogatory analysis of promise-making since they are forms of free giving motivated by good intention toward another person. However for Hobbes even these motivations are considered as 'self-interested' and

of this article originate in that debate, although here in more detail and often in a different form. Benn (2014) joined this debate later (see below).

[2] I intentionally avoid taking a stand on the widely discussed issues of the nature of promises—the conditions of their validity, the excuses or justifications for not keeping them, the conditions of their authenticity, their relation to intending or being determined, the performative ways in which they are created, and whether their normative force is natural (Hume and Rawls) or conventional (Raz and Scanlon). What I do assume is that promises involve the undertaking of an obligation.

hence cannot be really associated with supererogation. Nevertheless, in free gifts, says Hobbes, "I am enabled to merit only by the benignity of the giver" rather than by my own normative power to which I have been authorized to exercise in the contractual setting (95–6). Or in other words, in contracts one party can expect the other party to *make* a promise to do her share in the agreement, while in promises as free gifts the promisee has a claim only over the *keeping* of the promise (after it is freely made). Hobbes does not therefore deny the possibility of promises as free (supererogatory) acts of kindness.

Hume is equally utilitarian as Hobbes with regard to the practice of promising. Promising is an artificial human invention which aims at human cooperation. In mutual interpersonal relationship one cannot rely on gratitude and reciprocity which are the natural responses to being the beneficiary of some good. Promises therefore serve to reinforce human cooperation. In a world of generous and decent people, promises would be superfluous. But note that Hume speaks of promises exclusively in the context of agreements and ignores what Hobbes referred to as unilateral gifts. Promises are always 'interested' with the expectation of some return. "Interest is the *first* obligation to the performance of promises" (Hume, 1960, 523). The making of a promise is a matter of interest and once the promise is made the motivation to fulfil it is the natural fear of being shunned by society. Hume may be correct when he argues that promise-keeping cannot be accounted for in natural terms, which proves that the obligation to keep a promise must be a matter of social convention; but that does not answer the question about the reason for making promises which is definitely not a matter of duty. Indeed, Hume does not deny that there are acts of kindness and free giving motivated by love and friendship, but he does not include promise-making in those acts (521–2). He is too concerned with showing that promise-keeping is a function of an artificially invented convention that he fails to notice the natural motivation to make promises, what we might call a supererogatory sentiment.

But beyond these utilitarian considerations which mostly preoccupied Hobbes and Hume, promises are given unilaterally and are motivated by the will to give the promisee some assurance, to alleviate her anxiety, to express a sincere intention to do something in favor of another, to reinforce oneself against the possibility of weakness of the will, or to make a gesture of good will. In what way do these motivations decide the deontic status of promise-making? Given that it is obligatory to keep promises, is promise-making permitted, prohibited, obligatory, or supererogatory?

Surely, promising is usually permitted. Promising is a common way of exercising our autonomy and managing our interpersonal engagements. Although promises are often given for self-interested reasons, as Hobbes and Hume hold, they are perfectly legitimate. Even if they have no moral value, they are not morally wrong.

This does not mean that making a promise is always permitted. Promise-making may be prohibited. Making a promise to perform and immoral act is surely forbidden; promising to do an act which one knows to be impossible to perform is equally wrong (being a kind of deception); giving a promise which one knows would conflict with another duty or obligation is morally problematic; promising something that one does not sincerely intend to do is manipulative. Fox (1993) proposes a radical and counterintuitive thesis: promise-making is *always* immoral and hence forbidden. For if it is always obligatory to keep a promise then all other conflicting obligations and duties would have to be left unfulfilled (even when they are morally overriding); and if promise-keeping is only prima facie obligatory then the promisee would never be able to *rely* on the promise which would make it void and the assurance of the promisee illusory. Fox's conclusion is surely hard to accept since not only would it imply the abolition of the universal practice of promising but also entail that contracts and indeed any undertaking or commitment should be prohibited since in all these practices there is a chance that conflicting duties which are more urgent would arise. Yes, there is a chance that despite having promised to help you move your furniture to a new home it turns out later that I have to take my father to the hospital at the same time; but that equally applies to the taxi driver to whom you paid in advance and who for pressing reasons must cancel the ride. So although there are some cases in which promises should not be made, the practice of promising cannot be forbidden altogether.

A more interesting question is whether promise-making can be obligatory. On the face of it the act of promising is a matter of free choice. But one can think of cases in which one might have a duty to make a promise. Think, for example, of a father who planned to take his son to the zoo but due to the weather the zoo was closed. The child is very disappointed. A caring father may have a duty to compensate the child by promising to take him to the zoo once it reopens. When the cause of the child's disappointment is the father's more urgent business, the duty to promise the child an alternative date is even stronger. Another example is an offender's request for forgiveness by the offended party which commonly involves a promise not to repeat the offense in the future. A third example is a politician seeking the vote of citizens: she is expected to make promises to carry out certain policies so as citizens can rely on her declared intentions. Forgiveness and political appeal to voters are structurally similar to Hobbesian contracts: if you want me to forgive you, or vote for you, or strike a deal with you, you *ought* to undertake the obligation of promising never to offend me again, to carry out certain political agendas, or do your fair share in our deal.

However in those cases the duty to make a promise is dependent on another duty (to compensate one's child for not taking him to the zoo, or to assure the offended that the offense would not be repeated). However, Enoch (2015) suggests that there are cases in which there is a duty to undertake an obligation which is independent of any prior duty. His main aim is to show that one can take responsibility for acts for which the agent has no responsibility, but as a reinforcing analogy he considers promise-making that might be a duty even when there is no previous duty or obligation. If I have a dying friend who is concerned about the future of his small children, and none of his immediate family or close friends are willing to take responsibility for the children, I can be said to have a duty to promise my dying friend to do so despite having no independent duty to him (or to his children). This example is problematic. There cannot be a duty to make a promise unless there is some previous duty concerning the content of the promise. In Enoch's example, it is either a duty toward the dying friend or toward the children. It definitely might be the case that if no one of his closer acquaintances is prepared to assume responsibility for the children, it might be my turn do so as a matter of duty. But again it is not the promising that is my duty but the care of the children. I can of course *volunteer* to take care of the children but then it would be a typically supererogatory choice rather than a duty to make a promise.

Another way of treating promise-making as obligatory is to consider it as an *imperfect duty*. Even if no particular promise is obligatory, never promising is morally wrong. This is typically true in close relationships of marriage, friendship, or parenthood. Since, as we shall see, promising is a considerate act of assuring the promisee that she can rely on certain help or favor, the spouse, friend or parent ought to grant such an assurance from time to time (depending on the time available to her or other circumstances). A promise is like a gift: a mother does not owe her daughter a surprise present at any particular time, but if she never ever surprises her daughter by some act of giving, she does not live up to her (imperfect) parental duty. One cannot be considered a virtuous person without ever making a promise.

Cholbi (2017) is one of the very few philosophers who offer a supererogatory analysis of promise-making. His starting point is the critique of utilitarian justifications of the practice of promises (such as Hobbes' and Hume's). If promises promote social or individual well-being, the utilitarian must claim that promises are obligatory. This is not restricted to promise-keeping but equally applies to promise-making. Promising is valuable because it serves to give the promisee the guarantee that a certain act (in her favor) will be done in the future. Furthermore, this guarantee is important even when the promise made is to carry out an act which is itself obligatory; it is an added value which reinforces the reliance on the promisor. But if that is the case then for the utilitarian we ought to make promises whenever this is possible. Cholbi argues that this conclusion is absurd and hence that the utilitarian account fails. And the possible objection of the utilitarian that it is often rational to make a promise does not seem convincing to Cholbi since in many cases there is nothing rational in making a promise and the promise can be explained only in purely moral terms of a favor or a free gift. Cholbi's conclusion is that promise-making is supererogatory, at least sometimes (Cholbi, 2017, particularly 268–9). It is a laudable but non-obligatory practice which beyond its social value is an expression of good will motivated by altruism.

The idea that promise-making is supererogatory calls for a closer examination of the *value* of promises. Their social value as enabling contracts and covenants is uncontroversial, but as we have seen it is an instrumental rather than a moral value since promises to keep mutual agreements are in the rational interest of the parties concerned. There is nothing supererogatory in making a promise when it is in my interest to do so since I am about to gain from it directly. Supererogatory acts are by definition done with the intention of benefiting another person. They may even be irrational from the point of view of the agent's interests.

The moral value of promise-making is considered by most philosophers as the assurance of the promisee that she can expect the promisor to carry out her intention. Intention and resolution of an agent are not sufficient to enable the other party (the beneficiary) to *rely* on the actualization of the intention. An early articulation of the value of assurance is a passage in Thomas Aquinas' chapter on vows. In contrast to vows, which are made to God who does not need human assurances, promises are made to fellow humans who stand to benefit from them: "We promise something to a man for his own profit, since it profits him that we should be of service to him, and that we should first assure him of the future fulfilment of that service" (Aquinas 1947, question 88, article 4, in the second part of the second part). Thomas Aquinas and others note that a promise must be accepted by the promisee to be valid, since only by such acceptance is the benefit of the promise guaranteed and the goodness of promising secured.

Promising is a strong sign of commitment by the promisor who binds himself by an obligation from which only the recipient of the promise can release. Assurance is often quite valuable since it relieves the recipient's anxiety about future circumstances. And beyond this psychological benefit, it has a practical coordinative advantage of saving the recipient from taking alternative steps to achieve some practical goal. If you promise to drive me to the airport tomorrow night, I can sleep well without worrying about tomorrow's transportation arrangements and I am free from the need to make a taxi reservation. This peace of mind is not achieved by the mere knowledge that you are a friend and that you intend to offer me such a drive when the time comes tomorrow night since intentions might change due to changing circumstances. Assurance can be the added value of a promise to do what an individual is under an obligation to do. If I owe you money and you do not know on which day of the week I am going to pay you my debt (or are not sure if I remember my debt or intend to fulfil my obligation), a promise to return the money by Tuesday is a kind of assurance that is a welcome benefit to you. Duties of friendship or parenthood are often vague in their scope and application and a promise to comply with them is a clear announcement of the commitment to fulfil out a specific duty.

David Owens and Joseph Raz suggest a different source of the value of promises. The benefit to the promisee is constituted by the transfer of *authority* over the promised act from the promisor to the promisee. By undertaking to do the promised act, the promisee becomes normatively speaking the controller of the relevant part of the promisor's future action. The promisor is not free to break her promise and the promisee gains the power to release the promisor from the promise or insist on it being carried out. Owens quotes Kant when he says that "in accepting a promise you 'take possession of my choice'" (Owens, 2006, 71).

But is the authority over the promisor's behavior a major interest of the promisee? Indeed, we might benefit significantly from the authority over other people's actions in political contexts or in cases in which we have a plan which requires the cooperation of other people. But promises are usually limited to particular actions which give us a fairly limited control over the promisor's behavior. When we are given a promise to be given a ride to the airport our main interest is the convenience in which we are going to get to the airport and the assurance that we can rely on that promise; it is not the sense of control over the behavior of our good willing friend. This is particularly the case when the promise made to us is supererogatory in its content (thus, involving sacrifice by the promisor and a demonstration of a strong commitment to our well-being). When you promise to help me out of a deep financial crisis, I am touched by your extreme altruism and feel no extra satisfaction in being in control of your financial decisions. In such cases unless we are "control freaks" it seems strange that the authority over our benefactor is conceived as our benefit. For Owens, the reluctance of the promisor to bind himself by an obligation is due to her hate of losing control or being controlled by another person (Owens, 2006, 70). But is that common in the case of promising? We hesitate before making a promise because we are not sure we can deliver, or because we fear we could encounter more urgent business to take care of, or because we are afraid of disappointing the other party—rather than because we hate losing normative control. But even if Owens is right that promises serve the promisee's interest in gaining authority over the life of the promisor, what his account lacks is the explanation of the interest of the promisor in making the promise to begin with.

Here Raz (1977) seems to offer a more convincing account. In a way he reverses Owens' authority model of promises: in Owens' model it is the *authorized* party of the promise whose welfare is the source of value, while in Raz's analysis it is the *authorizer's* free choice in undertaking the obligation thus creating new normative relations between her and the promisee which is the source of value. Unlike the promisee, whose interest lies primarily in the content of the promised act and the assurance added by the promise that this act will actually take place, the promisor is motivated by the will to grant another person's claim against the promisor a 'peremptory force'. The promisor's second-order exclusionary reason to ignore first-order reasons which might arise is justified by the special interpersonal bonds created by such authorization (Raz, 1977, 228). Raz does not argue here for the value of these bonds, but we can think of social bonding and trust that such renunciations of normative power create as intrinsically good.[3] The intrinsic value of supererogatory acts lies both in the exercise of personal autonomy and in the creation of interpersonal relationships which transcend those of mutual respect.[4]

We can end this section regarding the deontic status of promise-making with the following conclusions. Due to their focus on the political context of promises, Hobbes, Hume, and Rawls are more interested in the duty of promise-keeping than in the reasons for promise-making; the obligation to keep a promise is a constitutive part of creating agreements and contracts. The justification of the institution of contracts (which consists of promise-making) is prudential or utilitarian. But this analysis ignores the much more common practice of promising, the personal gesture to a friend or a stranger to do something in her favor, the unilateral attempt to give assurance about one's binding commitment to benefit the other. Unlike the mutual assurance in contracts, without which the promises of the two parties do not bind

---

[3] I have argued that the supererogatory value of forgiveness and toleration lies exactly in the relinquishment of a right. Tolerating my neighbor's noisy party in the middle of the night is typically expressed by refraining from calling the police. Not insisting on my rights is a form of giving beyond the call of duty (Heyd, 2021, 92).

[4] Raz warns us that promises are not necessarily good. They can be deceptive, manipulative, immoral in their content, and impossible to keep. But these cases demonstrate that the practice of promises which is valuable in itself can be abused rather than qualify the value of the very practice. Raz admits that an undesired promise can be refused by the promisee which is one way of narrowing down the number of cases in which particular promises have negative value.

them, the unilateral character of personal promises cannot be exhaustively explained in prudential terms.

So what is the deontic status of promise-making? Surely promising is *permitted* unless it is deceptive in motivation, immoral in content, and known to be impossible to carry out. The suggestion that promises should be *forbidden* because there is no guarantee that they would always be kept (due to unpredicted circumstances or the appearance of a more stringent obligation) seems to go too far. In every moral relationship, commitment or undertaking there is such a chance and the recipient must take that into account when relying on them. Apologies and compensation may be required in such forced violations of promises but not the abolition of the practice as a whole. Finally, can promises be considered *obligatory*? We saw that they might sometimes be, but only when there is a prior independent obligation or a duty such as in the case of the failure to perform a certain act which created disappointment in another person or as compensation for unforeseen circumstances which made the fulfilment of the promise impossible. As we have seen, promise-making can be obligatory as an imperfect duty: a person who never makes a promise to his spouse, friend or children does not behave as he is morally expected. But again, these cases of obligatory promise-making are relatively rare. This leads us to the final conclusion that promise-making is in the majority of cases supererogatory. These cases consist of intentional altruistic valuable acts which are not obligatory. The sources of the value of these acts are both the content of the promised act and the assurance that the act be actually performed as well as the free exercise of the agent's autonomy.[5]

3 Can One Promise to Supererogate?

The very act of promising is often supererogatory, but what about the *content* of the promise? The deontic status of the promise-making and that of the promised act is not necessarily the same. We should distinguish between the undertaking of an obligation and the object of this undertaking. As we noted, it is on the one hand obligatory to return a loan, but promising to do so tomorrow (or on time) may be an extra gesture of reassurance which is supererogatory; but on the other hand it could be obligatory for a parent to make a promise to take one's child to the amusement park which to the disappointment of the child was closed, though taking the child to the park is not itself a parental obligation. But of course there may be cases in which both the making of a promise and the promised act are supererogatory. But here we encounter a problem: despite the supererogatory nature of both the giving of the promise and the act itself once the promise is made, it is *obligatory* to perform the (supererogatory) act.

[5] There is a growing literature on *suberogatory* acts. The case mentioned in the context of promise-making as an imperfect duty could be also analyzed as suberogatory, i.e., bad actions which are not wrong or prohibited: never making a promise is not contrary to duty (either perfect or imperfect) but is still bad. I shall not examine here which of the two accounts fares better—that of imperfect duty or that of suberogation.

This raises the question whether there can be a promise to perform a supererogatory act. This question was first discussed by Jason Kawall who argued that such a promise creates a paradox (Kawall, 2005, 389–90). To take Kawall's own example, imagine that you made a solemn promise to yourself on New Year's Day that you are going to do at least one supererogatory action every month of the coming year. Now a few days later, your friend Alicia tells you that she is worried about the reliability of taxi services to the airport from which she has a scheduled flight at 4 a.m. in the morning. You promise Alicia to drive her to the airport which you do with pleasure despite the inconvenience and your busy schedule. Can you fulfil your New Year's promise to yourself? On the one hand it seems that you can, since by driving Alicia to the airport you do a supererogatory act and that is the act you promised yourself to do once a month. On the other hand it seems that you cannot fulfil your New Year's promise since once you made the promise you are under an obligation to keep it and driving Alicia becomes obligatory, thus failing to fulfil the promise to do a supererogatory act. Another way to put it is to ask whether the act of driving Alicia is supererogatory. Here again, on the one hand it is, since waking up at 4 a.m. to drive a friend to the airport is definitely beyond one's duty. On the other hand it is not, since once you promised to yourself (and then to Alicia) to do the act, the act becomes obligatory.

Kawall's example is unfortunate because two elements in it obfuscate the logic of supererogatory promising. First it involves a duty which is similar to an imperfect duty in not specifying the recipient of the promised act (to do once a month a supererogatory act). It could be anybody. Hence, since the promise was not addressed to Alicia, Alicia must feel grateful for the act of driving her to the airport. She had no claim on her benefactor. If to be valid a promise must be accepted by the promisee, Alicia was not part of the original New Year's promise and its fulfilment is not due *to her*. She cannot release the promisor from the New Year's promise. Kawall's example does not distinguish between the New Year's promise (made to oneself) and the promise made *to Alicia* on the day before the drive to the airport (over which she does gain normative power). Secondly, the idea of promising to oneself is problematic since one cannot release oneself from such a promise (the promisor and the promisee are the same person). And even if one can have duties to oneself, can one act supererogatorily, beyond the call of duty, to oneself? Can I take care of my health more than I am required by some standard of duty to myself?[6] For those who do not believe in promises made to oneself Alicia's mental act on New Year's Day is not of promising but of deciding or resolving, and that does not raise any problem of supererogation. No obligation is created by making a decision, including a decision to supererogate.

I do not wish to examine here these questions any further and prefer to discuss the question of supererogatory promises in less controversial contexts. So why not think of a New Year's promise *to* Alicia that you are going to perform a supererogatory act *for her* at least once every month (e.g., bring her a present every month). If she accepts this promise, she gains control over your behavior and only she can release

[6] For a different view, see Alfred Archer's article in this volume.

you from the obligation to keep your promise. In this straightforward example the natural analysis is that you made a promise to do a certain act every month and now you keep the promise. The promise itself was supererogatory but its fulfilment now is obligatory. This is no more puzzling or paradoxical than promising to give you a gift for your birthday and then keeping the promise. In terms of the gratitude you owe me, despite having now the right to be given the birthday present you ought to thank me for the promise to give it to you or better for the fulfilment of the promise to give it to you. Once we understand that the act of promise-making is itself supererogatory the air of paradox vanishes. The only difference between Kawall's original example and the birthday example is that the former is a multiple-step promise while the latter is a single-step promise. We can treat the multi-step promise as *one* single promise which involves 12 separate stages for it to be completely fulfilled.

Things seem to get more complicated when the promisee is not the party which directly benefits from the promise. If I promise to Alicia's *mother* that I will drive Alicia to the airport, the mother gets normative control over my behavior. Alicia has no claim against me although she is the one who benefits most from the content of my promise. But again, there is no real puzzle in this case: there is just one supererogatory act. For the mother it is the giving of the promise (and the peace of mind she now enjoys) while for the daughter it is the actual deed of being driven to the airport (the convenience and the saving of costs). The split between the two beneficiaries of the supererogatory promise serves us to illustrate the distinction between the *deontic* status of the execution of the promise and the *content* of the promised act: while the execution of the promise is obligatory the content is supererogatory. This distinction explains how the fulfilment of the promised act can be obligatory and supererogatory at the same time.[7]

4 The Individuation of a Promised Action

Claire Benn claims that I have not noticed that Kawall's paradox involves *two* supererogatory acts—the making of the promise and its performance (Benn, 2014, 57). This raises the question of the individuation of promised actions which has not been given sufficient attention. If I drive you to the airport without a previous promise I do a supererogatory act. If I drive you after having promised you to do so, this has an extra value of the overall act of driving you to the airport but it is not necessarily—as Benn proposes—a *further* supererogatory act. For the act of promising is

[7] Benn (following Gregory Mellema) notices that an act can be both obligatory and supererogatory (Benn, 2014, 58). This is true for Benn's suggested case like promising to give one hundred dollars and then proceeding to give two hundred dollars. However Alicia's case is not just a case of "over-subscription" (giving more than is owed). The ride to the airport is both obligatory and supererogatory in a different sense—supererogatory in content but not in deontic status. Benn argues that "[t]he promise to supererogate is a promise to perform an action that remains supererogatory after the promise is made" (59). This is true only if we take the fulfilment of the promise as the independent content of the act rather than the deontic context of its performance.

supererogatory only if it is executed, i.e., promising in itself is not a supererogatory act. The only way in which we can speak of two independent supererogatory acts is if I decide that beyond driving Alicia to the airport (as I promised) I help her carry her luggage from the parking lot to the terminal, which was not part of my promise (cf. Benn, 2014, 60).

Promises in general and promises to supererogate in particular can be individuated in two ways. One could take the making of the promise and its (later) fulfilment as *one* act performed in two steps or as *two* acts separated by time, motivation, and deontic status. According to the first way there is one promised action; according to the second there is one act of promise-making and another act of promise-fulfilment. These are two theoretically valid analyses of promising and I shall not decide which makes better sense. But I will propose that neither analysis raises a paradox or a puzzle in promising to supererogate.

4.1 A single action

Take the single-action account first. If I intend to walk to the grocery the intention and the walk to the grocery are elements of a single action—the intended walk to the grocery. If I intend to walk to the grocery and do not rise from my sofa, then there is something incomplete in the act of intending; it may be an action which was aborted or an infelicitous case of intending. Unless I can provide some reason or explanation for this incompleteness we can say that there was no real act of intending. The same applies to promising: if I make a promise and do nothing to fulfil it, one can say that I did not really make a promise. When promises are not kept, some explanation or excuse must be given. Making a promise without intending or trying to carry it out is vacuous in the same way as intending without acting on that intention. Levi Eshkol, the Israeli finance minister in the 1960s, used to retort to politicians who complained to him for not having received the funds which he promised them: "indeed, I promised, but I have never promised to keep my promise." The jocular absurdity of this response lays bare the impossibility of separating promise-making from promise-keeping.

If a promise consists of one act, then the actual performance of the promise is a constitutive part of the promising act. Think of the analogy to forgiveness, which in my view is in typical cases supererogatory (Heyd, 1982, 154–164). If forgiveness means the abandonment of resentment, then if the forgiver still feels resentment toward the offender a year later then one may say that he has not really forgiven. The act of forgiveness has failed by not being actualized. Similarly, lacking any excuse for the non-performance of the promised act, we may say that there was no genuine promise in the first place. The act of promise-making was miscarried.

The duty of gratitude for a supererogatory act supports the single-action analysis: Alicia owes gratitude to her benefactor who following his promise drove her to the airport; but strictly speaking she is expected to thank the driver only once or only for one action; she does not have thank him twice, for two separate actions. For if the driver does not show up in the morning she does not owe him thanks for the act of making the promise on the previous day, complaining 'only' about its non-fulfilment. Although we tend to thank the promisor after receiving the promise, the gratitude is *conditional* on the execution of the promised act. The same applies to praise: we do not praise the supererogatory promisor once for making his generous promise and then again for fulfilling it.

An objection can be made to this conclusion on the basis of what we have shown to be the *double* benefit of a promised act. As proposed above, beyond the benefit of not having to rely on an unknown taxi service which costs a lot of money Alicia enjoys the peace of mind that accompanies the promise, the benefit of assurance. Does this show that there are two actions rather than one, each having its own benefit? Not really. The single act can yield two benefits, saving costs and giving the peace of mind, but the one is dependent on the other: without the actual drive the benefit of assurance is worth nothing. The double benefit results only from the completion of the single promised act. Imagine that I pop up spontaneously at 4 a.m. in Alicia's home and offer her a ride to the airport. This is surely a supererogatory act, although there was no prior promise made. If I promise her a day earlier and perform the same service this is also a (single) supererogatory act, but it has the extra benefit of assurance and the peace of mind.

It is therefore unnecessary to ask whether it is the promise-making *or* the performance of the act which makes the promised act supererogatory. Indeed making promises is supererogatory but driving someone to the airport is equally so. The fact that after having promised Alicia to drive her to the airport I am under an obligation to do so does not detract from the supererogatory nature of the *content* of my promise, namely, waking up early in the morning and driving to the airport. When Alicia thanks me when I come to pick her up and I say that I only did what I was under an obligation to do (keeping my promise) I refer at most to one step in the performance of the promised action. I cannot deny that my action *as a whole* (doing the promised act) is supererogatory. There is no use in evaluating the deontic status of each *step* in the overall action since that would mean that Alicia would have to thank me also for having set my alarm clock at 3 a.m. or for every further mile in driving her to her destination (rather than deciding mid-way that I am too tired and turn back home). If without having any prior duty I promise to support you next year by giving you a certain sum of money, it does not make any difference if I give you the money all at once or in installments. Each installment is not a separate supererogatory act. And although being polite you thank me for every installment, it is of course absurd to say that I deserve more praise or gratitude than if I had given you all the money at once.

A limiting case is a promise to make a promise. If I can promise to do for you some supererogatory favor next month, can the supererogatory favor be the making of a promise (which as we have seen is a supererogatory act)? This sounds somewhat artificial or even superfluous. I can choose to take you to the airport for your flight next month, but then there is very little use in promising you now to promise you again next month to take you the airport. I simply should announce when the time comes that I am going to drive you the next morning as I promised in my original promise. For that reason, Kawall's example of the promise to do *some* supererogatory act for someone in the future is of little value due to the little assurance value it has. Promising to promise has little value as is the case of promising to give a present: for in the first case the promisee does not know the content of the promised promise and in the second case the promise is superfluous because the promisee is interested almost exclusively in the content of the promise (the surprise present) rather than in the assurance secured by the promising.

4.2 *Two actions*

We turn now to the second option of individuating a promised action, namely, taking it as *two* actions: the making of the promise and the execution of the promised action. This analysis seems natural due to the time gap between the giving of the promise and its fulfilment. Furthermore, the two actions can be naturally separated by the way they benefit the promisee—the assurance of the act of promising and the material benefit of the deed itself. In addition, although a broken promise means that the second action (being driven to the airport by the promisor) was not performed, the act of promising did take place and the assurance was given (despite not being justified in retrospect).

However, the double-action account must explain the special relation between the two actions. As we have shown, the broken promise often makes the promise-giving an infelicitous act: it could be insincere, deceptive, and manipulative or simply forgotten due to being lightly taken. In that case the issue of individuation of the action of promising does not arise, since there is no second action (or completion of the single action) in the first place. But promises which have been broken could also be due to external circumstances over which the promisor has no control and in those cases the promise-giving act cannot be said to have been infelicitous. According to the two-action analysis, the fact that a promise was broken does not mean that it was not an action, possibly a complete action like any attempt which failed. Like failed attempts, broken promises can be evaluated as admirable and praiseworthy independently of their success or fulfilment. A promise which due to *force majeure* could not be kept is still supererogatory and the promisee may acknowledge that despite her disappointment and often also the material costs of the non-fulfilment. But we should also take into account that the separate value of the assurance granted by the promise-giving may be set off by the disappointment in its non-fulfilment, thus making the on-balance value of the supererogatory promise-giving negative (Alicia missed her flight). Still, if we define supererogatory action as that which is *intended* to benefit another person, then the actual negative value should not detract from its supererogatory nature.

The two-act account of promising can also solve our original puzzle: the act of promise-giving is supererogatory but the act of keeping the promise is obligatory.

Driving Alicia to the airport is in itself supererogatory but doing so after having promised to drive her to the airport is obligatory. There is no contradiction between the free undertaking of an obligation and the binding force of the obligation itself. In our scenario there is no tension between the supererogatory nature of the first act and the obligatory nature of the second. The first act morally necessitates the second.

In the single-act analysis of promising we had one act which can be described as having a deontic *form* (the undertaking of the obligation to do something) and the *content* or object of that undertaking. Without the content of the promise being realized the undertaking is incomplete, vacuous, or infelicitous. In the two-act analysis we have the first act which can be separately identified and evaluated and a second act which stands in a normative relation to the first.

5 Conclusion

Can one promise to supererogate? Both Jason Kawall and Claire Benn believe that the answer is no. By fulfilling a promise we cannot fulfil the obligation to do a supererogatory act, because once the promise is made its fulfilment cannot be supererogatory. I have argued throughout the article that there is no problem in making a supererogatory promise to take Alicia to the airport tomorrow night. The obligatory nature of the drive itself does not mean that the act as a whole (according to the single-act account) is not supererogatory, or that the promise-giving is not a supererogatory act the nature of which is the obligation to drive Alicia to the airport (according to the two-act account). By both the single-act and the two-act analyses the apparent paradox is resolved. The apparent paradox only arises when we take as our basic example Kawall's New Year's Day promise *to myself* to do one supererogatory action every month (to whomever it is). The difficulty can be overcome by distinguishing between promising which does not apply reflexively and undertaking a commitment which is typically reflexive. A second possible source of confusion is the awkward example of making a promise *to Alicia* to do a supererogatory act in her favor at least once a month. Again, I see no particular problem here in the supererogatory analysis of promises: the promise is supererogatory and the execution of the promise is obligatory although it comes in 12 steps (each one of them being obligatory).

Even if we take a third possible example which looks perplexing, namely, promising to promise, my proposed analysis can account for it. Suppose I promise Alicia that I will make a promise to her next month. The promise is somewhat artificial since although promises are supererogatory in benefiting the promisee they do not come without content and in this case Alicia does not know and cannot rely on the original promise except that on one of the days of next month she will be given some promise. This in itself gives her only little benefit because it does not give her assurance of any particular action. Still, conceptually there is no problem in such a contrived case: the original promise-giving is surely supererogatory (as Benn proposes) while the second promise-making becomes obligatory by the first promise. And of course the execution of the second promise (e.g., giving her a ride to the airport) is obligatory too. The fact that the second promise is more valuable to Alicia than the first vague promise does not affect the deontic status of the three acts or stages in this scenario. I fulfil my obligation by making the second promise in exactly the same way as I do my obligation in keeping any standard promise that I gave. A gift is supererogatory; a promised gift is obligatory. Similarly, a promise is supererogatory; a promised promise (and its execution) is obligatory. Again, we should avoid double counting: there is only one supererogatory act in this example, one action which deserves praise and gratitude. The sense of puzzle is created only by the pointlessness of promising to promise.[8]

References

Aquinas, T. (1947). *Summa Theologica* (trans. Fathers of the English Dominican Province). Benziger Bros.

Benn, C. (2014). What is wrong with promises to supererogate? *Philosophia, 42*, 55–61.

Cholbi, M. (2017). A plethora of promises—or none at all. *American Philosophical Quarterly, 51*, 262–272.

Enoch, D. (2015). Being responsible, taking responsibility and penumbral agency. In U. Heuer & G. Lang (eds.), *Luck, value, and commitment: Themes from the ethics of Bernard Williams* (pp. 95–131). Oxford University Press.

Fox, R. (1993). The immorality of promising. *Journal of Value Inquiry, 27*, 81–84.

Heyd, D. (1982). *Supererogation: Its status in ethical theory.* Cambridge University Press.

Heyd, D. (2005). Supererogatory promises a comment on Kawall's 'promising and supererogation.' *Philosophia, 32*, 399–403.

Heyd, D. (2021). The mutual independence of toleration and liberalism. In J. Drerup & G. Schweiger (eds.), *Toleration and the challenges liberalism* (pp. 79–96). Routledge.

Hobbes, T. (1996). *Leviathan.* In R. Tuck (ed.), Cambridge University Press.

Hume, D. (1960). *A treatise on human nature.* In L. A. Selby-Bigge (ed.), Oxford University Press.

Kawall, J. (2005). Promising and supererogation. *Philosophia, 32*, 389–398.

Owens, D. (2006). A simple theory of promising. *Philosophical Review, 115*, 51–77.

Rawls, J. (1971). *A theory of justice.* Harvard University Press.

Raz, J. (1977). Promises and obligations. In P. M. S. Hacker & J. Raz (eds.), *Law, morality and society* (pp. 210–228). Clarendon press.

Scanlon, T. M. (1998). *What we owe to each other.* Harvard University Press.

[8] And if the story is that I promise 'to myself' that I will do one supererogatory act every month next year (as Kawall and Benn present it), then—as I claimed—we do not have a promise to make a promise because the first 'promise' has no addressee, no specific promisee who is given normative control over my behavior except myself. Alicia does not have any expectation and cannot release me from the promise. Driving her to the airport is a purely supererogatory act because I did not promise *her* anything, or if I made the second promise (to drive her to the airport, *this* was a supererogatory act rather than obligatory in fulfilling of my promise to myself).

What's the Use of Non-moral Supererogation?

Alfred Archer

Abstract While moral philosophers have paid significant attention to the concept of moral supererogation, far less attention has been paid to the possibility that supererogation may also exist in other areas of normativity. Recently, though, philosophers have begun to consider the possible existence of prudential, epistemic, aesthetic, and sporting supererogation. These discussions tend to focus on aspects of our practices in these areas of normativity that suggest an implicit acceptance of the existence of supererogation. In this chapter, I will offer a different kind of defense of non-moral supererogation. I will begin by considering a particular kind of argument made in support of moral supererogation. According to this line of argument, we should accept the existence of moral supererogation because a moral code which makes room for supererogation is likely to be more effective at promoting morally desirable behavior than a moral code which leaves no room for the supererogatory. I will begin by outlining this argument. I will then develop a similar line of argument for prudential, epistemic, aesthetic, and sporting norms.

Keywords Supererogation · Aesthetic supererogation · Epistemic supererogation · Non-moral supererogation · Sporting supererogation

1 Introduction

Many moral philosophers accept the existence of acts of supererogation.[1] Those who give up their lives to save the life of another, dedicate their lives to serving the needy, or who simply go out of their way to help others when they have no special reason to do so, all seem to be acting in ways that go beyond their moral duties. While

[1] Though it is worth noting that some moral philosophers deny the existence of supererogatory acts. See Hale (1991), Baron (2016), and Fitts and Miller (2020).

A. Archer (✉)
Tilburg University, Tilburg, Netherlands
e-mail: A.T.M.Archer@tilburguniversity.edu

D. Heyd (ed.), *Handbook of Supererogation*,
https://doi.org/10.1007/978-981-99-3633-5_14

the concept of moral supererogation has been extensively explored by philosophers,[2] non-moral forms of supererogation have only recently started to receive philosophical attention. The basic idea driving these investigations is that if morality allows for the possibility of acts that are beyond the call of duty, then perhaps other forms of normativity do so as well. This guiding idea has led philosophers to explore whether supererogation exists in relation to prudential (Benn & Bales, 2020; McElwee, 2017), etiquette (McElwee, 2017), epistemic (Hedberg, 2014; Li, 2018; McElwee, 2017), aesthetic (Archer & Ware, 2017), and sporting (Archer, 2017) norms.

Those seeking to defend the existence of non-moral supererogation typically do so in one of two ways. The first way is to point to features of the normative practices associated with these domains of normativity that suggest that these are domains in which supererogation exists. Brian McElwee (2017; see also McElwee's contribution to this volume), for example, has argued that two features need to be present in order for supererogation to exist in any given normative domain. First, that domain must include an evaluative scale for actions which ranks the actions from better for worse. Second, that domain must involve a deontic categorization of acts into the obligatory, the permissible, and the forbidden, where the notion of obligation does not simply equate to an act being top of the evaluative scale. McElwee then argues that the domains of prudence, etiquette, and the epistemic meet both conditions and so we should accept that supererogation exists in these normative domains.[3] Similarly, Benn and Bales (2020) argue that just as it would be too demanding to require perfection in the moral domain, so too would it be too demanding to demand perfection in the domain of prudence.

The second, related, way in which people have defended the existence of non-moral supererogation is to point to cases that intuitively appear to be praiseworthy in relation to the domain of normativity but would not be fitting subjects of blame had they not acted in that way. Trevor Hedberg (2014), for example, argues for the existence of epistemic supererogation by presenting a series of cases where an act appears to be epistemically praiseworthy but does not seem to be epistemically required of them. I have presented a similar kind of argument in support of sporting supererogation (Archer, 2017).[4]

In this chapter, I will offer a different kind of defense of non-moral supererogation. I will begin by considering a particular kind of argument made in support of moral supererogation. According to this line of argument, we should accept the existence of moral supererogation because a moral code which makes room for supererogation is likely to be more effective at promoting morally desirable behavior than a moral code which leaves no room for the supererogatory. I will begin, in Sect. 2, by outlining this argument. In the sections that follow I will then develop a similar line of argument

[2] See, for example, Grigoletto (2019), Heyd (1982), and Mellema (1991) for monographs dedicated to the topic. For overviews of the literature on moral supererogation see Heyd (2011) and Archer (2018a).

[3] Archer & Ware (2017) use the same strategy to argue for the existence of aesthetic supererogation.

[4] Though see Borge (2021) for a critical reply to this argument.

for prudential (Sect. 3), epistemic (Sect. 4), aesthetic (Sect. 5), and sporting (Sect. 6) norms.

2 Moral Supererogation

In Urmson's (1958) influential essay 'Saints and Heroes', he provides several different kinds of argument in support of supererogation. He begins by making the case that our intuitive reactions to some of the exceptional deeds performed by saints and heroes are incompatible with viewing these actions as morally required. For example, a soldier who dives onto a grenade and sacrifices his life to save the lives of his comrades performs an act that is praiseworthy. However, the soldier would not be blameworthy if he had not dived on the grenade, and no one could reasonably demand that he act in this way. If we think that a moral duty is the kind of act which we can demand of each other and which people can be blamed for failing to perform, then we should accept that the soldier's act was not obligatory. This means that the soldier's act is morally praiseworthy but not morally required.[5] In other words, it is beyond the call of duty.

In addition to this argument, Urmson also argues that moral codes that make room for the supererogatory will be preferable to those that do not. Urmson provides five reasons in support of this claim, though I will focus only on two of these.[6] The first argument Urmson makes is the following:

> It is important to give a special status of urgency, and to exert exceptional pressure, in those matters in which compliance with the demands of morality by all is essential [...] while life in a world without saints and heroes would be impoverished, it would only be poor and not necessarily brutish or short, as when basic duties are neglected. (1958: 70)

Urmson's point is that there is a certain level performance that is needed for society to function. For example, if people harmed other people whenever they wanted to, then we would have to live under the constant threat of being attacked by others. It is important, then, to do what we can to ensure that people perform these basic acts. An important way of ensuring this is to give these actions the special status of duties and to exert pressure on everyone to act in fulfil these duties. There are other morally worthy acts that improve our lives but are not essential to societal functioning. A society in which no one dedicated their lives to helping others would certainly be less desirable than one in which some people do act in this way. However, a society with no saints would still be capable of functioning in an adequate way.

The upshot of this is that a distinction between acts of duty and acts of supererogation can help to ensure that a special effort is made to ensure that they perform those acts needed for society to function. The concept of a moral duty which includes the idea that people are required to act in certain ways and can be legitimately targeted

---

[5] For a more detailed discussion of this argument see Archer (2016).

[6] See Crisp (2013) for a critical discussion of Urmson's argument.

with demands and blame in order to ensure they do so gives these acts this special urgency. These are then distinguished from supererogatory moral acts which can be praised and admired but for which demands and blame are inappropriate. A society which includes the distinction between duty and supererogation will, then, be better placed to ensure that people perform the most basic and essential moral acts.[7]

The next reason Urmson offers for the superiority of moral codes that make room for supererogation is the following[8]:

> If we were to represent the heroic act of sacrificing one's life for one's comrades as a basic duty, the effect would be to lower the degree of urgency and stringency that the notion of duty does in fact possess. The basic moral code must not be in part too far beyond the capacity of the ordinary men on ordinary occasions, or a general breakdown of compliance with the moral code would be an inevitable consequence; duty would seem to be something high and unattainable, and not for 'the likes of us'. (1958: 70)

The point here is that a moral code according to which the morally best act is always the act we are morally required to perform will be one in which moral criticism and moral demands lose their force. With such a code in place, most people will regularly violate their moral obligations. When most people get used to violating their obligations, they will become accustomed to ignoring the moral rules. As a result, the general level of moral behavior will be lower than it would be under a moral code in which duties are less demanding.

In support of this claim Urmson gives the following example:

> The prohibition laws asked too much of the American people and were consequently broken systematically; and as people got used to breaking the law a general lowering of respect for the law naturally followed; it no longer seemed that a law was something that everybody could be expected to obey. (1958: 70)

While this is an example of a legal rather than a moral rule, Urmson takes this to show the danger that arises from imposing unrealistic rules on people. When people find that obeying the law is too difficult, then they get used to breaking the law. This erodes the respect that they have for the law and makes it more likely that they will break other laws as well. This means that overly demanding legal rules may lead to behavior that is further away from what the behavior that the legislators are aiming for than rules requiring less demanding but more achievable standards of behavior. Urmson claims that the same is true for moral rules. If the standards for moral duty are too high so that most people will fail to fulfil their duties, then people become accustomed to violating their moral duties. This, then, would lead to a lower average level of moral behavior than a less demanding and more realistic set of moral duties. It is preferable, then, for moral duties to be set at a level that is lower than moral

[7] John Stuart Mill (2001: 59) offers a similar argument claiming that "Justice is a name for certain classes of moral rules which concern the essentials of human well-being more nearly, and are therefore of more absolute obligation, than any other rules for the guidance of life." In other words, we accept a set of basic moral rules that people will comply to in order to secure the most essential conditions for well-being. These rules can be enforced with punishment and blame. Other moral actions are good but not 'absolute obligations' in the same way.

[8] The discussion of this part of Urmson's argument closely follows Archer (2018b).

perfection and when moral duties do not require moral perfection there will be room for acts that are morally better than what morality demands.

Benn (2018) offers a similar argument in defence of the practical value of the distinction between the obligatory and the supererogatory. According to Benn, overly high standards for moral duty would create large psychological costs:

> [B]eing required to do the best takes a psychological toll. Knowing that even the smallest deviation from a particular course of action will lead to the legitimating of sanctions is likely to have an effect on an agent. Even if you are motivated to do what is right, having a moral gun to the head (so to speak) can make doing so much more difficult psychologically speaking. (Benn, 2018: 343)

In support of this claim Benn draws a parallel to psychological perfectionism. Psychological perfectionists set overly high standards for themselves and subject themselves to excessive self-blame when they fail to meet these standards. This can have a paralyzing psychological effect, as a fear of failing to live up to these standards often leads to anxiety and even depression. This, is turn, often leads to avoidant behavior, such as procrastination, abandoning tasks before they are finished, or a failure to begin tasks in the first place. The upshot is that demanding perfection from oneself is often counter-productive. People would achieve more if they demanded less from themselves. As the saying goes, the perfect can often be 'the enemy of the good' (Benn, 2018).

Benn argues that views that hold that we are always required to do what is morally best are forms of moral perfectionism. These views, then, are also likely to be counter-productive in the same way that perfectionism more generally is counter-productive. Those who hold themselves to maximally demanding moral duties are likely to avoid situations where they may be able to act morally, for fear of failing to meet their overly demanding moral standards. This means that if we want people to be morally good, we should not demand that they be morally perfect. Instead, we should accept a less demanding set of moral duties, which is not set at the level of moral perfection. Such a set of moral duties will leave room for supererogation by allowing that people can perform acts that are morally better than what is required to merely fulfil these duties.

In summary, one approach that some philosophers have taken to defending the existence of moral supererogation is to point out that a moral code which leaves room for acts of supererogation will be morally preferable to one which does not. The reason for this is that it is important to ensure that as many people as possible reach some basic minimum level of moral performance to secure the pre-conditions for a functioning society. Moral duties which are enforceable through demands and punishment help to ensure that people succeed in meeting this basic level. If these duties were to demand too much of people, especially if they were to demand that people be morally perfect, then they would be less effective, and the average level of moral achievement would be lowered. While several criticisms have been made to this line of argument,[9] my primary aim here will not be in evaluating it but rather in showing that this same argumentative strategy can be employed to support the practical value of *non-moral* supererogation.[10]

3 Prudential Supererogation

The first form of non-moral supererogation that I will consider is prudential supererogation. I understand prudence as the area of normativity concerned with self-interest. The prudentially best act will be the act that is best from the point of view of promoting an agent's self-interest. As McElwee (2017: 509) points out, we often use deontic terms when discussing prudential normativity. We say that people really ought to take better care of their health, that they ought to exercise more, or that they ought to take their studies more seriously. We also use distinctive forms of criticism when people fail to act in these ways, describing them as 'foolish' or 'imprudent' (McElwee, 2017: 509).[11] As McElwee points out, though, these forms of criticism tend not to be used whenever someone fails to act in the best way possible from the point of view of self-interest. Rather, they are reserved for those who to those who fall below a basic level of prudential behavior. This means that there is room for acts of prudential supererogation, actions that exceed this basic level.

McElwee's argument is designed to show that our prudential practices implicitly allow room for prudential supererogation. My interest here is not in evaluating whether we are implicitly committed to the existence of prudential supererogation but rather to argue that an approach to prudence that makes room for supererogation will be better at promoting prudential behavior than an approach that leaves no room for the prudentially supererogatory.

The starting point for this argument is Urmson's claim that it is useful to have a set of moral norms that help to ensure that we perform the most essential moral acts. A set of basic moral duties which can be enforced through demands and blame helps to ensure that people perform the actions that are needed to allow the basic functioning of society. In the same way, it is useful to have a set of prudential norms to help to ensure that people perform the most essential prudential actions. These

---

[9] Eg. Baron (2016), Crisp (2013), Hale (1981). It is also worth noting that this is not the only kind of consideration offered in favor of the value of a moral code that leaves room for the supererogatory. David Heyd (1982: 178–181), for example, argues that moral codes should leave room for supererogation to allow room for acts of voluntary altruism which help to promote social cohesion.

[10] It is worth noting that the origins of the concept of supererogation lie in Catholic Theology where a distinction is made between precepts (commandments) and counsels (recommendations) see Heyd (1982 Ch.1). This theological distinction could also be justified in similar terms, as a way of effectively ensuring people perform their most essential religious duties.

[11] McElwee (2017: 509) argues that this special form of disapproval gives us reason to distinguish prudential duties from self-regarding moral duties. See Muñoz (2020) for a discussion of self-regarding moral duties. For a discussion of self-regarding acts of moral supererogation see Kawall (2003).

will be the acts for which people will seriously threaten their self-interest when they fail to perform them. For example, we might think that is prudentially required for those with an adequate but modest salary to prioritize one's spending in such a way that one will have enough money to pay for the basic essentials of life, such as food, throughout the month. Spending all of one's salary on expensive clothes and whisky so that one no longer has enough money left over for food at the end of the month would violate this prudential requirement. On the other hand, saving as much money as possible each month to make wise future investments may count as prudentially supererogatory. Similarly, for those in a position to do so, making sure that one is eating a reasonably healthy diet may count as a prudential requirement. However, to spend hours each week tracking nutritional details to ensure one is eating the healthiest diet possible would be prudentially supererogatory in the case where it really does advance ones interests or foolish or silly if it does not do so.

While it is valuable to have a set of basic prudential requirements to ensure that we are performing the most essential prudential actions, it is also important that these prudential requirements are not set at too high a level. There are two reasons for this. First, it may undermine the important role that these prudential requirements are supposed to play. We have all sorts of prudential reasons to perform all sorts of actions. Having a basic set of prudential requirements, though, can ensure that we feel a special force to perform the most essential actions, those that are needed to secure a basic level of well-being. If we were to view every act that we have most prudential reason to do as prudentially required, then these prudential requirements would lose their force. In such a situation, most people would quickly find themselves regularly violating their prudential requirements. This in turn would lead to a lowering of respect for these prudential rules, as people no longer feel any special compunction to ensure they act in line with these requirements. This would be likely to bring about a lower level of prudential performance as people fail to ensure that they are performing the actions needed to secure the most basic level of self-interest.

The second worry with viewing ourselves as subject to excessively demanding prudential demands is that someone who demands of herself that she always do what is best from the prudential point of view is a clear example of a perfectionist. Such a person will face demands that she always perform as well as possible in her job, has a maximally nutritional diet, a perfect exercise regime, spends the exactly right amount of time with her friends and no more, and has exactly the right number and selection of hobbies. As we saw in the previous section, perfectionism of this kind has a damaging psychological impact. It often leads to a fear of failure, which in turn motivates avoidant behavior such as procrastination, abandoning tasks before they are completed, and avoiding tasks altogether (Benn, 2018; Shafran et al., 2002). This clearly has a negative impact on our self-interest.

Prudential requirements, then, have an important role to play in our lives but setting these requirements too high may lead them to fail to perform this function and lead to damaging forms of perfectionism. As a result, if we want to promote our self-interest then we should not make excessive prudential demands of ourselves. Rather, we should accept a set of less demanding prudential requirements that secure our basic self-interest. That does not mean that we should aim only to achieve this basic level, we can and should aspire to do more than simply fulfil our prudential duties. However, we should not view ourselves as having a prudential requirement to always do what is perfect from the prudential point of view. Accepting this gives us a view of prudential requirements which makes room for the prudentially supererogatory. Prudential requirements help to secure a basic level of self-interest but there will generally be many ways to go beyond these requirements and perform an act that is prudentially better than what is prudentially required.

4 Epistemic Supererogation

The second form of non-moral supererogation that I will consider is *epistemic* supererogation. This is an area of normativity concerned with what we ought to believe. A belief is good from the epistemic point of view insofar as it tracks the truth. Those who form beliefs in a responsible way may be praised by others, while those who fail to do so are likely to be blamed by others.[12]

Consider a trivial example, suppose a group of friends are going to the cinema to see the new Nicholas Cage film. Dave confidently tells the group that he knows that the local arthouse cinema will be showing the film. The group walk to the cinema only to discover that the film is not being shown there. The group turn to Dave and ask him why he was so confident the film would be showing there. He replies that he knew they showed the previous Nicholas Cage film, so he assumed that they would be showing this one as well. The group roundly criticize him for making such a confident assertion based on so little evidence and for wasting everyone's time.[13]

This example seems to show that we do engage in blame when someone fails to form their beliefs in a reasonable and trustworthy way.[14] As McElwee (2017: 513) argues, when someone forms beliefs in an irresponsible way, as Dave did, it can be appropriate to blame them and perhaps even exclude her from the group of people whose testimony we are willing to trust. The next time the group of friends is looking for a cinema showing a film they want to see, it would be reasonable for them not to listen to Dave's claims about where it is showing, or at least to ask him to explain how he formed this belief. There seems good reason, then, to think that there exist epistemic duties that people may rightly be blamed for failing to fulfil.

However, as several philosophers have argued, the epistemic duties to which we hold each other do not require epistemic perfection (Hedberg, 2014; Li, 2018;

---

[12] See Boult (2021) for an overview of the literature on epistemic blame.

[13] This example may be thought to involve both a moral and an epistemic wrong. Dave deserves criticism not only for his belief-forming process but also for ruining his friends' plans. Importantly, though, even if he is a fitting target of moral criticism in this case, he is also a target of epistemic criticism. He should be criticized not only for ruining his friends' plans but also for the way in which he forms beliefs and for his willingness to confidently make factual assertions for which he has so little evidence.

[14] Though see Matheson and Milam (2021) for an argument against the existence of non-moral blame.

McElwee, 2017).[15] While we might reasonably expect Dave to check the cinema's online schedule before asserting that he knows the film is showing there, it would go beyond the call of duty for Dave to also phone the cinema to check that their online schedule is correct. Taking this extra step to make sure his belief is justified is praiseworthy but is not required and Dave would not be blamed if he simply relied on the cinema's online schedule.[16] There seems good reason, then, to think that our epistemic practices implicitly make room for epistemic supererogation.

My interest, though, is in arguing that a set of epistemic norms that make room for epistemic supererogation will do a better job at advancing our epistemic goals than one that requires epistemic perfection. I will do so by applying Urmson's argument once more, this time to epistemic normativity.

The reason that a set of epistemic norms that makes room for supererogation is preferable to one that does not is that it is likely to do a better job of ensuring that people meet an essential, basic epistemic standard in their belief-forming processes. A set of basic epistemic duties which can be enforced through demands and blame helps to ensure that people form their beliefs in a way that meets basic epistemic standards. For our epistemic practices of relying on testimony and social deliberation to be effective, we need to make sure that people are forming their beliefs in a reasonably responsible way. For example, someone who is never willing to reconsider their beliefs in the light of reasonable criticism, fails to form their beliefs in a responsible way (Hedberg, 2014). If we were to continue to trust this person's testimony, then we would run a high risk of accepting beliefs that are not justified and that are likely to be false. By blaming such a person for their belief-forming processes we can hope to discourage them and others from forming beliefs in this way in the future. In this way, they will hopefully change their behavior and begin to reevaluate their beliefs when facing reasonable criticism. If they do not do so, then we may have to respond in a more drastic way by excluding them from the group of people we are willing to trust and whose testimony we will depend upon.

While it is important to ensure that people are meeting basic epistemic standards in their belief-forming processes, it is also important that these epistemic requirements do not demand too much from people. If these requirements were too demanding, then they may undermine the important role that these requirements are supposed to play in promoting epistemic values, such as the discovery (or acquisition) of truth. Most of the time, there will be a whole host of actions we could take to make our beliefs more justified. To return to the cinema example, as well as double-checking

---

[15] See Li (2019) and Siscoe (2022) for a discussion of whether epistemic supererogation can provide support for epistemic permissivism, the view that there can be more than one rationally permissible belief which one may hold in relation to some body of evidence.

[16] Those unconvinced by this example may consider the following example suggested by Nado (2019): "[S]uppose that I check ten separate sources before resting content with my belief that Mogadishu is the capital of Somalia. It seems obvious that I did not need to do that; surely one or two would have been enough (indeed, in actual fact I only checked one while writing this paragraph). But it also seems obvious that I have improved my epistemic position at least a little by the additional checking - though, all things considered, my time might have been better spent in other ways. The epistemic action of decuple-checking was epistemically supererogatory.".

the cinema times by phoning the box office, Dave could triple-check by looking up the times in the local newspaper. He could also ask people he knows if they have ever heard of the cinema making a mistake in their schedule. While taking each of these steps may increase the justification Dave has for believing that the cinema will show the film, it would go beyond what we could reasonably expect from people to ensure that they try to achieve this level of justification for all their beliefs. A set of epistemic norms that were this demanding would quickly lose their force. In such a situation, most people would soon become accustomed to violating their epistemic requirements and would lose their respect for the norms. This in turn would lead to a lower level of epistemic performance as people begin to ignore the most essential steps that they need to take to form beliefs responsibly. By having a less demanding set of epistemic requirements, we can help to protect the special force of the most essential epistemic norms that need to be followed to ensure that our epistemic practices can continue to function.

A maximally demanding set of epistemic norms may also lead to problems of perfectionism. Someone who strives always to ensure that their beliefs are as well-supported as possible will face a very difficult life. They will have to double and triple-check their beliefs whenever they make any factual claims to others or even rely on such beliefs in their own reasoning processes. They will spend so much time checking their beliefs that they are unlikely to ever have much time for anything else. This form of perfectionism may also likely have a similarly damaging psychological impact to the other forms of perfectionism we have considered so far. Epistemic perfectionism is likely to lead to a fear of being found to hold a false or unjustified belief, to procrastination in relation to evidence gathering, and perhaps even to the avoidance of forming beliefs altogether, where possible. While it is important to ensure that people meet some basic epistemic norms when forming their beliefs, it would be counter-productive to make these norms too demanding.

That is not to say that we will only ever demand from people that they meet these most basic norms. Some roles that people hold may be accompanied by a more demanding set of epistemic duties. Those serving on a jury in a legal trial, for example, may be expected to take greater care than normal about the extent to which their beliefs are justified. Similarly, people who are regarded as experts by others or whose testimony is likely to have a large audience might reasonably be expected to be more careful in the ensuring that any factual statements that they make in public are well-justified. This, though, is fully compatible with the argument I have made which holds that it is useful for there to be a set of epistemic requirements that ensure people meet some basic standards (which may be higher for some people than others) in their belief-forming processes. For the purposes of promoting epistemic values, though, it is also important to ensure that these standards are not too demanding. It is desirable, then, for our epistemic norms to leave room for epistemic supererogation.

5 Aesthetic Supererogation

The next form of non-moral supererogation to consider is *aesthetic* supererogation. While the existence of aesthetic requirements is not universally accepted,[17] several different kinds of example have been offered in support of the claim that these are a recognizable feature of our aesthetic practices.[18] I will focus on three. First, Howard Press (1969: 525) argues that there are obligations to appreciate what is beautiful. We ought, for example, to take the time to appreciate the beauty in the world around. As Archer and Ware (2018: 115) argue, we might think there is something blameworthy about someone who never makes the effort to appreciate the beauty in their local surroundings. For example, if someone lives next to a beautiful forest and has never taken the time to visit for no other reason that they cannot be bothered to do so, we might think that there is something to be criticized in this indifference to aesthetic value (Archer & Ware, 2018: 116).

Second, it seems plausible to think that there are obligations not to destroy things that are aesthetically valuable, at least when there is no good reason to do so. As Yuriko Saito (2007: 214) has argued, people who destroy beautiful things, such as the natural environment, are often subject to fierce criticism. To illustrate this kind of case, Archer and Ware (2017: 111) give the real-life example of a local official who responded to the vandalism of a Scottish beauty spot by saying that the vandals "have very ugly minds." Here again it seems as if a distinctive form of aesthetic blame is being addressed to those who destroy beauty.

Third, there may be aesthetic requirements to meet particular aesthetic standards on certain occasions. Saito (2007: 213) gives the example of the Northwestern University lacrosse team who were criticized for attending a function at the White House while wearing flip-flops. This, Saito argues, is a clear case where people are deemed blameworthy and disrespectful for violating an aesthetic norm. Similarly, we might think people can be criticized for dressing informally at a wedding or a funeral.

These three kinds of cases are supposed to show that aesthetic requirements are a recognizable feature of our aesthetic practices. As Archer and Ware (2017, 2018) point out, in all these cases the aesthetic requirements are ones that can be surpassed. While someone may be criticized when they make no effort whatsoever to appreciate beauty, most people do not expect others to dedicate their entire lives to the appreciation of beauty (Archer & Ware, 2017: 109). Similarly, someone who dedicates their life to the protecting areas of great natural beauty seems to surpass the duty not to engage in the destruction of beauty (Archer & Ware, 2018: 120). Finally, while there are certain aesthetic standards for clothing that people expect others to meet on certain occasions, these too can be surpassed. Wearing flip-flops to the White House may violate such standards but someone who spends a large portion of their savings

[17] For arguments against the existence of aesthetic obligations see Dyck (2021) and Matheson and Milam (2021).

[18] We might also think that the existence of aesthetic obligations follows from other aspects of aesthetic normativity such as the existence of aesthetic dilemmas (Eaton 2008).

on buying a new outfit for the occasion would likely be seen as going beyond what is required here.

These examples give us reason to think that aesthetic supererogation is presupposed by our aesthetic practices of holding people to certain aesthetic standards.[19] My interest here, though, is in defending the claim that a set of aesthetic norms that makes room for supererogation is preferable to one that does not.

As with the domains of normativity that I have considered so far, it is important to have certain basic aesthetic norms we can demand that people comply with. This point should not be overstated. These basic aesthetic norms do not seem to have the same level of importance as the basic moral or epistemic norms. Urmson's claim was that society could not function without a basic set of moral duties that people can expect others to comply with. Similarly, I argued that our practices of relying on other people's testimony and social deliberation depend upon people being able to count on others fulfilling certain basic epistemic duties. Nothing quite so serious seems to hold in the aesthetic case. Nevertheless, it is certainly valuable to be able to rely on others to meet some basic standards. We expect that when others invited us round to make food that they have prepared, that they will have made some kind of effort to try and make food that is aesthetically pleasing. Without being able to rely on people to meet this basic aesthetic standards we might be much less willing to accept dinner invitations. Similarly, in many places, areas of natural beauty can be made publicly accessible without a major security presence, as most people can be relied upon not to destroy this beauty. It is also useful to be able to rely on others to comply with aesthetic norms that apply in particular occasions, such as dressing appropriately for weddings, funerals, or visits to the president. It is useful to be able to depend on others to make some basic steps to appreciate, promote, or protect aesthetic values and aesthetic requirements can enable us to do this.

While it is important to ensure that people are meeting basic aesthetic standards, it is also important that these aesthetic requirements do not demand too much from people. If these requirements were too demanding, then they may undermine the role they play in enabling us to depend on others to make some basic efforts to protect aesthetic value. On most occasions there will be a wide range of actions we can take to promote aesthetic value. People could dedicate their lives to appreciating works of art, preserving natural beauty, or dressing as well as possible. While there would certainly be something aesthetically valuable about these projects, it would not be helpful to demand this level of aesthetic dedication from everyone. Most people are not willing to dedicate their entire lives to these kinds of aesthetic projects and so would find themselves regularly violating their aesthetic duties. Moreover, these various aesthetic projects are likely to be incompatible, so even people wishing to dedicate their lives to aesthetic projects are likely to find themselves violating aesthetic demands regularly. A maximally demanding set of aesthetic norms, then, would quickly lose its force, as people become accustomed to violating their aesthetic duties. This would be likely to lead to lower levels of aesthetic performance. An overly

[19] See Kubala (2018) for a defense of the claim that some of these obligations can be grounded in a self-directed promise.

demanding set of aesthetic norms may be counter-productive in relation to promoting and protecting aesthetic values.

Maximally demanding aesthetic requirements are also likely to lead to problems of perfectionism. If we demand what is aesthetically best from everyone at all times, then people will quickly find themselves facing an impossible set of tasks. They must make sure that their houses are as beautiful as possible, that they are dressing as well as they possibly can, that they are making as much effort as possible to appreciate art and the aesthetic value of nature, while developing whatever aesthetic talents they possess as much as possible. Aesthetic perfectionism is likely to be psychologically damaging in similar ways to the other forms of perfectionism we have considered so far. For one thing, aesthetic perfectionism is likely to lead to procrastination in aesthetic production. Aesthetic perfectionism may lead people to be less willing to take the kinds of risk involved in true artistic creativity or even to avoid artistic creation altogether. While it is useful to be able to depend on others to comply with some basic aesthetic norms, it would be counter-productive to make these norms too demanding.

Another reason to accept a clear distinction between the aesthetically required and the aesthetically supererogatory can be found by looking at how certain aesthetic norms function in sexist societies. Naomi Wolf (1990) famously argued that sexist beauty norms have an oppressive impact on women's freedom. These norms make involve highly demanding aesthetic standards for personal appearance for women but not for men. Archer and Ware (2018) argue that a major problem with these norms is that they treat what should be considered aesthetically supererogatory approaches to aesthetic standards of personal appearance for women as if they were aesthetically required.[20] They argue that the concept of aesthetic supererogation can play an important role in highlighting what is going wrong with these demands. By making it clear that aesthetic perfection is not an aesthetic requirement, the concept of aesthetic supererogation can help women resist the pressure to strive for aesthetic perfection in their personal appearance.

6 Sporting Supererogation

The final form of non-moral supererogation that I will consider is sporting supererogation. Elsewhere I have argued that there is good reason to accept the existence of acts of sporting supererogation (Archer, 2017). We can see an initial case for this by considering two examples.

In an English Premier League football match between Arsenal and Liverpool in 1997, the referee awarded a penalty after Liverpool's Robbie Fowler appeared to be fouled by the Arsenal goalkeeper. Surprisingly, Fowler responded by saying that he had not in fact been fouled and so the referee should not award a penalty to his team.

[20] See Widdows (2018) for an alternative analysis of this problem according to which aesthetic values are transformed into *ethical* demands.

While most players would simply have accepted a referee error that benefitted their team, Fowler made the effort to try to persuade the referee to correct the mistake. Fowler was widely praised for his sportsmanship and received a special UEFA, the governing body of European football, for his actions.

A similar case of sportsmanship can be found in another English Premier League football match between West Ham United and Everton in 2000. Everton's goalkeeper, Paul Gerrard collapsed just as West Ham crossed the ball into the Everton box. Paulo Di Canio was left with an open goal and an easy chance to put his side ahead in the match. Rather than shoot at goal, though, Di Canio caught the ball in his hands, pointing to the injured goalkeeper to signal that the match should stop to allow Gerrard to receive treatment. Like Fowler, Di Canio was widely praised for his sportsmanship. In recognition of his act of good sporting behavior, he received the annual fair play award from FIFA, the World's governing body for football.

These two cases suggest that our sporting practices implicitly make room for acts of sporting supererogation. While Fowler and Di Canio were both widely praised for their acts of fair play, they would not have been widely criticized for taking advantage of the goalscoring situations they were presented with. This contrasts with other acts of fair play that are demanded from athletes. Football players are subject to criticism for deliberately misleading referees by simulating fouls. Golf players would be fiercely criticized for moving their golf balls with their hands to gain an advantage during a match. These acts would not only be criticized but criticized in a way distinctive to sport. These players would be accused of unsporting behavior, having no sense of fair play, and perhaps even of being a disgrace to the game. It appears, then, that there are basic norms of fair play that are demanded of athletes. There are also acts, like Fowler's and Di Canio's, that are praiseworthy from a sporting point of view for which athletes would not be blamed if they chose not to perform. There seems good reason, then, to accept the existence of sporting supererogation.[21]

There is also good reason to think that a set of sporting norms which allows for supererogation would be preferable to one that does not. There is good reason to want people to follow some basic norms of fair play when engaging in sport. As Bernard Suits (1978) has argued, sports are a subclass of games that involve physical skill and exertion. Sports and other games are activities that involve trying to achieve a particular aim, such as putting a ball in a goal or basket, by following rules that constrain what means you can take to achieve that aim. In football, for example, outfield players may not touch the ball with their hands, while in basketball players cannot hold the ball and run with it. It is these rules that make the game possible, and they allow athletes to challenge themselves and to test the limits of their physical abilities (Loland, 2002: 10; Simon et al., 2014: 47). We need to make sure these rules are followed then, in order to make these challenges possible. Referees are introduced to enforce these rules, but players are also expected to meet basic standards of fair play. This involves not only following the official rules of the game but also complying

[21] Though see Borge (2021) for a response to these arguments.

with informal norms such as kicking the ball out of play when a player is injured in football.

We need, then, a set of basic norms of sporting conduct that athletes can be expected to comply with in order to make the valuable challenges of sport possible. However, there is also good reason not to make these norms too demanding. There is good reason, for example, not to make it a norm of football for players to always attempt to change the referee's mind when they believe the referee has made a mistake which favors their team. One problem that would arise from such a norm is that players who are able to look at the game in an impartial way would be more likely to call for the referee to overturn decisions than those who view the game in a more biased way. This norm, then, may undermine the aim it is seeking to promote, namely, the fair application of the rules to both teams.

In addition, this norm would also require players to view the game more from the perspective of a referee rather than the perspective of a player. This is a problem, as these are different ways of viewing sport that require different focuses of attention. An official should attend to whether the rules are being broken, while a player should attend to how to how they should respond to the play and what their next move should be. A norm that players should always seek to overturn referee calls in their favor that they disagree with would require players to change their attentional focus so that it incorporates an official's view of the game. In doing so, it may lower the general level of sporting performance. Just as there is good reason to ask athletes to comply with some basic norms of sporting behavior, there is also good reason not to make these norms demand that athletes act in the most sporting way possible.

While there is good reason not to demand that athletes act in a maximally sporting way, there is also good reason to praise those athletes who exceed the basic level of sporting behavior expected of athletes. Asking players to make their own judgement about every referee decision may be damaging for sport, but it is still praiseworthy for players to ask referees to overturn mistakes benefitting their team in the case of particularly clear errors, such as Fowler's case. A set of norms for sporting behavior that leave room for sporting supererogation will then be preferable to a set of norms that leaves no room for the supererogatory.

7 Concluding Remarks

In this chapter, I have examined why a set of non-moral norms that make room for non-moral supererogation may be preferable to norms that require perfection in that normative domain. I began by considering Urmson's and Benn's arguments for the claim that a set of moral norms that leaves room for supererogation will produce better moral results than norms requiring moral perfection. I then argued that similar arguments could be given for a range of non-moral norms. Some basic prudential, epistemic, aesthetic, and sporting norms may also play an important role in our lives but if we want to promote these various kinds of value we have good reason not to require perfection in any of these areas.

Before finishing, it is worth noting the limitations of this line of argument. Urmson's and Benn's arguments rest on somewhat speculative empirical claims. While both provide plausible reasons for thinking that moral norms that require perfection would lead to a lower level of moral performance, this is nevertheless an empirical claim which could turn out to be false. That is not to say that this is an empirical claim that would be easy to test. Perhaps we could examine the effect of more demanding norms in some localized area of life but a wholesale study of the effect of making moral perfection a moral requirement on a societal level is unlikely to ever be feasible, nor likely to be approved by an ethics committee! The effects of demanding non-moral perfection are also likely to remain uncertain. While I have offered some reasons to think that requiring perfection could be damaging in these areas too, these considerations are a long way from a conclusive proof of the truth of this claim. Nevertheless, they do give us reason to be cautious about making non-moral norms more demanding, as this may frustrate the values that these norms are intended to promote. Moreover, the argument for non-moral supererogation that I have outlined in this chapter is not the only form of argument that may be offered for this position (see, for example, McElwee's contribution to this volume). So even if my argument is found wanting, there may nevertheless be good reasons to accept the existence of non-moral supererogation.

Finally, my discussion has focused on forms of non-moral supererogation that have already been discussed in the philosophical literature. It would be interesting for future work to consider whether similar arguments could be made in support of supererogation in other areas, even if these areas do not constitute a separate area of normativity from the moral, prudential, epistemic, or aesthetic. For example, in most workplaces it is useful to have a set of basic norms of conduct that employees can be expected to comply with that help advance the interests of their employers and their co-workers. Making these basic norms too demanding, though, could potentially undermine the values they are designed to promote. Demanding perfection may not be the best way to improve performance.

References

Archer, A. (2016). The supererogatory and how not to accommodate it: A reply to Dorsey. *Utilitas, 28*(2), 179–188.

Archer, A. (2017). Sporting supererogation and why it matters. *Journal of the Philosophy of Sport, 44*(3), 359–373.

Archer, A. (2018a). Supererogation. *Philosophy Compass, 13*(3), e12476.

Archer, A. (2018b). The problem with moralism. *Ratio, 31*(3), 342–350.

Archer, A., & Ware, L. (2017). Aesthetic supererogation. *Estetika, 54*(1).

Archer, A., & Ware, L. (2018). Beyond the call of beauty: Everyday aesthetic demands under patriarchy. *The Monist, 101*(1), 114–127.

Baron, M. (2016). A Kantian take on the supererogatory. *Journal of Applied Philosophy, 33*(4), 347–362.

Benn, C. (2018). The enemy of the good: supererogation and requiring perfection. *Utilitas, 30*(3), 333–354.

Benn, C., & Bales, A. (2020). The rationally supererogatory. *Mind, 129*(515), 917–938.
Borge, S. (2021). A critical note on sporting supererogation. *Journal of the Philosophy of Sport, 48*(2), 247–261.
Boult, C. (2021). Epistemic blame. *Philosophy Compass, 16*(8), e12762.
Crisp, R. (2013). Supererogation and virtue. *Oxford Studies in Normative Ethics, 3*, 13–34.
Dyck, J. (2021). There are no purely aesthetic obligations. *Pacific Philosophical Quarterly, 102*(4), 592–612.
Eaton, M. M. (2008). 'Aesthetic obligations. *The Journal of Aesthetics and Art Criticism, 66*(1), 1–9.
Fritts, M., & Miller, C. (2020). Must we be perfect?: A case against supererogation. *Inquiry*, 1–30.
Grigoletto, S. (2019). *Only through complexity. Morality and the case of supererogation.* Padova University Press.
Hale, S. (1991). Against supererogation. *American Philosophical Quarterly, 28*(4), 273–285.
Hedberg, T. (2014). Epistemic supererogation and its implications. *Synthese, 191*(15), 3621–3637.
Heyd, D. (1982). *Supererogation: Its status in ethical theory.* Cambridge University Press.
Heyd, D. (2011). Supererogation. In E. N. Zalta (ed.), *The Stanford encyclopedia of philosophy* (Winter 2011 Edition). http://plato.stanford.edu/archives/fall2008/entries/supererogation/
Kawall, J. (2003). Self-regarding supererogatory actions. *Journal of Social Philosophy, 34*(3), 487–498.
Kubala, R. (2018). Grounding aesthetic obligations. *British Journal of Aesthetics, 58*(3), 271–285.
Li, H. (2018). A theory of epistemic supererogation. *Erkenntnis, 83*(2), 349–367.
Li, H. (2019). How supererogation can save intrapersonal permissivism. *American Philosophical Quarterly, 56*(2), 171–186.
Loland, S. (2002). *Fair play in sport: A moral norm system.* London & New York: Routledge.
Matheson, B., & Milam, P. (2021). The case against non-moral blame. *Oxford Studies in Normative Ethics, 11*, 199–222.
McElwee, B. (2017). Supererogation across normative domains. *Australasian Journal of Philosophy, 95*(3), 505–516. https://doi.org/10.1080/00048402.2016.1253760
Mellema, G. (1991). *Beyond the call of duty: Supererogation, obligation and offence.* State University of New York Press.
Mill, J. S. (2001). *[1861]) Utilitarianism.* Hackett Publishing.
Muñoz, D. (2020). The paradox of duties to oneself. *Australasian Journal of Philosophy, 98*(4), 691–702.
Nado, J. (2019). Who wants to know? *Oxford Studies in Epistemology, 6*, 114–136.
Press, H. (1969). Aesthetic obligation. *Journal of Philosophy, 66*(16), 522–530.
Saito, Y. (2007). *Everyday aesthetics.* UK: Oxford University Press.
Shafran, R., Cooper, Z., & Fairburn, C. G. (2002). Clinical perfectionism: A cognitive–behavioural analysis. *Behaviour Research and Therapy, 40*(7), 773–791.
Simon, R. L., Torres, C. R., & Hager, P. F. (2014). *Fair play: The ethics of sport.* Westview Press.
Siscoe, R. W. (2022). Rational supererogation and epistemic permissivism. *Philosophical Studies, 179*(2), 571–591.
Suits, B. (1978; 2005). *The grasshopper: Games, life and utopia.* Broadview Press.
Urmson, J. O. (1958). 'Saints and Heroes' Reprinted in *Moral Concepts* Joel Feinberg (ed.) (1969) Oxford: Oxford University Press.
Widdows, H. (2018). *Perfect me. Beauty as an ethical ideal.* Princeton University Press.
Wolf, N. (1990). *The beauty myth.* Vintage.

Going Above and Beyond: Non-moral Analogues of Moral Supererogation

Brian McElwee

Abstract Apparent analogues of moral supererogation can be found in other normative domains, such as the prudential domain and the epistemic domain. Vindicating moral supererogation requires a convincing response to the challenge of the 'paradox of moral supererogation': if some act would be morally best, why would it not be morally required? Vindicating putative non-moral types of supererogation requires responding to analogous challenges: if some act would be best by the lights of some normative domain, why would it not be required by the lights of that domain's standards? I argue that the key to responding to such challenges involves giving a substantive account of what requirement is within the domain in question. The most promising type of account, I suggest, is what I call the Critical Reaction Account.

Keywords Supererogation · Non-moral supererogation · Non-moral obligations · Prudential supererogation

1 Introduction

The morally supererogatory is a familiar normative category. We speak quite readily of actions that go above and beyond what morality requires of us. Most striking are cases of heroism and saintliness. Someone risks their life for a stranger, or devotes themselves relentlessly to helping people in severe need. More prosaically, there are cases where a simple act of politeness or helpfulness, far from heroic or saintly, may go beyond what morality requires. Someone asked for directions takes the time to draw a map, rather than just pointing in the right general direction.

But there are contexts in which we speak of going above and beyond in ways that seem not to be a moral matter at all. A small but growing recent literature considers examples of non-moral analogues of moral supererogation.

B. McElwee (✉)
University of Southampton, Southampton, UK
e-mail: B.M.Mcelwee@soton.ac.uk

D. Heyd (ed.), *Handbook of Supererogation*,
https://doi.org/10.1007/978-981-99-3633-5_15

Here are some examples:

Epistemic: Someone reads an extra article on a philosophical question even after reading enough to form a justified view; reads an encyclopedia in the evenings instead of watching mindless TV; reads articles on quantum physics rather than just relying on the conclusions of experts. (Hedberg, 2014).
Etiquette: 'You must at least wear a shirt and jacket, but wearing a suit would be better' (McElwee, 2017).
Aesthetic/Vocational: An artist (e.g., Dmitri Shostakovich) continues to strive to produce great art at a time when many artists are being arrested or killed for non-conformity (Hills, 2018).
Prudential rationality: Someone shows exceptional self-discipline in resisting a short-term temptation in order to secure a larger long-term benefit; e.g., an athlete does extra training in awful weather to maximize chances of success (McElwee, 2017). An ambitious student of mathematics resists the temptation to go and build a snowman so that he can study for longer (Benn & Bales, 2020).

There are also cases which have a moral element, in that some practical good is done for the sake of another person, yet where some quite specific set of standards appears to be in play, distinct from generic moral standards, which the agent exceeds. For instance:

Neighborliness: You ask your neighbor to pick up your mail while you're on holiday. Unasked, they also water your plants in the dry weather.
Professional: A plumber goes beyond fixing the problem he was contracted to, and provides an extra service free of charge to an elderly customer.
Collegiality: An academic colleague consistently helps out when others are under pressure, takes on extra marking, volunteers for after-hours events, gives detailed feedback on colleagues' papers, etc.
Sportsmanship: A footballer urges the referee to reverse a decision to give his team a penalty, knowing it's been wrongly awarded. Or a player unilaterally stops play when he has a chance to score against an injured goalkeeper (Archer 2017).

In all these cases, we might well be inclined to say that the agent in question has gone above and beyond. They've gone beyond what they *must* do, by the lights of some set of standards.

In this chapter, I aim to do three things: first, to offer an overview of non-moral supererogation; second, to argue that a positive account of what obligations are in a given domain is necessary in order to establish that there's a type of supererogation in that domain; and third, to argue that what I call the Critical Reaction Account is the most promising account outlined to date.

In Sect. 2, I offer a structural account of how *moral* supererogation is best understood. Section 3 indicates how this structure can be instantiated in other normative domains. Section 4 fills this out by developing the Critical Reaction Account of obligation within a domain, arguing that this is the most promising way to vindicate variations of supererogation. In Sect. 5, I discuss Claire Benn and Adam

Bales' recent account of *prudential* obligation, as an example of an account of non-moral supererogation that doesn't say enough—I argue that it needs to be supplemented with the Critical Reaction Account or something similar. In Sect. 6, I discuss some challenges for the Critical Reaction Account emerging from Benn and Bales' discussion.

2 Moral Supererogation

Morally supererogatory acts are commonly characterized as ones that are morally good but not morally required.[1] This suggests that the category of moral supererogation emerges from the combination of two ways of morally classifying acts: (a) triaging acts into the *deontic categories* of morally required, the morally forbidden, or the morally optional (neither morally required nor morally forbidden) and (b) an *evaluative ranking* of the acts available in particular circumstances from the morally best to the morally worst.

Morally supererogatory acts are ones which (i) fall into the deontic category of the morally optional, and which (ii) figure sufficiently high up the evaluative scale of moral goodness.

This provisional characterization prompts two questions.

Firstly, how do the two types of classification interact? Intuitively, morally permissible acts (whether obligatory or merely optional) will generally, perhaps always, figure higher in the evaluative ranking than morally impermissible acts. Morally permissible acts are *morally better* acts than their morally impermissible alternatives.[2]

The second question is: How high in the evaluative scale is 'sufficiently high' to count as supererogatory? It might seem that the obvious answer is that it must be sufficiently high to count as 'morally good'. We might think this answer follows from the colloquial description of morally supererogatory acts as ones which are morally good but not morally required. But this isn't quite right. In some situations, I may have two morally permissible options—call them A1 and A2—which are equally morally good, and a range of further options all of which are morally impermissible. For instance, I am in a position to save one of two people in danger of drowning at no risk to myself, but cannot save both. In such cases, performing action A1 (saving the person to my left) is morally good, but not morally required, since I may permissibly omit A1 and do A2 (saving the person to my right) instead. Yet we would not describe A1 as morally supererogatory, since I do not *go beyond* what duty requires; I do not exceed the bare minimum that duty requires.

---

[1] In this chapter, I use 'requirement', 'obligation' and 'duty' interchangeably.

[2] Though see Pummer (2016) for a class of cases where there may be some temptation to dispute this.

To be supererogatory then, acts (as well as falling into the deontic category of being morally optional) need to be better than some permissible alternative—i.e., to figure higher on the evaluative scale than some morally permissible alternative act.

Examples such as these, as well as showing why the 'good but not required' formula is not quite right, also illustrate why a condition sometimes proposed as necessary for supererogation is mistaken. Trevor Hedberg, for example, following Gregory Mellema, says that in order for an act to be morally supererogatory, it is necessary that 'performing the act fulfills no moral duty'. (Hedberg, 2014, p. 4) But consider now a case similar to the one just described, but where one of the two morally permissible options is clearly morally better than the other. For instance, I am in a position to save one person at no risk to myself (call this B1), or three people, at a different location, at very significant risk to myself (call this B2). I cannot save all four people.

In this case, I have a disjunctive duty to do either B1 or B2. Doing B2 would be supererogatory, given the high degree of risk. But in doing B2, I *do* fulfill a moral duty; doing B2 is one way of discharging my disjunctive moral duty.[3]

One further issue should be mentioned: it is sometimes suggested that a necessary condition for an act's being morally supererogatory is that it is morally praiseworthy, or worthy of moral admiration. I will assume here that this is not a necessary condition of the core concept.[4]

Let us accept then the following definition:

Morally supererogatory acts: Acts which (i) fall into the deontic category of the morally optional, and which (ii) figure higher up the evaluative scale of moral goodness than some other morally permissible alternative.[5]

3 Supererogation in Other Normative Domains

We can use this definition of *moral* supererogation as a starting point for thinking about analogues in other normative domains. In light of the foregoing, the following schema looks promising:

X-wise (e.g., epistemically) supererogatory acts are ones which (i) fall into the deontic category of the X-wise (e.g., epistemically) optional, and which (ii) figure higher up the evaluative scale of X-wise (e.g., epistemic) goodness than some other X-wise (e.g. epistemically) permissible alternative (McElwee, 2017, p. 506).

---

[3] Note that one doesn't have a duty to do B1 here, since if one did B2 instead, one wouldn't have violated any duty by failing to do B1. So the duty is the disjunctive one to do B1 *or* B2. There are other cases, of course, where one has a duty to perform some specific act, and where some further, supplementary act is supererogatory. For instance, the Good Samaritan may fulfill a duty to provide immediate assistance to the person in need, before performing the supererogatory act of paying the innkeeper for further care.

[4] See Archer 2016 for argument for this conclusion.

[5] Cf. Heyd 2015; Archer 2018; McNamara 2011.

One issue worth considering is whether any interesting normative category (one that might be described as a type of supererogation) results from a mix-and-match approach to normative domains:

Acts which (i) fall into the deontic category of the P-wise optional, and which (ii) figure higher up the evaluative scale of Q-wise goodness than some other P-wise permissible alternative.

(i.e., cases where it's not the case that P = Q, as in the original schema, where a single type of normativity, X, is in play).

There may be interesting and important instances of such mix-and-match categories. For instance, one might wonder whether there are such things as *legally* supererogatory acts. One barrier to accepting the existence of such acts is that, while the deontic categories are perfectly familiar in the case of legal normativity—legally permissible, legally impermissible, legally required—there's no obvious sense in which available actions can be given a comprehensive evaluative ranking from a legal point of view. We don't readily speak of some legally permissible acts being legally better than some other legally permissible acts. Nevertheless, we might be interested instead in a mix-and-match category of actions such as those that are *legally* optional, and *morally* better than some *legally* permissible alternative. For example, consider two corporations developing a site in a residential area. If one corporation, without being legally required to do so, conscientiously and open-mindedly consults local residents about their plans, and the other corporation does not, then we might be inclined to judge that while both acted in a way that was legally permissible, the first acted in a way that was clearly better than the other, *going beyond* what was legally required. This going beyond seems to be moral, rather than legal: the corporation that performs the consultation acts morally better than the one that doesn't. (If the consultation is morally optional, then performing it is also a standard case of moral supererogation—but whether or not it is morally optional, it has the interesting status of being morally better than some *legally* permissible alternative.)

A related issue is that the boundaries between the moral and non-moral may not be clear-cut. The plumber, the neighbor, and the footballer arguably act *morally* better than is required by the standards of professionalism, neighborliness, and sportsmanship. I don't aim to settle here whether in some cases the relevant norms may be subcategories of moral norms. What does seem clear is that these more specific standards do have a life and shape of their own: there are independent *ideals* of good sportsmanship, good neighborliness, etc., and the people in the examples act notably well by the lights of *these* ideals.

All these issues merit further exploration. For the remainder of this chapter, however, our primary focus will be on cases where just one type of normative is in play, where P = Q.

One further issue is important to note. The account of moral supererogation, and the generic account of its putative non-moral analogues, are *structural*. And this structure can be instantiated in cases where we think the norms are mistaken. Amongst the examples already cited, we might have reservations about etiquette, for example. Are its norms ones that we have genuine reasons to comply with? More strikingly, some norms are unequivocally mistaken, and yet they can instantiate the structure outlined here. One may go above and beyond in 'being a good Nazi'. The fact that a set of norms fits the structure does nothing to vindicate that set of norms.

Let us return now to the putative instances of non-moral supererogation offered at the outset. Should we take these at face value? Are there any reasons for skepticism about analogues of moral supererogation in other normative domains?

When it comes to vindicating the category of *moral* supererogation, it is seen as crucial to be able to offer a response to the so-called paradox of supererogation: *If some action were morally best, why would it not be morally required?* We require some *explanatory story* about why some actions are not correctly regarded as morally required even when they are the morally best available.

So likewise when it comes to vindicating normative categories of non-moral types of supererogation, we need a satisfactory response to an analogous challenge in each case: *If some action were X-wise best, why would it not be X-wise required?* We again require some explanatory story about why some actions are not correctly regarded as X-wise required even though they are the X-wise best available.

This sort of challenge has been the focus of some recent discussions of non-moral supererogation—in particular, McElwee (2017) and Benn and Bales (2020). Each of these discussions takes its cue from promising ways of responding to the paradox of moral supererogation as the starting point for how to respond to analogous paradoxes of non-moral supererogation.

In the remainder of the chapter, I argue that the key to establishing supererogation in some normative domain lies in giving a positive account of what obligations are in that domain. The McElwee (2017) account, which I call the Critical Reaction Account, takes up this challenge—this is the focus of Sect. 4. I then argue in Sect. 5 that the Benn and Bales (2020) account of prudential supererogation is incomplete as it stands, precisely because it fails to give any substantive account of prudential obligation. Section 6 defends the Critical Reaction account against some concerns raised by Benn and Bales.

4 The Critical Reaction Account

McElwee (2017) takes the main challenge as one of providing conceptual space for the category of supererogation. Within a given domain, we need to offer a clear account of what is to be required or obliged. It is essential that an act's being required in the relevant sense (morally, prudentially, epistemically, etc.) is different from the act's simply being optimal by the lights of the relevant domain.

The starting point in thinking about the substance of a requirement is to note the *force* of claims of requirement. If one is required, or obliged, to do X, then one *must* do X. A claim that X is the optimal option amongst the alternatives is quite clearly *not* to say that it *must* be done. Such a claim at most makes a recommendation. 'Must' has more force than 'best', but what more can we say about this force? A natural question to ask is: *Must or else what?*

In distinguishing moral requirements from non-moral requirements, we might expect the distinction to lie in how we answer this question. That's the approach taken by the Critical Reaction Account. To say that someone is required to do something is to say that they must do it, on pain of meriting some critical reaction. And to say that someone is X-wise (morally, prudentially, epistemically) required to do something is to say that they merit a *particular* critical reaction distinctive of the domain in question.

In the case of moral requirements, a strong candidate for the relevant critical reaction is moral blame. To say that someone is morally required to do X is to say that, absent excuse, they would merit moral blame for failing to do it. The punchy force of a claim of moral obligation on this account comes from the prospect of being morally blameworthy: meriting anger, cold-shouldering, distancing from others, and perhaps feelings of guilt on one's own part.

The Critical Reaction Account has the virtue of offering a clear response to the challenge of the paradox of supererogation. Some action may be morally best, yet not be morally required because someone who failed to do it would not thereby merit blame. Risking my life by diving into cold and choppy waters to save a stranger might be the morally best thing to do, but it's not morally required, in virtue of being an action that one wouldn't merit any blame for refraining from. Giving 50% of my income to help the needy might be the morally best thing to do, but it's not morally required, because giving some smaller portion is sufficient to make blame inappropriate.

Turn now to the putative non-moral analogues. There are surely very many sets of standards with clear and distinctive evaluative rankings. We speak freely of what is good, better, or best, from the perspective of self-interest/prudence, epistemology, sportsmanship, aesthetics, and so on. What might be less clear, however, is whether there is any substantive conception of *obligation* or *requirement* in all those domains in which there is an intelligible and distinctive evaluative ranking.

For each putative type of non moral supererogation, we should ask whether there really are meaningful *deontic* categories within that domain which are sufficiently independent of the evaluative scale in question. On the Critical Reaction Account, we can answer this by attending to those critical reactions that are distinctive of a particular domain. Insofar as we use strong deontic language in that domain ('must', 'required', 'obliged', 'forbidden', 'wrong'), how do we respond to those who do what they X-wise *must* not do, and who omit to do what they X-wise *must* do? And are there instances within each domain where an agent may fall short of what's optimal, yet clearly doesn't merit such forceful critical reactions? When we can answer this question in the affirmative, we have identified plausible cases of supererogation for the domain in question. If the agent acts in a way that is X-wise *better* than some other act that was sufficiently X-wise good to avoid meriting the relevant critical reaction, then they act X-wise supererogatorily. The reader of extra articles and the athlete doing extra training in our introductory examples seem clearly to fit this bill, so their actions are instances of, respectively, epistemic supererogation and prudential supererogation.

So this type of account looks well-placed to answer challenges of non-moral paradoxes of supererogation. It is not my aim in this chapter to argue that no alternative account is plausible, but I do want to insist that *some* account of the nature of requirements in a given domain is necessary to vindicate non-moral instances of supererogation. It's quite striking that in most discussions of non-moral supererogation (and many discussions of moral supererogation) very little if anything is said about what the requirements or obligations in play are supposed to be. To put it in terms deployed by Richard Yetter Chappell, we should ask: *What is the line between permissible and impermissible supposed to signify?* (Chappell, 2020).

The distinction between the permissible and the impermissible seems clearly a different distinction between the optimal and the suboptimal. But what distinction is it? What, for instance, does it mean to say that something is *prudentially* obligatory? But what is the force of this alleged prudential must? How is a prudential must distinct from the moral must? Having nothing whatsoever to say in response to these questions seems to me to leave one well short of vindicating any new distinctive type of supererogation.

To say that the quietist approach is unsatisfactory is not to say that the Critical Reaction Account is obviously correct. It faces two related challenges:

Challenge 1. Can we really identify a relevant critical reaction for each putative type of non-moral supererogation?
Challenge 2. Will it be extensionally adequate? That is, is it plausible that those cases where we're inclined to say that something is X-wise obligatory are just those cases where the relevant critical reaction would be warranted?

I say more about Challenge 2 in the final section. What about Challenge 1? If we return to the putative instances of non-moral supererogation at the start of the paper, it certainly seems as if there are salient candidate critical reactions—ones which are plausibly warranted in response to falling egregiously short of what's best by the lights of the domain, and which are plausibly warranted in just those cases where we would be inclined to say that one *must* act in some way or other:

Epistemic: One epistemically *must* not form beliefs in ways that are likely to lead you make false assertions, or else one merits exclusion from the epistemic community—from the community of people whose testimony we rely upon. (McElwee, 2017, Sect. 3.3).
Etiquette: Etiquette is a very disparate phenomenon. Candidate critical reactions include exclusion from social acceptance, and in some cases feelings of disdain.
Aesthetic/Vocational: This is a tricky one. What kind of must is involved when Shostakovich says 'I must pursue my art'? Or when F. Scott Fitzgerald feels that he must focus on writing great novels rather than more lucrative but less artistically valuable work?[6] Or when someone feels a similar sense of vocational obligation in academia, sport, or charity? The most plausible candidate may be some sort of internal, first-personal reaction: *I must, or else I can't rest easy; I can't be at*

[6] Hills discusses the case of Fitzgerald in unpublished work.

one with myself. It's less obvious in these cases that any critical response from a third-party would be apt. (And of course, we might deny in specific cases that the first-personal feeling really is fitting.)

Prudential: One prudentially must not act in egregiously imprudent ways, or else one merits a charge of foolishness; in Mill's words 'a sentiment the opposite of admiration… distaste, or in extreme cases, even… contempt' (Mill 1859 [1997]). I discuss this further below.[7]

Neighborliness: If one falls short of the obligations of neighborliness, one is liable to acquire a bad reputation, to be gossiped about, to be shunned and excluded from neighborly help in return.

Professional/Collegiate Obligations: In some cases, there will be formal penalties—disciplinary procedures in the workplace. In other instances, legal penalties for breach of contract. In other cases, the penalty is a bad reputation—through low customer satisfaction ratings or word of mouth. The consequent loss of custom may be a 'natural penalty' for failing in some professional obligations.

Sportsmanship: Candidate critical reactions here include formal penalties, such as fines or suspensions, codified in a sport's rules—or more frequently perhaps, shunning from one's fellow professionals, booing from fans.

Each of these domains would require detailed thought and discussion in order to fully vindicate the existence of a distinctive type of supererogation. However, it's worth noting a couple of general issues raised by the examples:

First, it seems as if in some domains, there are *multiple* more or less punchy responses to more or less egregious departures from what is X-wise optimal. What should we conclude from this? Is there one privileged reaction in each domain which, like moral blame on the above account of moral obligation, gives a determinate shape to talk of obligation or requirement in that domain?

Relatedly, are these critical reactions able to set a sufficiently determinate cut-off point between the permissible and impermissible in that domain? We might think that in response to at least *mildly* suboptimal actions, some *mild* critical response is warranted. But where the appropriate response is a mild one, we will be less inclined to think the must-y talk of obligation or requirement appropriate.

One option is to conclude that if *any* punchy response distinctive of the domain in question is warranted, then one X-wise must perform the action. In settling on what counts as sufficiently punchy to undergird an obligation, we can be guided by our disposition to use must-y language in cases that seem to warrant the response. Only *punchy* critical reactions (i.e., nothing mealy-mouthed of the order 'you could do better' or 'that's a bit silly') match up well with strong deontic language like 'must', 'required', 'forbidden', 'wrong', etc.[8]

---

[7] An interesting case is that of purported aesthetic obligations. Alfred Archer and Lauren Ware (2017) endorse the sanctions approach to obligations. They postulate a distinctive critical reaction they call aesthetic blame which picks out the force of aesthetic musts.

[8] I omit 'ought' from this list of strong deontic terms. While it is sometimes used as the verb form of obligation (which is a strong deontic term), it (as well as 'should') is sometimes used instead to articulate a weaker claim about what there's strongest reason to do. To say that someone has

5 Benn and Bales on Prudential Supererogation

To bring the issues into focus, let's turn now to Benn and Bales' account of prudential supererogation. In this section, I'll argue that their account is at best incomplete, since it fails to give an account of what prudential obligations are. In the final section, we'll turn to concerns Benn and Bales raise about the Critical Reaction Account.

Benn and Bales defend the existence of prudential supererogation and aim to give a positive account of the phenomenon. They too see as the key to establishing the existence of prudential supererogation successfully addressing the challenge of the 'paradox of supererogation': If some course of action is prudentially optimal, why would it not be prudentially required?

Benn and Bales state that 'Resolving the paradox in a way that allows for the supererogatory is harder in the rational [prudential] case than the moral.' (Benn & Bales, 2020, p. 920).

One reason for thinking this, they claim, is that 'the literature is dominated by optimizing views of rationality, according to which we are rationally required to do what is rationally best. As supererogation entails permissible suboptimality, optimizing views entail that no acts are rationally supererogatory. The prominence of such views might leave us pessimistic about the prospects for rational supererogation.' (Benn & Bales, 2020, p. 920).

However, Bales and Benn are perhaps led astray by their idiosyncratic terminology here. While they are surely right to say that there is no ideal label for the phenomenon in question (no label that has no misleading connotations), 'prudential supererogation' strikes me as much less misleading than the term 'rational supererogation' they settle on. And what dominates various academic literatures is an optimizing view of a more *general* rationality (one is rationally required to do what has the strongest overall reason to do) or an optimizing view of instrumental rationality (one is rationally required to do what best fulfills one's goals), *not* an optimizing account of *prudential* rationality (one is *prudentially* required to do what best promotes one's own interests).

Bales and Benn offer a second, more convincing reason why resolving the paradox may prove tricky in the case of prudence. A standard response to the paradox of *moral* supererogation seems, at first sight at least, to be unavailable in the case of prudential supererogation. The response in the moral case is to observe that the primary cases where we are inclined to deny that some action is morally required, even though it's morally best, are cases where there's a significant prudential cost to the agent in doing what's morally best. It seems *too demanding* in many such cases to morally require the agent to impose such a significant cost upon herself. In the prudential case, however, the prudentially optimal act *just is* that act that is best for oneself, so performing that act cannot involve imposing (overall) costs on oneself. It, therefore, seems that putative prudential obligations cannot be 'over-demanding' in the way that putative moral obligations may be.

---

strongest reason to do X is not to make a strong deontic claim: it does not have must-y force, and does not implicate any punchy critical reaction for failure to comply.

Benn and Bales propose, however, that we reinterpret talk of over-demandingness: 'Perhaps over-demandingness is not specifically about prudential sacrifice, but about the sacrifice of something that is valuable by the lights of another domain of normativity.' (Benn & Bales, 2020, p. 921)

A claimed prudential obligation then might be rejected as 'over-demanding' in some case where it would involve doing something that is suboptimal by the lights of morality (or the epistemic or some other authentically normative domain).

They thus propose: 'Just as with morality, [prudential] rationality is surely not so demanding that it always requires us to make sacrifices by the light of some other domain. For this reason, perhaps we are, at times, [prudentially] rationally permitted to act [prudentially] rationally suboptimally.' (Benn & Bales, 2020, p. 921).

A couple of comments are in order.

First, it is surely quite plausible that I can be justified in doing what's prudentially suboptimal when, and just because, the prudentially optimal act would involve sacrificing something of non-prudential (e.g., moral) value. But need this always be because prudential rationality would be too 'demanding' otherwise? This is a natural enough description of those cases where the non-prudential value is something that one cares about. A prudential demand or requirement that I sacrifice the well-being of my child or the success of a project I care deeply about whenever it would most benefit me does indeed sound too demanding. It sounds distinctly odd, by contrast, to describe a putative prudential requirement as *too demanding* on the basis that aiding a complete stranger whom I didn't care about would be the morally best thing to do—but some such cases (e.g., where the benefit to the stranger is great and the prudential cost low) seem to be amongst the cases where we'd be inclined to say one isn't prudentially *required* to do what's best for oneself.

A second worry is that it seems strange to be so confident that '[prudential] rationality is surely not so demanding that it always requires us to make sacrifices by the light of some other domain' in the absence of any account of what prudential requirements *are*. Suppose someone disputed Benn and Bales' treatment of these cases. They might insist that when what's prudentially best differs from what's morally best, one is still prudentially required to do what's prudentially best, while acknowledging that one may not have most overall reason (or may not be all-things-considered normatively required) to do what's prudentially best. Without an account of what prudential requirements are (or what difference their presence is supposed to make), it's difficult to choose between these treatments. By contrast, the Critical Reaction Account has something substantive to say. For an agent to be prudentially required to do A is for them to merit the relevant critical reaction for failing to do A. But one *wouldn't* merit any critical reaction if one did what was morally best instead of what was prudentially best in the sorts of conflict cases Benn and Bales have in mind, and so we *can* confidently reject the claim that there's a prudential obligation.

Take the following example offered by the authors:

> Kavya, a doctor, could volunteer to provide free cataract surgery to those in need. Yet, in terms of her own well-being, Kavya would be marginally better off if she instead went on a writing course she has been longing to attend. (Benn & Bales, 2020, p. 922).

Benn and Bales conclude that in this case, it's not only morally permissible for Kavya to go on the course, but it's prudentially permissible for her to instead do the volunteering:

> Prudential rationality is surely not so demanding as to require that she sacrifice something of substantial moral value for the sake of the minor prudential gain of attending the writing course. (Benn & Bales, 2020, p. 922).

Again, it seems odd to be so confident in this sort of judgment in the absence of any positive account of what prudential demands or requirements are. It might help us to get a grip on the authors' concept if they provided some examples where an agent *is* prudentially required to do something, but they don't do so. (Plausibly in this case Kavya has the strongest *overall* reason to volunteer, given the goods at stake, and it could be argued that *that's* why it's inappropriate to insist that she's required in any sense to instead go on the course.)

It strikes me as crucial to give a substantive account of what prudential requirement is, for at least three reasons.

First, any response to a paradox of supererogation challenge requires a clear distinction between what's best and what's required in the relevant domain. In the absence of what we mean by, for example, prudential requirement, we cannot be sure that such talk of requirements is not just carelessly expressed reference to what's optimal in that domain. It's an open question whether, in some domain, there are any meaningful strong deontic categories to be had; perhaps all that's available is a ranking of actions from best to worst by the lights of that domain.

Second, we want to know how prudential obligation differs from moral obligation (which it presumably does, if there is some distinctive normative category of prudential supererogation). Moral obligation is a quite familiar category (even if its nature is not transparent to us). But prudential obligation is not. Consequently, it is incumbent on theorists to give some positive account. It's just not clear, absent further philosophical commentary, what we're talking about by bandying around the term 'prudentially required'. Distinguishing moral requirement from prudential requirement seems to me best done by appeal to the differing force of each—the first in terms of warranted blame, the second in terms of some distinct critical reaction (to be discussed further in Sect. 6). Those who disagree owe us an alternative account.

Thirdly, it is difficult to get any sense of why we should *care* about the existence of prudential supererogation in the absence of any clear grip on the concept of prudential obligation. Why is it important that some prudentially optimal acts are not properly regarded as prudentially required? What is it about the nature or function of prudential requirements—the role they play in our normative lives—that makes it wrong to so classify some prudentially optimal actions? (See Alfred Archer's chapter in this volume for relevant discussion.)

So it seems to me that Benn and Bales' account, which may be plausible as far as it goes, actually requires filling out with either the Critical Response Account of prudential obligation, or else with some alternative conception that addresses these three issues. The point here is not that the examples that Benn and Bales offer are implausible instances of prudential supererogation. And we need not deny their central claim that instances of prudential supererogation are to be found at points of 'normative conflict' (where what is optimal by the lights of one normative domain, such as morality or prudence, differs from what is optimal by the lights of another normative domain). The issue rather is that these explanations are incomplete, and need supplementing with a positive account of prudential obligation.

Our focus has been on prudential requirements, but just the same issues can be raised about putative requirements in any other normative domain. In order to vindicate the existence of supererogation in some normative domain, we will need to give some account of what talk of requirement, permission and prohibition amounts to within that domain. For instance, we cannot vindicate the category of the epistemically supererogatory until we give some positive account of what epistemic obligation is. The characterization of epistemic obligation I favor is in terms of what merits exclusion from the epistemic community (the community of people whose testimony we rely upon). We have specifically epistemic obligations to refrain from belief-forming mechanisms that are liable to lead us to form false beliefs and consequently make false assertions. We *epistemically must* refrain from this, on pain of meriting epistemic exclusion. Alternative characterizations may be possible, but quietism about what epistemic obligation is seems to me a non-starter.

6 A Challenge to the Critical Reaction Account

Toward the end of their discussion, Bales and Benn offer a critique of the Critical Reaction Account of prudential obligation and prudential supererogation defended in McElwee (2017). Ultimately, they conclude that if the Critical Reaction Account works, then so much the better: it's a way of establishing the existence of prudential supererogation that need not be in conflict with their own. This seems correct. However, for the reasons cited above, it strikes me that their account actually *requires* supplementation with the Critical Response Account, or else with some alternative substantive account of what prudential requirements are.

In any case, the challenges they raise in this section are good ones, whose lessons apply in thinking about any type of supererogation.

Here is how Benn and Bales frame the McElwee (2017) account of prudential obligation:

'an action is rationally [prudentially] obligatory only if failure to perform it merits the 'distinctive prudential criticism' of 'foolishness, or simple imprudence' (McElwee, 2017, p. 509). Further, McElwee claims, this charge is appropriate only when a person acts in a 'grossly deficient' manner with respect to their prudential interests (McElwee, 2017, p. 210)...

(P1) It is only fitting to charge someone as foolish or imprudent if she acts in a manner that is grossly deficient at promoting her well-being;

(P2) If it is not fitting to charge someone as foolish or imprudent for making a decision than it's rationally [prudentially] permissible for her to make that decision.' (Benn & Bales, 2020, p. 933).

In response to this, they worry about equivocation in the depiction of the relevant critical reaction as a charge of 'foolishness or imprudence', and present a nice dilemma. The root of the problem is that a charge of foolishness is clearly punchier than a mere charge of imprudence.

'Suppose that a friend goes for a walk on a warm spring day without taking water, knowing that she won't get lost and that the weather won't get oppressively hot. Such behavior does not involve gross deficiency in self-care but is nevertheless imprudent. Gross deficiency is not required for a charge of imprudence.' (Benn & Bales, 2020, p. 933).

This seems quite plausible. Consequently, (P1) should really read 'It is only fitting to charge someone as [prudentially] *foolish* if she acts in a manner that is grossly deficient at promoting her well-being'. However:

'Now take (P2): It is plausible that if someone acts impermissibly then it must be fitting to charge her with imprudence. However, it's far less clear that the same is true of foolishness: plausibly, one can act rationally impermissibly without deserving the extremely harsh judgment that they acted foolishly. Indeed, it seems perfectly coherent to say, for example, that there are times when it would be rationally [i.e., prudentially] impermissible to forgo sunscreen when going for a walk but where it would be a bit much to call someone a fool for doing so.' (Benn & Bales, 2020, p. 933).

The charge is that P1 only holds if the stronger charge of foolishness is in play, while P2 only holds if the weaker charge of imprudence is the operative one. So the McElwee (2017) view only looks appealing because it unfairly plays on equivocating between weaker and stronger critical responses in the blurry phrase 'foolish or imprudent'.

How should the Critical Reaction Account respond to this?

First, the basic point about ambiguity is well-taken. A charge of foolishness does sound harsher than a charge of imprudence, so we shouldn't equivocate when aiming to set a boundary for prudential impermissibility.

Second, the key for the Critical Reaction Account is to pick out *some* critical reaction that's appropriate in response to deficiencies in prudence sufficiently serious that we'd be inclined to say that one *must* not act that way. There may not be a word that perfectly articulates that attitude in all contexts. As noted in McElwee (2017), John Stuart Mill, in his famous and very eloquent discussion distinguishing moral criticism from prudential criticism, seems to be fishing around trying to find the right description for the attitude in question, without perhaps being able to nail it down:

'If he is eminent in any of the qualities which conduce to his own good, he is, so far, a proper object of admiration... If he is grossly deficient in those qualities, *a sentiment the opposite of admiration* will follow. There is a degree of folly... which, though it cannot justify doing harm to the person who manifests it renders him necessarily and properly *the subject of distaste, or in extreme cases, even of contempt*' (*On Liberty*, 100–1, emphasis added).

Third, 'foolishness', the harsher charge, is a better candidate term than 'imprudence' for picking out the threshold of acts that are sufficiently imprudent that we're inclined to say one must not perform them.

What Benn and Bales say about imprudence seems correct. What they say about foolishness appears much less convincing, for just those reasons cited in the previous section.

Again, they say: 'plausibly one can act rationally impermissibly without deserving the extremely harsh judgment that they acted foolishly. Indeed, it seems perfectly coherent to say, for example, that there are times when it would be rationally impermissible to forgo sunscreen when going for a walk but where it would be a bit much to call someone a fool for doing so.' (Benn & Bales, 2020, p. 933).

In this case, 'foolish' does indeed seem too strong. But, contrary to Benn and Bales, I'd be ill-inclined to say they've done anything impermissible. At worst, they've been mildly imprudent. It seems to me that those acts where it's appropriate to say that one *mustn't* do them (for prudence-related reasons) are just those acts that are so imprudent that they'd be *foolish* things to do. Of course, my intuitions, and linguistic and attitudinal dispositions, might just differ from those of Benn and Bales. But the crucial point is that such differences cannot be settled in the absence of any positive characterization of prudential obligation whatsoever. Reference to a charge of 'foolishness' is the best attempt to pick out the *attitude*, common enough in the course of our everyday lives, that we have in response to especially egregious cases of prudentially suboptimal behavior. We have no such reaction, and we would never call 'foolish', acts that are only minimally prudentially suboptimal, where the adverse consequences for the agent are really quite minor.

The attitude in question, and a label such as 'foolish', help give determinate content to the concept of prudential obligation: content which distinguishes the concept both from the concept of the prudentially optimal, and from the concept of moral obligation.

Again, the lacuna in Benn and Bales' account is that, in offering no positive elucidation at all, it provides no guide to the distinctive content of the concept.

7 Conclusion

We speak readily of actions that go above and beyond in ways that are not moral. I have argued here that the key to vindicating any putative non-moral analogue of moral supererogation lies in giving some positive account of what obligations or requirements are in the domain in question. For now, I have suggested, the Critical Reaction Account's template looks most plausible.

Many interesting issues remain concerning supererogation in non-moral domains. If, as I have proposed, there are critical reactions distinctive of particular domains, we might ask likewise whether there are laudatory reactions distinctive of each domain. Are there different types of admiration, praise, or other positive reactions, that are appropriate in response to impressive degrees of prudence, epistemic fastidiousness, and so on? These intriguing questions should be addressed in future work.[9]

References

Archer, A. (2016). Are acts of supererogation always praiseworthy? *Theoria, 82*(3), 238–255.
Archer, A. (2018). 'Supererogation'. *Philosophy Compass, 13*(3).
Archer, A., & Ware, L. (2017). Aesthetic supererogation. *Estetika, 54*(1), 102–116.
Benn, C., & Bales, A. (2020). The rationally supererogatory. *Mind, 129*(515), 917–938.
Chappell, R. Y. (2020). Deontic pluralism and the right amount of good. In D. Portmore (ed.), *The oxford handbook of consequentialism* (pp. 498–512). OUP.
Gibbard, A. (1990). *Wise choices, apt feelings*. Harvard.
Hedberg, T. (2014). Epistemic supererogation and its implications. *Synthese, 191*(15), 3621–3637.
Heyd, D. (2015). 'Supererogation'. *The Stanford Encyclopedia of Philosophy* online. https://plato.stanford.edu/entries/supererogation/
Hills, A. (2018). Moral and aesthetic virtue. *Proceedings of the Aristotelian Society, 118*(3), 255–272.
McElwee, B. (2017). Supererogation across normative domains. *Australasian Journal of Philosophy, 95*(3), 505–516.
McNamara, P. (2011). Supererogation, inside and out: Towards an adequate scheme for common-sense morality. In M. Timmons (ed.), *Oxford studies in normative ethics* (vol. 1). OUP.
Mellema, G. (1991). *Beyond the call of duty: supererogation, obligation, and offence*. Albany, SUNY.
Pummer, T. (2016). Whether and where to give. *Philosophy and Public Affairs, 44*(1), 77–95.
Skorupski, J. (2010). *The domain of reasons*. OUP.

[9] I presented some of the material discussed here to philosophy colleagues at the University of Southampton at a Departmental Research Day and at meeting of our Normativity Group. I'm grateful to the attendees for very helpful discussion. I'd also like to thank Alfred Archer, Jonathan Way, and David Heyd for very useful feedback on a draft of this chapter.

Feminist Perspectives on Supererogation

Katharina Naumann, Marie-Luise Raters, and Karoline Reinhardt

Abstract It is commonplace by now that moral philosophy has a long history of gender biases, not only regarding the pertinent moral issues but also regarding the development of concepts and theories. This insight from feminist philosophy, however, has not yet received sufficient attention in the debate on supererogation. That is not least surprising, since we all are familiar with the phenomenon that what is morally expected of an agent is not gender-neutral but at least to some extent relates to gender roles and social expectations. Therefore, the main aim of this chapter is to build on theoretical and methodological insights from feminist philosophy to carve out this research gap and to map the field for future research. It will be argued that feminist approaches can only be ignored within the debate on supererogation at the cost of philosophical rigor since they do not only give rise to novel perspectives and insights on questions already discussed, but also make matters visible that would have been overlooked otherwise. Thereby this chapter seeks essentially to contribute to a more extensive understanding of the social, political, and epistemic dimensions of supererogation that have hitherto been neglected.

Keywords Blame · Care ethics · Epistemic injustice · Feminist ethics · Feminist social epistemology · Heroism · Moral saints · Oppression · Praise · Post-heroic feminism · Resistance · Social expectations · Stereotyping · Supererogation

---

K. Naumann (✉)
Otto-Von-Guericke University, Magdeburg, Germany
e-mail: katharina.naumann@ovgu.de

M.-L. Raters
University of Potsdam, Potsdam, Germany
e-mail: mlraters@uni-potsdam.de

K. Reinhardt
University of Passau, Passau, Germany
e-mail: karoline.reinhardt@uni-passau.de

D. Heyd (ed.), *Handbook of Supererogation*,
https://doi.org/10.1007/978-981-99-3633-5_16

1 Introduction

> All the moralities tell them that it is the duty of women [...] to live for others; to make complete abnegation of themselves, and to have no life but in their affections.
>
> (Mill/ Taylor Mill, The Subjection of Women, 272f.)

It is commonplace by now that moral philosophy has a long history of gender biases, not only regarding the moral issues that take center-stage, but also regarding the development and justification of moral theories and their pivotal concepts (cf. Friedman, 2000; Held, 1998; Norlock, 2019). This insight from feminist philosophy, however, has not yet received sufficient attention in the debate on supererogation. Ever since the publication of James O. Urmson's seminal essay "Saints and Heroes" in 1958, the focus lies on the definition and justification of supererogation, the place it can be given in different moral theories and its application to particular moral issues. The neglect of feminist perspectives with regard to these questions is astonishing, since what is morally expected of an agent is not gender-neutral but relates to gender roles and social expectations, as has already been pointed out in various philosophical debates (cf. Baron, 1987; Calhoun, 1989; Held, 1998; Holroyd, 2021; Thomson, 1971). Furthermore, the impact of gender roles seems to be particularly striking regarding ascriptions of heroism and saintliness, the paradigm cases of supererogation. Therefore, a systematic analysis of supererogation from a feminist perspective constitutes a research gap, or so we will argue. Accordingly, we seek to build on theoretical and methodological insights from feminist philosophy to highlight this gap, to sketch the challenges and potentials from the outlined observation for the category of supererogation, and to map the field for future research. The article, thereby, contributes to a more extensive understanding of the social, political, and epistemic dimensions of supererogation.

In this article, however, we do not take a supererogationist or anti-supererogationist stance. Quite honestly the authors of this article do not agree on this matter. However, we do agree that feminism provides theoretical and methodological insights that can only be ignored at the cost of philosophical rigor. We are aware that there is no unified account of supererogation. Since the publication of "Saints and Heroes", the debate on supererogation has evolved in numerous directions, each putting emphasis on different aspects. We cannot do justice to each and every single text ever written on supererogation in this article; what we can and will do, however, is to provide good reasons for why feminist approaches would give rise to novel perspectives on, and insights into, the field of supererogation research.

A similar caveat is necessary regarding feminism, since there also is no unified account of it, neither in public debates nor in philosophy. Feminism surely is not only a political movement, but also an intellectual endeavor, and as such it can be understood as "a critical theory centered around the category of gender; its historical constitution, and the relations of oppression it legitimizes and reproduces" (Benhabib, 1990, 199). Feminist philosophy has developed into a complex field of research that cannot be outlined here. Feminist philosophers have been concerned with a multitude of topics and they have done so coming from a multitude of theoretical backgrounds.

Therefore, there is neither a unified account of what feminist philosophy is nor is there *the* feminist position on any particular topic. Yet, despite all the differences, broadly speaking there are two aims that are widely shared in feminist philosophy: Firstly, it describes patterns of gender-based ideas, role models, and structures of oppression. Secondly, it asks normative questions, such as where does gender-based oppression take place, what can be done against it, and whether different rights and duties for women* and men* can actually be justified.[1]

Against this background, we will show that feminist approaches to supererogation not only reveal epistemic, moral, social, and political injustices done to women*, but they also help to point out certain difficulties as well as potentials of the category of supererogation itself that have not yet been sufficiently explored. Thereby feminist perspectives make matters visible that would have been overlooked otherwise in the debate on supererogation. That is because they promise to give rise to novel perspectives and insights on questions that are already widely discussed. Thus, Sects. 2, 3 and 4 will each take up one of the main questions of the debate on supererogation. In Sect. 2, we will allude to the problem of defining and justifying the notion of supererogation focusing on the way gender stereotypes are entrenched in the typical examples that are brought to illustrate it. Section 3 asks whether feminist ethics can make room for the supererogatory. Section 4 focuses on the application of the concept to particular moral issues and critically discusses whether there are frequently gendered double standards involved regarding the supererogatory types of action such as beneficence, volunteering, and forgiveness. Our problematization of the concept of supererogation from a feminist perspective thus takes place on three levels: The examples and paradigm cases employed, the concept of supererogation itself, and its application in social practices of praise and blame. Furthermore, feminist perspectives not only give rise to such criticism but also to new questions and they are able to build bridges to other debates relevant for supererogation as we will highlight in Sect. 5. In this vein, we will focus on questions that highlight the social, political, and epistemic dimensions of supererogation that have hitherto been neglected, before summarizing our findings in Sect. 6.

2 The 'Saints and Heroes' of the Supererogation Debate: A Feminist Analysis

In this section, we argue that the supererogation debate is no exception regarding the widespread problem within (moral) philosophy that in terms of the examples deployed it is interwoven with (implicit) notions of who the agents are that we

[1] The authors' understanding of gender is non-binary and non-essentialist. In this article, we are, thus, using an asterisk when we employ gendered notions (e.g. man*, woman*, mother*) to indicate that we are referring to how people are perceived and categorized by others. We want to draw attention to the fact that this perception might or might not coincide with their self-identification. For hero/heroine, we use hero*ines with the same impetus.

are talking about. We will lay this out in detail regarding the examples Urmson provides in his seminal essay "Saints and Heroes" (Urmson, 1958). We will show that women*'s achievements are naturalized and thus marginalized, that the notions of 'saints and heroes' are linked to normalizing ideas of what can and cannot be expected of somebody, and that these notions are to a large extent tied to conventional and stereotypical gender roles, and that further assumptions, e.g. able-bodiedness, are not made explicit. In the remainder of this section, we will argue that these observations do not only hold with respect to Urmson, but at least to some extent also for the subsequent debate.

Following Urmson, a person can be called a saint, "if he does his duty regularly in contexts in which inclination, desire, or self-interest would lead most men not to do it [...] by exercising abnormal self-control" (ibid., 200), or "without effort" (ibid.), or "if he does actions that are far beyond the limits of his duty" (ibid., 201). Parallel to the saint, Urmson conceives of heroes as follows: We can call a person a hero, "if he does his duty in contexts in which terror, fear, or a drive to self-preservation would lead most men not to do it [...] by exercising abnormal self-control" (ibid., 200), or "without effort" (ibid.), or "if he does actions that are far beyond the bounds of his duty" (ibid.). The first two types of 'saints and heroes' respectively can be explained in terms of fulfilling one's duty no matter how difficult this is or seems to be. Whereas these are, according to Urmson, only "minor saints and heroes" (ibid., 201), the third stage is the saint or hero "par excellence" (ibid.) and it is precisely their 'heroic and saintly deeds' he considers to be paradigmatic cases of supererogation, cases that cannot be explained in terms of merely fulfilling one's duty "in difficult contexts" (ibid.).

To illustrate what he means by the different types of heroic and saintly deeds he gives a number of examples, all of which involve some kind of self-sacrifice. Before turning to these examples, let us first turn to the telling class of self-sacrificing acts that he explicitly excludes altogether from the discussion. What has nothing to do with supererogation for Urmson, or even with morality in general for that matter, are actions motivated by affection. Therefore, for him, the sacrifices that mothers* make for their children should not be counted as supererogatory. For Urmson these are cases of "natural affection" (ibid., 202). In contrast, he wants to examine only cases that have as little to do with feelings as any other moral action. Some achievements and extraordinary deeds, e.g. the sacrifice of a mother* for her child, are thus excluded from the discussion by emotionalizing and naturalizing them.[2] This often leads to a misjudgment of women*'s achievements (cf. Gilligan, 1982 and Pieper, 1998, 95 1982; and with reference to the supererogation debate: Baron, 1987, 253) and excludes some of their extraordinary deeds from the realm of the supererogatory. And not only is the category of supererogation not applicable to these deeds, but there is also a complete exclusion of themes and questions of female* life and

[2] Whereas the exclusion of certain cases has been frequently noticed by feminist philosophers (cf. Friedman 2000, 221 and Held, 1998 94), to our knowledge his examples and the examples employed in the subsequent debate on supererogation have not been analyzed in detail from a feminist perspective so far.

experience from the moral-philosophical debate, such as for instance motherhood (Held, 1998, 94f.). It is a well-known theme of feminist philosophy that through an androcentric perspective, women*'s achievements are either not perceived at all or they are perceived but naturalized and thus deprived of moral praise and recognition.[3] What follows from this is an unfair representation of the achievements and accomplishments of women* and consequently a lack of social recognition and reward, often interlinked with concrete socioeconomic consequences.

Before turning to the particular examples that Urmson discusses, a second issue worth noting is that androcentric assumptions take normalization as their starting point. The term normalism refers to the determination of prescriptive norms via 'normalities' (Foucault, 1975). These normalities can be based on data and use averages and margins, but they can also be based on an assumed general understanding that is used without being precisely defined or described. The latter is what seems to be going on in "Saints and Heroes". For Urmson, the decisive factor in determining what is saintly and heroic at least regarding what he calls minor saints and heroes is going beyond what "most men" would or wouldn't do. The conceptual limits are thus inferred from an assumed majority behavior. The notion "abnormal self-control" also appears repeatedly in this context. For Urmson, there is something like "normal behavior" in a particular situation, that is what he assumes most would do.[4] Urmson, furthermore, relies frequently on a normality that is based on the capabilities and experiences of able-bodied men* of a certain age with a certain standing in society, etc. Other examples for normalizing tendencies are his restriction to male* examples to define what goes beyond the bounds of duty, his bias for actions that exemplify virtues that have been traditionally perceived as 'masculine' (cf. Baron, 1987, 253), and his neglect of categories of female* action that could be defined as supererogatory, as we will show in the next paragraphs.[5]

Finally, addressing Urmson's examples for different types of saintliness and heroism in more detail will additionally reveal problematic stereotypes. For the "minor saintly and heroic deeds", that is acts that go beyond what most of us would do, he refers on the one hand to "the unmarried daughter [who] does the saintly deed of staying at home to tend her ailing and widowed father" (Urmson, 1958, 200), and on the other hand "the terrified doctor [who] heroically stays by his patients in a plague-stricken city" (ibid.) as well as "the soldier who heroically does his duty in the face of such dangers that would cause most to shirk" (ibid., 202). Heroes

[3] Urmson does in fact assume that the sacrifices of a mother* for her child do not deserve moral praise but are "admirable in some different way" (Urmson, 1958, 202).

[4] The main point is not that Urmson speaks of "most *men*", when defining what minor saints and heroes are. We understand this is in line with linguistic conventions in the 1950s. However, as has been argued repeatedly in feminist philosophy, these conventions were and are in fact shaped by our understanding of normality: It is not a mere coincidence that languages that can make this distinction tend to use the masculine form as the standard pronoun.

[5] The normalizing tendencies in Urmson's essay relate the discussion of feminist perspectives on supererogation to the wider debate on whether normality also plays a crucial role in determining the limits of our duties and thus Urmson's conception of supererogation – a criticism anti-supererogationists frequently voice against his position (cf. Naumann, 2020, 42).

"par excellence" are others: "[T]he doctor who [...] volunteers to join the depleted medical forces in that [plague-stricken] city" (ibid., 201f.) and the soldier who "does more than his superior officers would ever ask him to do" (ibid., 202), for example sacrificing "his life by throwing himself on the grenade in protecting his comrades with his own body" (ibid.).

Several things are striking about Urmson's choice of examples. On the one hand, we have the unmarried daughter as the only example that features a female* protagonist. All other examples explicitly refer to men*. Since she is an example for saintliness, the first thing to note is that among the examples for heroic actions, there is not a single example that has a female* protagonist: There are heroes, but no heroines*. As has often been pointed out in feminist philosophy and gender studies, if women* want to be outstanding they have to be saintly, ideally also virgins: From Joan of Arc to the virgin queen Elisabeth I. to Malalai Maiwand—to Urmson's "unmarried daughter", thus severely limiting the possibilities of expressions and ways of life conceptually open to women*. Secondly, the only female* protagonist we find in Urmson's text is associated with the domestic sphere. She does not go out into the world; she stays at home confined to the household. Her situation is entirely subordinate to the life of a man*. She does not take on this task for a stranger; it falls to her as part of her relationship to her father. Contrary to her male* counterparts she does not have a special education or special skills; she did not choose a profession that comes with certain risks—or at least it does not matter for the argument of the text. Thirdly, the saintly deed done by the only woman* in Urmson's text consists in what feminist theory has called care-work: she is tending to her father, who is as we are told widowed. She takes on a task that would have fallen to the wife. In the absence of other women*, she is in charge of the task of tending to her father. Finally, the unmarried daughter is for Urmson an example for somebody who does a saintly deed suppressing self-interest, inclination, or desire. One wonders what these desires might be. Since we are told nothing about her education or her career, the key to the answer seems to lie in the information that she is unmarried. Urmson seems to think that not getting married demands enough of female* self-control to be considered "abnormal".

On the other hand, we have the soldier and the doctor as male* protagonists. Both of them, in their hero-par-excellence form, sacrifice or risk their lives. What is striking here is, for instance, as gender studies have pointed out, that there seems to exist a troubling connection between masculinity and death in literature, film—and philosophy. As Allen J. Frantzen, who employs Judith Butler's account of gender performativity to reinterpret medieval tales about saints, notes: "Real men finally transcend the body altogether, a performative gesture of their own, a repeated and ceaseless denial of the male body" (Frantzen, 1993, 467). They do not just endure the situation they are in, as the doctor who stays by his patients or the daughter who takes care of her father do; they perform an act, an act of negating their lifes. Furthermore, the self-sacrificing soldier is a powerful image of masculinity and at the same time of bellicose heroism. After all, it is not a civilian who saves others in this example; it is a soldier. Note further that Urmson's soldier does not face the enemy in this example. He throws himself on a hand grenade, which is a piece of war technology.

The emergence of the admiration of soldierly sacrifice regarding technology is historically closely interwoven with military mobilization in times of industrialized wars. Earlier notions of the war hero* either fearlessly fighting *mano a mano* or bravely leading his troops into battle no longer applied, so increasingly since the beginning of the twentieth century, we have experienced the heroization of the soldierly death brought about without enemy contact but by war technology. The specific type of heroic soldier, which Urmson imagines here and elevates to the ideal of heroism (the hero *par excellence*), was part of mobilization strategies in industrialized wars, wars that had no place for traditional heroes* but still needed 'man power'. The mobilization tactic referred to virtues and actions that were preconceived as particularly 'manly'—though the conditions of their application had changed dramatically. Thus, it is unclear whether the ideal of heroism that Urmson refers to here is well-founded, or only based on historically contingent ideas that are questionable from the perspectives of gender and peace studies respectively.

One might want to object that the three problems mentioned above—naturalizing, normalizing, stereotyping—might just occur in "Saints and Heroes" and thus tell us little about the debate on supererogation. However, this objection does not hold, because we find these phenomena in the wider supererogation debate, too. To give but a few examples for stereotyping: Gregory Mellema opens his *Beyond The Call of Duty* (1991) with a comparison of resistance as a soldier in the Third Reich against the orders of a superior with Urmson's example of the soldier who sacrifices his life to protect others and throws himself on a hand grenade. And he concludes (uncritically): "It seems much less controversial to claim that the soldier in Urmson's example performs an act of supererogation" (1991, 6) and that this act clearly contains elements of heroism. Rebecca Stangl also uses the case of the soldier to render supererogation plausible within a virtue ethics perspective (2020, 68). Yogendra Chopra questions the claim that the deed of the unmarried daughter is less praiseworthy than the soldier, but still takes it to be her duty not to marry and stay by her father (Chopra, 1963, 161f.). Strikingly, as Marcia Baron puts it, " [t]he favorite examples of supererogatory acts tend to be those which exhibit a remarkable degree of some masculine virtue" (Baron, 1987, 253). The only woman* who plays a prominent role in the supererogation debate is Mother Theresa, a nun (Curtis, 1981; Driver, 1992; Hale, 1991; Hare, 1981; Mieth, 2012; Saunders, 2015; Tomasi, 1991). So, even though various criticisms of the examples used in the debate have been voiced, for example, their classification or whether they are decisive or not, to this day their gender stereotyping content is rarely questioned in the debate on supererogation. Regarding normalizing tendencies in the debate, since normalism can be based on an assumed general understanding that is used without being precisely defined or described, what needs further scrutiny are claims that something is "obviously", "naturally" or "clearly" supererogatory. Finally, the naturalization of female* deeds is a rather widespread phenomenon, too. In his *Supererogation* (1982), David Heyd, for instance, writes about mothers in much the same way as Urmson did. He excludes all acts "which pertain to the sphere of natural relationship and instinctive feelings (which lie outside morality)" from the realm of supererogation and thus, he further explains: "A mother who makes a great sacrifice for her child is not strictly speaking acting supererogatorily" (Heyd, 1982, 134). The extraordinary deeds of women* keep getting naturalized and marginalized. Though, in more recent contributions to the debate, the deeds of women*, for instance as mothers, are not as explicitly downplayed, they are instead not mentioned at all. At first glance, this makes the texts appear more 'gender-neutral' and thus less gender biased, but in fact, it simply hides female* realities of life.

To sum up, Urmson's paradigmatic cases as well as his examples set the framework for the subsequent debate. This is not to say that other examples cannot be found, but Urmson's examples still have an impact on the debate on supererogation (cf. Raters, 2022, 46).[6] Thus, the current debate still carries forward a problematic legacy of stereotyping genders, marginalizing certain socio-demographic groups and prescriptive concepts derived from gendered assumptions of what can reasonably be expected of a person. We are not arguing that these difficulties cannot be overcome, but they must at any rate be addressed. One could counter these worries in various ways. One could choose examples different from the ones discussed here. And a feminist analysis of standard examples and gender-neutral examples would certainly be a first and important step. It will not, however, be sufficient to tackle the above-mentioned challenges. Our worry runs deeper. We would argue that the issues discussed raise fundamental questions with regard to the category itself. Thus, one might wonder whether a 'gender-neutral' understanding of supererogation would be possible at all and whether a feminist critique undermines the validity of the concept of supererogation. To deal with this more fundamental worry, one might either argue for the abandonment of the concept altogether, or one might propose its refinement. In the next section, we will hint at possible directions that such a refinement might take.

3 Can Feminist Ethics Make Room for the Supererogatory?

A dedicated feminist theory of supererogation is still a desideratum. We will not be able to provide such a theory here. However, what we will do is to take stock of existing attempts to do so, evaluate them critically, and point to further questions that would need to be addressed. Prima facie, two routes seem to be promising in providing a feminist account of supererogation. On the one hand, one can take feminist ethics as a starting point for supererogationism and thereby ask whether a certain moral theory can make room for the supererogatory. On the other hand, one can take supererogationism as a starting point for feminist ethics and thereby ask whether a certain theory of supererogation can make room for feminist concerns.

[6] As Heyd (2019) notes, the discussion of specific cases is "methodologically important" because "supererogation is a concept the applicability of which is controversial. [...] Examples [...] can definitely help in revising the various definitions of the concept as well as make a case for one or another of its justifications." Yet, he seems to neglect that the choice of examples is always prone to unwittingly and uncritically inherit prevalent norms. For a critical discussion of the methodological importance of examples in the debate on supererogation see Naumann (2017).

Since one cannot only find different conceptions of supererogation in the literature but also different feminist ethical theories, we obviously can only go by way of adhering to a small subset of theories to make our point.

As far as we can see, there is but one example of a broadly speaking feminist moral theory that explicitly tries to incorporate the notion of 'supererogation' and that is Michael Slote's Care Ethics. Care Ethics has for a long time been considered to be the paradigmatic feminist ethical theory. The external starting point of this theory was the research of Lawrence Kohlberg on the moral development of children and adolescents (Kohlberg et al., 1977). The research had at least three problems. Firstly, male* subjects were at 78% significantly overrepresented (Haan, 1978, 287). Secondly, it was found that "black adolescents had higher interpersonal than formal levels; whites reversed this pattern". And thirdly, control studies of adult probands found that female* probands mostly seemed to argue on the (according to Kohlberg) lower level of context-sensitive moral reasoning, whereas most male* probands argued at the (according to Kohlberg) higher level of principle-oriented moral reasoning (ibid., 278). Following Norma Haan, Carol Gilligan asserts a distinct female* morality. According to Gilligan, in women*'s moral judgments the determining factors are contextual characteristics, individual responsibilities, and personal relationships, whereas men* are guided primarily by general moral principles of justice (Gilligan, 1982). It should be emphasized that Gilligan does not claim "two moralities" but more carefully distinguishes a "perspective of justice" from a "perspective of care" (ibid., 20). Moreover, it was shown early that African male* probands also tend to argue in a care-ethical manner, which Sandra Harding explains in terms of social oppression (Harding, 1987). Later research showed that it is not as much gender as educational level and occupation or profession that determines the moral reasoning of a person (Kohlberg, 1984, 341 ff.; Walker, 1984, Nunner-Winkler, 1989). Nevertheless, since Gilligan and Haan the so-called "ethics of justice" is discussed as distinct male* ethics and "care ethics" as distinct female* ethics.

Explicitly following Gilligan's findings, Slote's theory is based on the premise that "the caring motivation is based on and sustained by our human capacity for empathy with others." (Slote, 2007, 4). Accordingly, he claims that "[a]ctions are morally wrong and contrary to moral obligation if, and only if, they reflect or exhibit or express an absence (or lack) of fully developed empathic concern for (or caring about) others on the part of the agent." (ibid., 29). But it should be emphasized that Slote, like most care ethicists, explicitly makes a universal claim for his ethics. On the dividing line between supererogation and duty, he writes: "The person who demonstrates *more* empathy than most people with fully developed empathy ever show—can be said to act in a supererogatory fashion" (ibid., 34). However, this definition is prone to four problems: Firstly, as Jason Kawall points out, one could question whether there is actually room for the supererogatory within this theory, because "there are not higher levels of human empathic concern available to us (given that rightness requires fully developed human empathic concern)" (Kawall, 2009, 185). Secondly, there is also a critical discussion of the 'nearest-and-dearest' principle of care ethics, according to which an "act of helping strangers can be seen as morally wrong because it shows a lack of fully developed empathic concern, say, for one's own spouse, parents, or children" (Slote, 2007, 29). A third problem is the premise of moral hero*ines, who are supposed to be especially capable of supererogatory action in contrast to normal actors because of "an unusual and unexpected gift for empathy and for transcending or limiting their self-concern". Thus, the problem of double standards for two kinds of moral agents arises (ibid., 2007, 34). Fourthly, Rodney Roberts labels Slote's supererogationism as "a sort of sentimentalism that is largely concerned with only the appearance of sentiment possessed by moral agents; an ethics of care according to which individuals do not really have to care", because supererogatory acts merely 'express' a high level of empathy according to Slote instead of demanding that the empathy be truly felt (Roberts, 2014, 601). Despite these problems, which certainly must be addressed, Slote's theory of supererogation is highly interesting from a feminist perspective.

There are, however, other approaches in feminist ethics. One example is Seyla Benhabib's discourse ethical approach. According to Benhabib, the contractarian theories of moral judgment presuppose an unrelated, impartial, autonomous (i.e., male*) self behind a veil of ignorance. However, because of their social role in society and family, women* are "not autonomous, independent, but by the same token, non-aggressive but nurturant, not competitive but giving, not public but private". Therefore "her identity becomes defined by a lack—the lack of autonomy, the lack of independence, the lack of the phallus" (Benhabib, 1986, 409f.). Thereby an "entire domain of human activity, namely nurture, reproduction, love, and care, which becomes the woman*'s lot in the course of the development of modern, bourgeois society, is excluded from moral and political considerations, and relegated to the realm of 'nature'" (ibid., 407f.). Against this background, she adds the "standpoint of the Concrete Other" to the moral "standpoint of the Generalized Other" in contractarian theories. But this standpoint and other forward-looking approaches of feminist ethics have not yet found any attention in the supererogation debate.

Whereas Slote has adapted feminist ethics for a theory of supererogation, conversely, there are to our knowledge two adaptations of theories of supererogation for feminist ethics. These are not worked out in detail, however. Therefore, these certainly would deserve further attention in future research as well as analyzing the potential of other theories of supererogation in this regard. A first example is Andrew Flescher who refers to Susan Wolf's essay "Moral Saints" (1982)—albeit only in a note—to support his thesis that "certain feminists, appropriate the category of supererogation as a way of advancing their conviction that positive duty is not always what should be overriding in an agent's conduct" (Flescher 2000, 14). Instead, feminists would "favor a strong sense of self-regard and question whether or not the disposition to behave altruistically is as worthy as has generally been supposed" (ibid.). Indeed, a feminist supererogationism could in this way use the category of supererogation to argue against traditional roles of women* in families or (religious or public) institutions and emphasize the moral value of self-regard and self-respect.

A second example is indirectly asserted when Gertrud Nunner-Winkler takes up a part of Heyd's book *Supererogation* in the section *Application Problems* in her

German anthology *Weibliche Moral* (engl. *Female Morality*; cf. Heyd, 1982, 172–183 in Nunner-Winkler 1991, 257–270). Under the heading *The Autonomy of the Individual*, Heyd here defends the premise that "individuals have the right to pursue their own ends, to satisfy their wants, and to try to realize their personal ideals" (Heyd, 1982, 172). Although this premise could certainly be made fruitful for a feminist plea against the confinement of female* agents to caring for others, Heyd obviously neither has this in mind, nor does he defend a feminist notion of 'supererogation' when he justifies a theory of supererogation based on this premise. The first negative justification is that "considerations of social good" do not necessarily have "more weight" over "considerations of individual good" (ibid., 173); the second positive justification stresses the "intrinsic value" of supererogation as "purely optional" moral actions (ibid., 174); finally, the "social aspects" are emphasized with the indication that "supererogatory morality adds love of one's fellow-beings to the duty of respect for persons" (ibid., 179). Although the capacity for such love is often considered to be a specific female* capacity, Heyd's justifications nevertheless does not allow the conclusion that he is attentive to the gendered perceptions of this capacity. Therefore, it would certainly merit further attention whether this justification carries some problematic notions with it or whether it could rather be appropriated from a feminist perspective. But there has been no debate as to whether there should be a feminist theory of supererogation, and if so, what conception of 'supererogation' it should be based on, if any, and what problems would need to be discussed.

Whereas so far we have focused on the ways naturalizing, normalizing and stereotyping often enter the philosophical debate on supererogation via the examples discussed (Sect. 1) and via biased conceptual analysis (Sect. 2), the next section will focus on the reverse: the application of the concept in and to our moral practice with a special emphasis on the lived experiences of women*. For that purpose, we will primarily look at some of the more mundane types of actions that are often assumed to be supererogatory such as beneficence and forgiveness (cf. Heyd, 1982).

4 Beyond the Call of Duty: Feminist Deconstructions of Supererogatory Types of Actions

From a feminist perspective, it seems obvious that members of marginalized groups are often held up to higher standards of beneficence than others; that they are rather expected to forgive wrongs done to them; and, finally, that they are expected to be especially grateful, if someone treats them well, especially if it goes beyond what they could usually expect. The main idea here is that oftentimes what they do or do not do is judged by different standards. Hence, even if they act in the same way as members of privileged social groups, unjustified double standards lead to the problem that their actions are less likely to be judged as supererogatory. Against this background we will sketch how deconstructing such practices does not merely reveal unjust structures of moral evaluation, that is unfair distributions of praise and blame, but also how this contributes to sustaining problematic gender roles and enforcing oppressive structures. For that purpose, we will assume that supererogatory acts are roughly those that are praiseworthy to perform but not blameworthy not to perform. Although in the philosophical debate on supererogation, it is controversial whether praiseworthiness is a necessary condition of supererogatory acts, we assume that even opponents of this view would concede that it is an important common feature of many supererogatory acts.[7] Therefore, the following considerations are structured along some relevant exemplary problems regarding our social practices of distributing praise and blame, starting with the latter.

One way of spelling out the problem of unfair distribution of blame along gendered hierarchies is to draw directly on the so-called 'argument of supererogation', which is used when a moral agent claims that she does not have to perform a certain act because it is not her duty (cf. Raters, 2022). In the first step, reclaiming this argument from a feminist perspective amounts to revealing double standards in its application. In the second step, this argument can then be used to reject unjustified gendered moral demands. Revealing these moral double standards will not only make unfair distributions of blame visible, but often also the severe social and political consequences resulting from them. The struggle for abortion rights is but one example for this. In her paper "A Defense of Abortion", Judith Jarvis Thomson highlights that far more is required of women* with regard to the legal treatment of pregnancy and the ban on abortion than most jurisdictions would require of any other group.[8] This injustice in the law shows, as she claims, "that the groups currently working against liberalization of abortion laws, in fact working toward having it declared unconstitutional for a state to permit abortion, had better start working for the adoption of Good Samaritan laws generally, or earn the charge that they are acting in bad faith" (Thomson, 1971, 63f.). Prima facie, this rather seems to be a side note by Thomson, since she is not actually concerned with the state of the law on abortion but rather with the moral status of abortion. Yet, what Thomson alludes to are moral double standards that might seem to be justified since they are legally endorsed. As such they are normalized and run the risk of becoming invisible.

Against this background she then uses the 'argument of supererogation' to argue for the moral claim that women* don't have the moral duty to save the life of their unborn children at any cost and that doing so is more than can be required of anyone; thus, undergoing an abortion should not be blameworthy. Her argument is based on the premise "that no person is morally required to make large sacrifices to sustain the life of another who has no right to demand them, and this even where the sacrifices do not include life itself; we are not morally required to be Good Samaritans" (ibid., 64). Thus, Thomson claims that there can be no duty for women* to carry a pregnancy to

---

[7] For an understanding of supererogation along these lines see for example and Mellema (1991) and Urmson (1958), whereas others deny that there is a necessary connection between supererogation and praiseworthiness, see for example Archer (2015) and Heyd (1982).

[8] *Nota bene*, the paper was published in 1971, two years prior to the Supreme Court decision Roe vs. Wade, conferring the right to have an abortion, yet today Thomson's observation surely is topical again after the Supreme Court has overturned Roe vs. Wade in 2022.

term if they have to make great personal sacrifices to do so.[9] Against the objection that doing so is a duty of rescue, she concedes that it would be "self-centered and callous, indecent in fact" not to save a person's life when one can do so with trivial effort, "but not unjust" (ibid., 61). Of course, Thomson's argument is far from uncontroversial, but there surely are further issues to which the 'argument of supererogation' could be asserted to reject unjustified gendered moral demands. Yet, in feminist applied ethics, it has only been taken up infrequently. It has for example been used regarding the problem that the (predominantly female*) nursing staff in the hospitals during the pandemic era was applauded empathetically but not paid for the many overtime hours under severe circumstances (cf. Cox, 2020 and Raters, 2021). Yet, it could for example also be employed regarding the question whether there is a duty to provide home care, which is also until today still predominantly done by female* relatives worldwide, to show that there can be no duty to do so if the agent would have to sacrifice her own life plans or overburden herself psychologically and physically. Thus, there seems to be much to do in further research in this regard.

Another way to spell out the problem of unfair distributions of blame is to allude to the "logic of moral language" (Calhoun, 1989, 403f.), thereby elucidating the reverse problem that members of privileged groups are often exempted from blame for their wrongdoing against members of non-dominant groups. There is a long-standing debate within feminist thinking about moral responsibility, on whether oppressors are responsible for oppressive wrongdoings in light of the following two problems. First of all, their wrongdoing "is in many ways 'normalized' such that participants and even victims become oblivious to it", and, secondly, "it is institutionalized and part of the very structure of society, which means that often it is carried out independently of anyone's harboring any bad attitudes toward persons in subordinate groups" (Superson, 2020).[10] Yet, the connection to judgments of supererogation has rarely been drawn explicitly. A notable exception to this is Cheshire Calhoun who points to the above-mentioned problem in "the logic of moral language", which she elaborates as follows: "'X is obligatory' means, 'Unless there are exceptional excusing conditions you are blameworthy and reproachable for not doing X.' And 'X is supererogatory' means, 'You are not blameworthy and reproachable for failing to do X and deserve special praise for doing X.' Unfortunately, this logic breaks down in abnormal contexts where individuals are routinely rather than exceptionally exempted from blameworthiness, and hence from reproach, for failing to do the obligatory." (Calhoun, 1989, 403f.). What is specific to abnormal contexts, following Calhoun, is that moral ignorance is not an individual failure but the norm. And in fact, one could argue that this gives us a good reason for excusing wrongdoers or at least gives rise to an expectation to forgive them, since it seems that they could not have

[9] Thomson's paper has been discussed rarely within the debate on supererogation. Yet it has been taken up prominently by Driver who takes the problem of abortion to shed new light on the category of 'subererogation', that is acts that are "bad to do, but not forbidden" (Driver, 1992, 286).

[10] For an overview over the varying positions on that question to which we cannot do justice here see Superson (2020). For a comprehensive study of moral responsibility from an intersectional perspective, analyzing the way power asymmetries result in unfair distributions of blame and praise, see Ciurria (2020).

known or only hardly could have known better. Yet, this entails the danger that those who go above the normal standard and do the right thing despite the widespread ignorance will easily conceive themselves as acting "supererogatorily, and hence deserving gratitude" (ibid. 403). Thereby they are prone to the same kind of ignorance as wrongdoers: they are impeded in seeing this as obligatory. The resulting problem then is that this contributes to the reinforcement of oppressive structures since in light of this "women have trouble sustaining their sense of what is owed them and find themselves feeling grateful when given their due. The logic of moral language dooms any attempt to sustain or convey the obligatoriness of X while simultaneously excusing most failures to do X" (ibid. 404).[11]

Against this background, let us now turn to the distribution of praise which *prima facie* may seem to be less problematic. Praise, as it seems, doesn't hurt anyone, so it doesn't come as a surprise that the debate on moral responsibility has so far been mainly concerned with blame (cf. Holroyd, 2021, 2f.). Yet, the preceding analysis already suggests that unfair distributions of blame can go hand in hand with unfair distributions of praise and especially that the latter are far from irrelevant in normalizing and reinforcing oppressive structures. As Jules Holroyd has recently pointed out, there is not only a "praise gap" in philosophical research, but also in the world. The examples she points to are familiar and show how stereotypes inform our social practices of praising, which therefore are "not insulated from social and political dynamics. Men are praised when women—especially women of colour—are not; white people are praised when people of colour are not" (ibid., 8): (i) with regard to parenting, men* are praised for their effort, whereas women* are not (ibid., 4); (ii) men*'s engagement in altruistic workplace behavior is praised and figures positively on their performance record, whereas women*'s is not, and withholding such behavior figures negatively in women*'s evaluations, although not in the case of men* (ibid., 6); (iii) we are quick to praise members of privileged groups for their participation or mere declaration of allyship in social movements fighting against oppression, whereas members of the oppressed group at stake are hardly praised if at all, even if their effort is bigger. And despite the fact that their participation in such movements is genuinely riskier for them. These cases might suggest that the main problem is merely one of undeserved praise or inappropriate withholding of praise which results in an unfair allocation of praise—which might be already problematic enough since it can lead to "forms of psychological oppression—including low self-esteem—that oppressed people may suffer" (ibid., 11). Yet, as Holroyd argues convincingly, the picture is much more complex: There may well be contexts in which women* or other members of oppressed groups receive more praise than others, oftentimes though due to the fact that they are held to lower expectations which are affected by social group membership. The received praise then "doesn't reflect the fact that [...] women were *really* better or (actually, unqualifiedly) meritorious" (ibid., 13). Accordingly, she concludes that "[i]t is the conjunction of the role of stereotyped expectations, and the asymmetric attributions of appraisal respect and esteem, favouring those in positions of privilege, that is problematic in the cases described " (ibid., 16). Holroyd does not

[11] For a pertinent discussion of the problem of misplaced gratitude see Card (1988).

relate her analysis to the idea of supererogation, but if we take our prejudiced and thus unreliable practices of praising as a sign of praiseworthiness and thereby at least in many cases as a sign of supererogation, two pertinent questions arise: Should we be more reluctant towards Urmson's idea that the "facts of morality" clearly point to a category of supererogation and hence that any moral theory that does not make room for it must be rejected? Or should we rather scrutinize how our practices of praising as well as our theorizing about supererogation can contribute toward overcoming oppressive structures? As has become clear by now, feminist philosophy holds valuable intellectual resources for this endeavor with regard to the examples employed in the debate, the concept of supererogation itself, and its application in social practices of praise and blame.

5 New Hero*ines? Insights from Feminist Epistemology, Social and Political Philosophy

In the three preceding sections, we took up the main questions of the debate on supererogation—the definition and justification of the concept of supererogation, the question whether certain moral theories can make room for the supererogatory and finally the application of the concept to particular moral issues. The main aim of this section is to show that feminist philosophy does not only have the potential of contributing to existing questions but also brings new questions about supererogation to our attention. For that purpose, we shall sketch two interrelated topics which to our mind deserve closer attention in this regard. On the one hand, the relevance of our epistemic situation for delineating the limits of moral duty. On the other hand, the question whether social and political progress calls for a heroic or for a post-heroic feminism.

"Feminist theory begins with epistemological critique", as Benhabib (1990, 200) once noted. In the preceding section we have already pointed to the relation between what we know and what we owe with a focus on the moral psychology involved, that is with a focus on the prevalent ignorance of privileged subjects and its broader implications for our moral judgments. Now, we turn to the question of whether our social position bears on our epistemic and moral duties and their respective limits. Feminist philosophers often contend that our knowledge is situated, i.e., social contexts or more specifically our social positions within these contexts shape our lived experience as well as our thoughts. Taking up this assumption, it is often claimed that members of oppressed groups are in a privileged epistemic situation with regard to the understanding of their oppression (cf. Harding, 1991).[12] This is not to say that they necessarily are in a better position or that it doesn't take an effort for them to understand their own situation, but that on average they do have an epistemic advantage *vis à vis* members of groups that are socially privileged.

[12] Obviously, there are different readings of this claim. For a recent discussion which also discusses the respective moral implications, see Dror (2022).

Ann Cudd, however, once noted, "[a]ll resistance begins with the recognition that there are serious injustices that can be addressed, and then must proceed to mitigate or at least protest the material and psychological harms" (Cudd, 2006, 221). With this in mind, one might wonder whether the epistemic situatedness has implications for epistemic and moral duties of the victims of oppression. Like Cudd, Ashwini Vasanthakumar argues that the oppressed do in fact have a duty to resist and that this is a duty against other victims of oppression. More specifically, she claims "that this duty is an instantiation of their duty to assist. In virtue of their experience of injustice, victims are epistemically privileged [...]. They are thus uniquely positioned to initiate and motivate resistance efforts [...]" (Vasanthakumar, 2018, 465f.). Others have argued, on the contrary, that the duty to resist is a duty to oneself (cf. for instance Hay, 2011).[13] The claim that victims of oppression have a duty to resist gives rise to an obvious objection, namely that such a duty would be too demanding to be morally required and runs the risk of 'victim-blaming'. Thus, many argue against such a duty, whereas at the same time, it is rarely denied that resistance would be morally desirable (cf. Vasanthakumar, 2020, 5). Yet, surprisingly the literature on this question lacks a sufficiently nuanced account of the supererogatory status of resistance either as an other-regarding supererogatory behavior[14] or possibly as a self-regarding supererogatory act.[15]

Furthermore, even if one assumes that the oppressed have an epistemic privilege regarding the knowledge of their own oppression, it seems that it is nevertheless not only hard to acquire this knowledge but that the oppressed also face severe disadvantages in sharing it. Therefore, we can easily connect the problem discussed above to epistemic injustice, which according to Miranda Fricker is "a wrong done to someone specifically in their capacity as a knower" (Fricker, 2007, 1).[16] Fricker distinguishes two forms[17] of epistemic injustice and both can help to illuminate further connections to supererogation: The first form is testimonial injustice, which occurs when a speaker "receives a credibility deficit owing to identity prejudice in the hearer" (Fricker, 2007, 28). Against this background, it may seem that even if the oppressed may have some kind of expert knowledge regarding their own oppression, it often doesn't receive sufficient uptake by the privileged due to widespread prejudices and the self-serving refusal of members of dominant groups to engage

[13] For an overview on the question whether the oppressed have a duty to resist and if so, whether this is a duty against oneself or against others see Superson (2020) and Vasanthakumar (2020).

[14] Cudd (2006, 199) is a notable exception since she discusses the application of Heyd's account of supererogation to the case of victim's resistance.

[15] The concept of self-regarding supererogatory acts was introduced by Kawall (2003) and surely deserves closer attention—not only in this regard.

[16] Even though Fricker coined this term and introduced the phenomenon into mainstream philosophy, it is important to note that there is a long history within feminist thinking of the problematic exclusion of certain epistemic agents from the practices of knowledge production. Especially black feminists and decolonial feminists have relentlessly pointed to forms of "epistemic violence" and "epistemic oppression" (cf. Collins 2000, Lorde 1984, Lugones/Spelman 1983, Spivak 1988).

[17] Yet, others have argued that there is a need for further distinctions to make sense of the complexity involved, see for example Dotson (2011).

with the experiences of the oppressed (cf. Pohlhaus, 2012). The second form is hermeneutical injustice wherein someone has "some significant area of their social experience obscured from collective understanding owing to a structural identity prejudice in the collective hermeneutical resource" (Fricker, 2007, 155). This is a possible explanation for the claim that having a privileged epistemic position has to be understood as an achievement. At the same time, it makes the idea plausible that sharing experiences among victims is important. One possible route from here would be to suggest that gaining sufficient knowledge about oppression is, in fact, a case of epistemic supererogation, that is to say, that it is epistemically praiseworthy but not epistemically required (cf. Hedberg, 2014, 3623). Whereas recently some authors have already taken an interest in epistemic supererogation (see Hedberg, 2014; Li, 2018 and McElwee, 2017), the main interest thus far, however, is to define and justify the introduction of this category in light of a supposed structural analogy between ethics and epistemology.[18] Yet, the moral implications have hitherto been ignored, but promise to be of interest—particularly regarding the aforementioned problems: Would it follow from the claim that gaining knowledge about one's own oppression is epistemically supererogatory, that acting accordingly, i.e., resisting oppression, is also morally supererogatory and especially praiseworthy, but not blameworthy to omit? And what would be the consequences regarding the above-mentioned debate on the moral responsibility of oppressors for oppressive wrongdoings: Should it also be regarded as epistemically supererogatory if the privileged gain sufficient knowledge about oppression and hence morally supererogatory if they counteract oppression, or could they nevertheless be seen as culpable for their own ignorance and thus blameworthy for omissions?

Pushing the analogy between moral and epistemic supererogation further one could also wonder whether there is a need for the category of 'epistemic heroism'. This notion has recently been introduced by Medina (2013) to better capture the social impact of figures like Rosa Parks and Sor Juan Inés de la Cruz, especially their role in overcoming epistemic injustices. For Medina, "epistemic heroes" are "extraordinary subjects who under conditions of epistemic oppression are able to develop epistemic virtues with a tremendous transformative potential" (Medina, 2013, 186), because "they come to epitomize the daily struggles of resistance of ordinary people" (ibid., 187). Therefore, they have the potential of facilitating social and political progress. Yet what is crucial for Medina is that success is genuinely dependent on "social networks and daily practices that *echo* that performance" (ibid.). So far, this approach has so not received attention in the debate on supererogation, but it would certainly be worthwhile to pursue it further.

The topic of epistemic heroism leads to the question whether we should strive for heroic feminism not only from an epistemic but also from a social and political standpoint, or rather abandon the idea of heroism from a feminist perspective: Dieter Thomä, on the one hand, is aware that heroism has been made a "male domain" (Thomä, 2019, 21). However, he does not see this as a reason to call for a post-heroic feminism. Rather, one should aim for an inclusive heroism (ibid., 22), since

[18] On epistemic supererogation see also Archer in this volume.

he believes that the democratic project is very much in need of hero*ines. What he wishes for is therefore that gender clichés would be overcome and heroism re-figured in an inclusive manner (ibid.). He sees particular potential for novel hero*ines in the great figures of the women*'s rights movements (ibid., 83). Thomä thus calls for a heroic feminism. Ulrich Bröckling, on the other hand, argues that though heroism has been associated with combat and war and, thus, appears for historically contingent reasons as a primarily male* domain (Bröckling, 2020, 36), stories about heroes* often negotiate "disparate notions of masculinity" (ibid.). Nevertheless, the gender order of heroic narratives is ultimately "binary and leaves little room for intermediate positions" (ibid., 37). Women* are often assigned the role of the (passive) admirer. Exceptions would, where they exist at all, from the Amazons to Brunhilde to Joan of Arc and Lara Croft, only confirm the "masculine hero code" (ibid., 38). In persons such as Greta Thunberg or Carola Rackete, he sees another type of heroine*. For him, they embody the hero*ines of virtue (ibid., 224). However, their fate is unfortunate as well, because they can only fail the standards that are applied to them (ibid., 223). Therefore, for Bröckling the main goal must be to leave behind any form of heroization. Neither the political nor the moral hero*ines pass his judgment, and he calls for a post-heroic feminism. Despite their deep disagreement on this matter, both Thomä and Bröckling observe a tendency to the heroization of victims in current debates. Does the heroization of victims generate a further type of hero*ine and what would be the basis for this: silent endurance or courageous, albeit perhaps futile, resistance? On the one hand, a closer look at victims might help in stripping heroism of its masculine*, bellicose, and elitist elements and might provide the basis for a different story of heroism. One that is much more inclusive than the notions that we find in, for instance, the supererogation debate thus far. On the other hand, this very approach is also criticized for upholding the notion of women* and marginalized groups as victims and thus for reinforcing problematic stereotypes. Therefore, it becomes clear once more, that feminist perspectives on supererogation are not bound to a supererogationist or anti-supererogationst stance but can be approached from different angles and taken in different directions. Feminist analysis can lead to abandoning or appropriating the concept of supererogation.

6 Concluding Remarks

Starting from the observation that what is morally expected of an agent is not gender-neutral, we have illustrated that there is a research gap in the philosophical debate on supererogation. Based on insights from feminist philosophy we have mapped the debate for further research: We have outlined that feminist philosophy can help to reach a fuller understanding of the category of supererogation and its philosophical as well as practical implications and that it contributes to a more substantial understanding of the social, political, and epistemic dimensions of supererogation. In particular, we have focused on the potential of a feminist analysis of the 'saints and heroes' of the supererogation debate, that is the examples that are employed in the debate on supererogation to make the category itself plausible; the question whether feminist ethics would be able to appropriate supererogation; a feminist deconstruction of types of action that are perceived in the debate on supererogation as particularly clear cases for supererogation; and whether perspectives from feminist epistemology, as well as social and political philosophy, would lead to different kinds of hero*ines. We have thus highlighted problems on three levels: the examples, the concept itself, and its application. We have argued that the current debate on supererogation still carries forward a problematic legacy of stereotyping gender and marginalizing the lived experience of women*. Gender is, of course, just one of many dimensions that impact what we describe as exceptional, saintly, or heroic, as we have pointed out throughout this article. We have focused on gender here but believe that feminist perspectives can help make implicit assumptions visible regarding other dimensions and their respective intersections as well. Concepts like intersectionality (Crenshaw, 1989), as developed in feminist legal theory and philosophy, appear to be especially promising and should also be pursued regarding the concept of supererogation. The extension to further dimensions and their intersections may again raise new questions and difficulties. Thus, further scrutiny is needed to address the topics sketched in this article regarding the whole spectrum of the examples, conceptual foundations, and action types discussed in the debate as well as the social and political consequences of their application. We have not taken a stance on whether our criticism would ultimately motivate the abandonment or rather a refinement of the category of supererogation and we are not arguing that the difficulties discussed in this article cannot be overcome. What we have shown, however, is that they have to be addressed as a matter of philosophical rigor as well as a matter of justice.

References

Archer, A. (2016). Are acts of supererogation always praiseworthy. *Theoria, 82* (3), 238–255.

Baron, M. (1987). Kantian ethics and supererogation. *The Journal of Philosophy, 84* (5), 237–262.

Benhabib, S. (1986). The generalized and the concrete other. Visions of the autonomous self. *Praxis International, 5* (4), 402–424.

Benhabib, S. (1990). The debate over women and moral theory revisited. In: H. Nagl-Docekal (Ed.), *Feministische Philosophie.* Wien/München: Oldenbourg, 191–201.

Bröckling, U. (2020). *Postheroische Helden.* Berlin: Suhrkamp.

Calhoun, C. (1989). Responsibility and reproach. *Ethics, 99* (2), 389–406.

Card, C. (1988). Gratitude and obligation. *American Philosophical Quarterly, 25* (2), 115–127.

Chopra, Y. (1963). Professor Urmson on 'Saints and Heroes'. *Philosophy, 38* (144), 160–166.

Ciurria, M. (2020). *An intersectional feminist theory of moral responsibility.* London: Routledge.

Collins, P. (2000). *Black feminist thought: Knowledge, consciousness, and the politics of empowerment.* London: Routledge.

Cox, C. L. (2020). 'Healthcare Heroes': Problems with media focus on heroism from healthcare workers during the COVID-19 pandemic. *Journal of Medical Ethics, 46* (8), 510–513.

Crenshaw, K. (1989). Demarginalizing the intersection of race and sex. *University of Chicago Legal Forum,* (1), 139–167.

Cudd, A. (2006). *Analyzing oppression.* New York: Oxford University Press.

Curtis, B. (1981). The supererogatory, the foolish and the morally required. *Journal of Value Inquiry, 15* (4), 311–318.
Dotson, K. (2011). Tracking epistemic violence, tracking practices of silencing. *Hypatia, 26* (2), 236–257.
Driver, J. (1992). The suberogatory. *Australasian Journal of Philosophy, 70* (3), 286–295.
Dror, L. (2022). Is there an epistemic advantage to being oppressed?, *Noûs*, 1–23, preprint.
Flescher, A. M. (2003). *Heroes, saints, and ordinary morality.* Washington, DC: Georgetown University Press.
Foucault, M. (1975). *Surveiller et punir: Naissance de la prison.* Paris: Gallimard.
Frantzen, A. J. (1993). When women aren't enough. *Speculum, 68* (2), 445–471.
Fricker, M. (2007). *Epistemic injustice. Power and the ethics of knowing.* New York: Oxford University Press.
Friedman, M. (2000). Feminism in ethics: conceptions of autonomy. In: M. Fricker, & J. Hornsby (Eds.), *The Cambridge companion to feminism in philosophy.* Cambridge: Cambridge University Press, 205–224.
Nunner-Winkler, G. (Ed. 1991). *Weibliche Moral.* Frankfurt a.M./New York: Campus.
Gilligan, C. (1982). *In a different voice.* Cambridge, MA: Harvard University Press.
Haan, N. (1978). Two moralities in action contexts: Relationships to thought, ego regulation, and development. *Journal of Personality and Social Psychology, 36* (3), 286–305.
Hale, S. (1991). Against supererogation. *American Philosophical Quarterly, 28*(4), 273–285.
Harding, S. (1991). *Whose science? Whose knowledge? Thinking from women's lives.* Ithaca: Cornell University Press.
Harding, S. (1987). The curious coincidence of feminine and African moralities. Challenges for feminist theory. In: E. Kittay, & D. Meyers (Eds.), *Women and moral theory.* Totowa/ New York: Rowman & Littlefield, 296–315.
Hare, R. M. (1981). *Moral thinking: Its levels, method and point.* Oxford: Oxford University Press.
Hay, C. (2011). The obligation to resist oppression. *Journal of Social Philosophy, 42* (1), 21–45.
Hedberg, T. (2014). Epistemic supererogation and its implications. *Synthese, 191* (15), 3621–3637.
Held, V. (1998). Feminist Reconceptualizations in Ethics. In: J. A. Kourany (Ed.), *Philosophy in a feminist voice. Critiques and reconstructions.* Princeton: Princeton University Press, 92–115.
Heyd, D. (1982). *Supererogation. Its status in ethical theory.* Cambridge: Cambridge University Press.
Heyd, D. (2019). "Supererogation". In: E. N. Zalta (Ed.), *The Stanford Encyclopedia of Philosophy.* https://plato.stanford.edu/archives/win2019/entries/supererogation.
Holroyd, J. (2021). Oppressive praise. *Feminist Philosophy Quarterly* 7 (4), Article 3.
Kawall, J. (2003). Self-regarding supererogatory actions. *Journal of Social Philosophy, 34* (3), 487–498.
Kawall, J. (2009). Virtue theory, ideal observers, and the supererogatory. *Philosophical Studies, 146* (2), 179–196.
Kohlberg, L. (1984). *The psychology of moral development. Vol. 2: Moral stages and the idea of justice.* San Francisco: Harper & Row.
Kohlberg, L. et al. (1977). Logical operational foundation of moral judgement. *Genetic Psychology Monographs* 95, 97–188.
Li, H. (2018). A theory of epistemic supererogation. *Erkenntnis, 83* (2), 349–367.
Lorde, A. (1984). *Sister outsider: Essays and speeches.* Trumansburg: Crossing Press.
Lugones, M. C., & Spelman, E. V. (1983). Have we got a theory for you! Feminist theory, cultural imperialism and the demand for 'the woman's voice'. *Women's Studies International Forum 6* (6), 573–581.
McElwee, B. (2017). Supererogation across normative domains. *Australasian Journal of Philosophy, 95* (3), 505–516.
Medina, J. (2013). *The epistemology of resistance. Gender and racial oppression, epistemic injustice, and resistant imaginations.* Oxford: Oxford University Press.

Mellema, G. (1991). *Beyond the call of duty Supererogation, obligation, and offence*. Albany: State University of New York Press.
Mieth, C. (2012). *Positive Pflichten*. Berlin/Boston: DeGruyter.
Mill, J. S., & Taylor Mill, H. (1869). The subjection of women. In: J. M. Robson (Ed.), *Collected works of John Stuart Mill, Vol. XXI: Essays on equality, law, and education*. Toronto: University of Toronto Press/London: Routledge and Kegan Paul, 260–340.
Naumann, K. (2017). Paradigmatische Fälle? Methodologische Überlegungen im Rahmen der Supererogationsdebatte. *Zeitschrift Für Praktische Philosophie, 4* (2), 141–168.
Naumann, K. (2020). *Die Kraft des Exempels. Eine kantische Perspektive auf das Problem der Supererogation*. Berlin/Boston: DeGruyter.
Norlock, K. (2019). Feminist ethics. In: E. N. Zalta (Ed.), *The Stanford Encyclopedia of Philosophy*. https://plato.stanford.edu/entries/feminism-ethics/.
Nunner-Winkler, G. (1989). Gibt es eine weibliche Moral? In: M. Haller et al. (Eds.), *Kultur und Gesellschaft*. Frankfurt a.M./ New York: Campus, 165–178.
Pieper, A. (1998). *Gibt es eine feministische Ethik?* München: Fink.
Pohlhaus, G. (2012). Relational knowing and epistemic injustice: Toward a theory of willful hermeneutical ignorance. *Hypatia*, 27 (4), 715–735.
Raters, M.-L. (2021). Samariter*innen in Weiß? Pflichten, Supererogationen und Dilemmata in den Krankenhäusern einer pandemischen Zeit. In: R. Kipke et al. (Eds.), *ZusammenDenken*. Wiesbaden: Springer, 89–119.
Raters, M.-L. (2022). *Das muss ich nicht tun! Das Argument der Supererogation und die Grenzen der Pflicht*. Berlin/Heidelberg: J.B. Metzler.
Roberts, R. C. (2014). Supererogation in an ethics of care. *Philosophia, 42* (3), 597–602.
Saunders, B. (2015). Is procreative beneficence obligatory? *Journal of Medical Ethics, 41* (2), 175–178.
Slote, M. (2007). *The ethics of care and empathy*. London: Routledge.
Spivak, G. (1988). Can the subaltern speak? In: C. Nelson, & L. Grossberg (Eds.), *Marxism and the interpretation of culture*. Urbana: University of Illinois Press, 271–313.
Stangl, R. (2020). *Neither heroes nor saints: Ordinary virtue, extraordinary virtue, and self-cultivation*. Oxford: Oxford University Press.
Superson, A. (2020). Feminist moral psychology. In: E. N. Zalta (Ed.), *The Stanford Encyclopedia of Philosophy*. https://plato.stanford.edu/archives/sum2020/entries/feminism-moralpsych.
Thomä, D. (2019). *Warum Demokratien Helden Brauchen*. Berlin: Ullstein.
Thomson, J. J. (1971). A defense of abortion. *Philosophy & Public Affairs, 1* (1), 47–66.
Tomasi, J. (1991). Individual rights and community virtues. *Ethics, 101* (3), 521–536.
Urmson, J. O. (1958). Saints and heroes. In: A. I. Melden (Ed.), *Essays in Moral Philosophy*. Seattle: University of Washington Press, 198–216.
Vasanthakumar, A. (2018). Epistemic privilege and victims' duties to resist their oppression. *Journal of Applied Philosophy, 35* (3), 465–480.
Vasanthakumar, A. (2020). Recent debates on victims' duties to resist their oppression. *Philosophy Compass, 15* (2), 1–8.
Walker, L. J. (1984). Sex differences in the development of moral reasoning. *Child Development, 55* (3), 677–691.
Wolf, S. (1982). Moral saints. *Journal of Philosophy, 79* (8), 419–439.

Supererogation in Christianity

Dimitrios Dentsoras

Abstract The philosophical origins of the concept of supererogation can be found in medieval discussions of actions that deserve extraordinary merit. These discussions focus primarily on the evangelical counsels of celibacy, poverty, and obedience, which Christian tradition has recognized as non-obligatory and especially efficacious ways of reaching perfection and salvation, ever since its early centuries. This chapter will provide a history of supererogation and the related counsels, primarily within the context of the Roman Catholic Church. It starts with the New Testament, where the counsels first appear, and the commendation of the counsels by the early Church Fathers. It then moves to St. Thomas Aquinas and discusses the main features of his understanding of supererogation, keeping an eye on how Aquinas' account relates to important theological challenges of his times. The chapter concludes by examining the Protestant rejection of the idea that some actions can possess superabundant merit, the Catholic response to this rejection, and the moral philosophical developments that came as a result.

Keywords Opera supererogationis · Evangelical counsels · St. Thomas Aquinas · Monasticism · Indulgences

1 Christian Origins of Supererogation

The first accounts of actions that are explicitly identified as possessing the two central features of supererogation—being non-obligatory and having exceptional moral value—appear in discussions of moral advice known as the *evangelical counsels* by early Christian theologians. The background for these discussions can be found in passages of the New Testament where Jesus identifies certain actions, such as giving away one's property or remaining celibate, as more secure and expedite ways of gaining salvation. The commandments-based moral framework of Christian ethics, which offers a fixed limit to the actions that count as obligatory, in addition

D. Dentsoras (✉)
University of Manitoba, Winnipeg, MB, Canada
e-mail: Dimitrios.Dentsoras@umanitoba.ca

D. Heyd (ed.), *Handbook of Supererogation*,
https://doi.org/10.1007/978-981-99-3633-5_17

to a seemingly open-ended view of moral value, which primarily consists in charity, seems to provide a good starting point for the existence of supererogatory actions. On the other hand, the discussions of counsels by the early Christian Fathers are often short and focus only on the specific actions that appear in the New Testament—poverty, celibacy, and obedience, which do not appear particularly representative of what most have in mind as supererogatory acts (and, to some extent, even morally praiseworthy acts).

A more extensive treatment of supererogatory acts appears in St. Thomas Aquinas' *Summa Theologiae.* Aquinas' discussion draws a clear distinction between the obligatory precepts of the Church and the optional counsels. It also gives some reasons why counsels are not obligatory, and a detailed listing of the various benefits they confer to those who perform them—and beyond. It is this last feature of the Catholic Church's *opera supererogationis*, the acts of virgins and martyrs, that make them distinct. The only true benefit of every act is related to its contribution to salvation. While keeping the commandments may lead to one's own salvation, performing supererogatory act creates a surplus of merit that can be used for the salvation of others—sinners who would otherwise be condemned. This special feature of the supererogatory acts of saints and martyrs was the basis of the Indulgences –the Catholic Church's contentious practice of distributing the accumulated merit of saintly actions according to its discretion, in exchange for money or participation in the crusades.

The selling of Indulgences was one of the central points of Luther's criticism during the Protestant Reformation. But Luther did not merely object to the excesses of the Catholic Church. Rather, he disputed the very idea of supererogatory acts, and the existence of actions that, once performed, merit salvation on their own (i.e., independent of God's grace), let alone possessing any extra merit. The Protestant objections to supererogatory acts were mostly focused on their religious aspects and the question of Salvation. But there is a broader philosophical point in the background, having to do with the idea of acts that go beyond the call of duty. If an action is beneficial, why would it be left out of the commandments? Moreover, why would God reward (and in a particularly generous way) actions that seem to be unnecessary and impracticable at a large scale, such as poverty and celibacy? Aquinas and other Catholic medieval theologians seem to be aware of these issues and concerned about them, to some extent. But their responses do not go far beyond defending the scriptural basis for the existence of counsels. As a matter of fact, the Protestant polemic against the selling of Indulgences, and the idea of supererogatory acts in general, seems to have brought about an abrupt conclusion to a move away from the monastic ethics of poverty and celibacy based on the counsels toward an emphasis on keeping the commandments, which had been taking place for a while during the medieval period.[1] While the existence of supererogatory acts based on the counsels and the scriptural authority of the doctrine was affirmed by the Catholic Church in the

[1] On this development, see Pinckaers (1995), 231.

Council of Trent,[2] the theological interest in the idea waned, as the Catholic Church gradually sought to move away from the excesses of the Indulgences, and, to some extent, the ethics of self denial that were intimately linked with the New Testament counsels. By the time the discussion of supererogatory acts attracted philosophical interest, in our period, it was divorced from the religious and scriptural associations that framed the original introduction of supererogation into moral discourse.

The following sections will attempt to offer a historical account of supererogation, from late antiquity to the early modern period. My focus will be on the Christian view of supererogation, starting with the New Testament in Section Two. Section Three will examine the treatment of the evangelical counsels, which provide the foundation of the Christian notion of supererogation, by early Church Fathers such as Ambrose, Tertullian and Augustine. The early Fathers focus primarily on the benefits of life-altering commitments, such as chastity, that formed the basis of the monastic movement and provided the canonical views that Thomas Aquinas leaned on in presenting his comprehensive treatment of the counsels and supererogatory acts in general. Section Four will give an outline of Aquinas' treatment of supererogation, and Section Five will address some of the philosophical challenges that the doctrine faces and attempt to draw some lessons from Aquinas' view on supererogation. Section Six will conclude with some comments on the Roman Catholic Church's response to the Protestant rejection of the doctrine of supererogation and its related practices, paying attention to the religious disagreements that exist in the background. While the Catholic doctrine of supererogation might appear too tied up to religious practice and quite foreign to our contemporary understanding of supererogation (for example, most of us do not think of virginity as a paradigmatic supererogatory act, or even as morally praiseworthy at all), looking at the debates it sparked could give us both a better understanding of the development of moral theory from the Middle Ages to the modern period and some interesting ideas about how to approach contemporary debates on supererogation.

2 The Counsels of the New Testament

The use of the term "supererogatory" for actions that are non-obligatory, yet contain exceptional moral value, can be traced back to the Roman Catholic Church's doctrine of *opera supererogationis* and the corresponding distinction between obligatory precepts or commandments (*praecepta*) and optional counsels (*consilia*). Etymologically, the term stems from the Latin translation of the New Testament and the parable of the Good Samaritan, in Luke 10:25–37.[3] After bandaging the wounds of a traveler that he found robbed and beaten, the story's Samaritan takes the unfortunate

[2] The Council of Trent was one of the high points of the Counterreformation. Although most the Council's decrees relate to other matters, there is a separate chapter on Indulgences at the very end of the Council's decrees (see Waterworth, 1848, 277).

[3] For a short account of the use of the term in the first Christian centuries, see Hruschka (1998), 94.

traveler to an inn. On the next day, the Samaritan pays for the stay and promises to come back at a later point and to reimburse the innkeeper for "anything you may require in addition" (*quodcumque supererogaveris*).

The term in the parable (it only appears once in the New Testament) does not have any moral meaning; it merely refers to the innkeeper's extra expenses. Moreover, the moral context of the parable concerns acts that are obligatory. The episode that Luke narrates starts with Jesus telling a legal expert that eternal life can be gained by obeying the Law and, more specifically, by keeping the two so-called *Great Commandments*: "Thou shalt love the Lord thy God with all thy heart, and with all thy soul, and with all thy strength, and with all thy mind; and thy neighbor as thyself." (Luke 10:27). Of course, one may read the parable as more than an illustration of a moral commandment and proclaim that the Samaritan's action went beyond what was required by duty and law, and that his coming back to pay the extra expenses is not something that could be expected, and is, in that sense, truly supererogatory.[4] Some have suggested that the parable's lesson is that moral worth goes beyond compliance with rigid rules, such as those that are contained in the Hebrew law that the parable's Priest and Levite (those who avoided any contact with the victim lying at the side of the road) adhered to.[5] But this may have to do with the open-ended nature of the command to love one's neighbor, and the ways in which the command can be translated into specific actions. In any case, there is no point in the parable that explicitly presents the Samaritan's actions as optional and non-obligatory.

The reference to *optional* morally praiseworthy action, on the other hand, appears in Matthew 19:16–21, which is the passage that has traditionally provided the base for the distinction between precepts and counsels in medieval theology. The passage's starting point is similar to that of Luke 10:25: Jesus responds to a young man's inquiry about how eternal life can be attained. Jesus' first answer is doing everything the Commandments require. But he does not stop there. After being incited by the young man ("All these things have I kept from my youth up; what lack I yet?" (19:20)), Jesus gives some advice on how to reach perfection: "If thou wilt be perfect, go and sell that thou hast, and give to the poor, and thou shalt have treasure in heaven: and come and follow me" (19:21).[6] Upon hearing this, the young man becomes disappointed

---

[4] See Derrett (1964), 30, who maintains that the Samaritan's obligation ends when he brings the victim to the inn and that his return to reimburse any extra expenses is above and beyond duty. Derrett's basis for the claim, and the analysis of the parable in general, is the Hebrew law at the time, although it is unclear that Matthew would have expected his readers to be aware of all the detailed legal requirements that might be applied to the case, or take them into account in learning something from the parable.

[5] On this thought, see Forbes (2000), who highlights the parable's emphasis on compassion and love, as the Samaritan's motivation.

[6] From the King James Bible. According to Aquinas (*Summa Theologiae* IaIIae, 108,4), in this passage Jesus presents not only the counsel of poverty, but also the counsel of obedience, which is implied by Jesus' exhortation that the young man follow him. One should distinguish between the obedience that Jesus requests when he asks the young man to follow him and obedience to the Commandments. The latter is obligatory and does not require any act of acceptance (since everyone is bound by the Commandments, which express the divine will on how human beings should act). The former is an optional and voluntary adoption of an attitude that goes beyond merely following

and goes away sorrowfully, obviously unwilling to give up his considerable wealth. This prompts Jesus to declare that "it is easier for a camel to go through the eye of a needle, than for a rich man to enter into the kingdom of God" (19:24).

Three features of Jesus' suggestion that the young man abandons his wealth and follow him—what became known as the counsels of poverty and obedience—stand out and provide the basis for the discussion of supererogation in Christian theology. The first is the optional character of the action ("*if* thou wilt be perfect"), and the indication that poverty does not merely lead to salvation but to something higher, namely perfection. This might give one pause: surely, none of the Commandments require that we abandon all wealth,[7] but doing so seems necessary for being perfect, according to Matthew 19:21. Is perfection something valuable on its own and separately from salvation? If that is the case, then poverty can be seen as a requirement for perfection, and therefore as obligatory, in a sense.

The Church's response to this was to deny a separate value to perfection and to insist that our only goal is the salvation of our soul and our eternal life, which can be reached *either* by adhering to the Commandments alone, *or* by actions that aim at perfection, in addition. But this opens a new question: why should one do something as radical (and difficult) as abandoning all material possessions, if the same result can be achieved in an easier way? The Church's answer to this, and the second important feature of the evangelical counsels, is that poverty makes salvation more secure and expedient. In that sense, Jesus' suggestion is nothing other than a piece of advice (thus the term counsel) on how one can secure salvation that becomes especially potent with the warning that money is, if anything, a hindrance to salvation ("a rich man shall hardly enter into the kingdom of heaven" (Matthew 19:23)). At the same time, Jesus notices that his advice is not for everyone. Some people are too attached to their wealth to abandon it completely.

The limited scope of the counsels and the fact that they are not for everyone is the third feature that Christian theologians point out. A clear expression of this appears in the passage that precedes the discussion with the rich youth, where Jesus discusses divorce (Matthew 19:3–12). In response to the clear prohibition of divorce for any reason other than fornication and of remarrying (Matthew 19:9; a central view of the Catholic Church and a clear departure from Mosaic Law), Jesus' students wonder if it is better, in that case, not to marry at all (presumably with the thought that, by marrying one binds themselves in a way that cannot be undone).[8] Jesus' response is rather enigmatic: "All men cannot receive this saying, save they to whom it is given. For there are some eunuchs, which were so born from their mother's womb,

commands. The person who chooses to adopt the counsel of obedience relinquishes his own plans and desires, and wholly places himself under the decisions of another (the one who is followed). In this sense, the counsel of obedience is an all-encompassing long-term life stance that largely eliminates individual choice and the pride and potential arrogance that might come with it.

[7] There are, of course, commandments against gaining wealth in immoral ways (e.g., against usury), and against amassing excessive wealth (e.g., against avarice). Also, there are obligations to share parts of one's wealth with others (almsgiving etc.). But these do not amount to adopting poverty, and do not intend to do so.

[8] On this point, see Plumptre (1879), 272.

[...] and there be eunuchs, which have made themselves eunuchs for the kingdom of heaven's sake. *He that is able to receive it, let him receive it*" (19:12). The Church interpreted Jesus' rather startling expression as a suggestion that a life of celibacy can both avoid the commitment of marriage and the obligations that come with it and also serve as a more expedient way to the kingdom of heaven. At the same time, Jesus recognizes that celibacy is not for everyone. This is not because universal celibacy would be disastrous for the human race. Rather, it is because few people have the will or desire to choose such a life (apparently, the general aversion to celibacy is also a natural and God-given attitude that ensures the survival of the species). A similar attitude is expressed by Paul in 1 Cor. 7:1–16. There, Paul claims that "it is good for a man not to touch a woman" and recommends celibacy, which he has chosen for himself. But he also clearly indicates that celibacy is not obligatory; rather it is only recommended, and (only) for those who have the "proper gift of God one after this manner, and another after that."

3 The Early Fathers of the Church

Despite its importance in the following centuries, the brief mentioning of the evangelical counsels of chastity, poverty, and obedience in the New Testament does not provide anything close to a moral theory of supererogatory action. One could, after all, simply place such actions within the more general framework of the New Testament's moral project, which is the abandonment of the stringent confines of the Mosaic law, and its replacement with the Christian Law of Liberty, which is based on the two Great Commandments of loving God and one's neighbor. In this case, poverty, celibacy, and obedience are ways of loving God and dedicating oneself to God.[9] These are open-ended duties that could be satisfied in many ways, and celibacy or poverty are two of the different ways of fulfilling the duty of loving God. More importantly, poverty and chastity were practiced by Jesus, who serves as the Christian moral paradigm.

With this in mind, it is not surprising that some members of early Christian communities saw the counsel as obligatory.[10] There is little evidence about how

[9] One might think that poverty can also benefit one's neighbors, since Jesus asks the young man to give his money to the poor, in Matthew 19:21. But, giving all that one has to the poor and then withdrawing from any economic activity is a rather inefficient way of helping others. If one wanted to help the poor, it evidently would be better to dole out smaller amounts over time, while keeping one's own source of income. In any case, the Matthew 19:21 passage is usually understood as advocating not merely helping the poor (itself a duty), but rather denouncing all worldly possessions and espousing a life of poverty (cf. Jesus' demand that his disciples do not possess any money, at *Matthew* 10:9). And such a life seems to have no positive practical consequences for others.

[10] For a discussion of the early Matthean communities with respect to property, see Nel (2015). Communities practicing poverty, or at least the communal sharing of limited property, as well as celibacy (in some occasions) had been existing in the Eastern Mediterranean for a while. Such where the Essenes, a religious group among the Jews of Palestine (together with the Pharisees and Sadducees), who resided in the Qumran region of the Dead Sea, and throughout Palestine. Philo of

widespread the practice of celibacy was among early Christians, but it must have been noticeable, since Paul mentions it as an issue that requires his attention, in 1 Tim. 4:3, where he criticizes those who forbid people to marry and turn celibacy into a duty.[11] It is not entirely clear why some overzealous early Christians might have thought this way.[12] Clearly, the idea that one should follow the example of Jesus has something to do with this. But there seems to be a further, and more controversial, thought behind the celibacy and poverty of early Christian zealots. This is the idea of a special, more strenuous, and demanding morality that applies to the initiated, or, at least, to those that have chosen a more devout version of Christianity. Such an understanding of morality has a parallel in rule-based ethical theory, including that of the Hebrew tradition, where moral precepts are adapted to one's position in life, which is itself the product of choice, as well as personal circumstances. According to this thought, Jesus's counsel "he that is able to receive it, let him receive it" (Matthew 19:12), does not introduce an optional action, but a duty that can (and should) be taken up by those who are worthy (in the case of early Christians, the most fervent devotees, which could have been Paul's target in 1 Tim. 4:3), and could be rewarded in an exceptional way, commensurate to the worthiness of those who take up such duties.[13]

This opens up an important question about the optional character of the evangelical counsels that can also be applied to contemporary cases of supererogation. Are the counsels "first order" optional, i.e., one may choose to follow them or not at any time, and they never become obligatory? Or are they "second order" optional, i.e., one may choose to make them obligatory for themselves (although they remain optional for others), since such actions are necessary for being a certain kind of

---

Alexandria draws a connection between the Essenes and similar earlier Greek philosophical sects, such as the Pythagoreans, although the similarities seem to be mostly limited to their practices (the main motivation behind the Essene insistence on purity had to do with the Scriptures and Jewish eschatology, rather than philosophical ideas, as with the Pythagoreans). The community of the Therapeutae in Egypt, also mentioned by Philo in his *De vita contemplativa*, is part of this group, too. For the Essenes, their practices and theological foundations, which have been greatly illuminated since the discovery of the Dead Sea scrolls, see Betz (1999). For the Therapeutae and their connection to the Essenes, Taylor and Davies (1998).

[11] Paul does not give any reasons for his repudiation. The most obvious response that celibacy, if universally adopted, would be catastrophic for humankind and society, would have held little appeal to the early Christian societies, which considered the end of the world to be imminent. If anything, Paul seems to think that the burden of celibacy would be too heavy for those who are not fit for it, and that such practices might discourage future converts.

[12] Part of the difficulty has to do with the unclear relationship between the communities Paul refers to and other similar groups that undoubtedly influenced them, such as the Essenes. Some aspects of the moral motivation of the early Christian groups seem comparable to the Essenes (for example, the idea that celibacy is compatible with the imminent end of the world). Some others might be more influenced by Neoplatonic philosophical ideas having to do with the nature of the soul and its detachment from earthly goods. For a discussion of the relationships between proto-monastic groups in Eastern Mediterranean, and corresponding debates in the Reformation, see Joest (2010).

[13] On the view that counsels are a special kind of duty (*officia perfecta*) that applies to ministers, see Ambrose, *De virginitate* 1.36, 3.10 (discussed in Hruschka 1998, 98).

person?[14] Similarly, is the extra moral value of the counsels attached to the actions themselves, or to being the kind of person that not only performs them, but, more importantly, would be able and willing to accept them as obligatory?

Both of these options seem to be supported by different readings of the evangelical counsels. Celibacy, for example, may not be obligatory to all, but it is for those who choose to become priests in the Catholic Church (in a way, "those who are able to receive it"), as is poverty for some monastic orders, whose increased importance for the doctrine of supererogation will be discussed further below. On the other hand, the recommendation of such acts of self-denial by the early Fathers of the Church focuses almost exclusively on the positive effects of the acts themselves, and not on the special value of taking them up as extraordinary obligations. In this respect, celibacy and poverty can function as "shortcuts" to a moral life and salvation, which might appear strenuous in the beginning, but pay off practical dividends.[15] The background thought in such a recommendation is the Neoplatonic idea that bodily pleasures and earthly possessions tie us up to this world and prohibit the spiritual progress toward God and salvation.[16] One can regulate such pleasures, which are in a way natural, but doing so might turn out to be more difficult than foregoing them altogether. Complete abstinence would be impossible with some pleasures, such as those related to food and sleep,[17] but not for sex and, to some extent, money. Seen this way, the evangelical counsels should function as prudential advice, which can ease one's way to salvation.[18]

---

[14] An example of such second order supererogation is being a donor (pledging a gift) and gift giving, or being a hero and sacrificing your life on battle. We may not be obliged to be donors or heroes, but donors and heroes are obliged to give gifts (and keep their pledge) and offer their lives to the country. Cf. Aquinas' indeterminate tone in, *Summa Theologiae* IIaIIae 185.6, in reference to bishops having money (and, indirectly, the counsel of poverty): "No one is bound to works of supererogation, unless he binds himself specially thereto by vow". Thinking of supererogatory acts as second order optional seems to fit well with how saints and heroes describe their own actions. According to them, they just did what "they had to do", although they do not think that the duties they took up should be universally required or even adopted.

[15] See Franks (2009), 118, about how poverty removes temptations, and could, therefore, make keeping the commandments easier. The idea of a shortcut to virtue, and of such an understanding of choices such as celibacy and poverty, seems to have roots in Hellenistic philosophy, especially Stoicism and Cynicism, which also sometimes recommend such acts. On this, see Dentsoras (2014).

[16] See 1 Cor. 7:32–33: "He that is unmarried careth for the things that belong to the Lord, how he may please the Lord. But he that is married careth for the things that are of the world, how he may please his wife." Tertullian *Ad uxorem* 1.3 interprets this as an indication that while marriage is acceptable, celibacy is preferable: "What, however, is better than this good, we learn from the apostle, who permits marrying indeed, but prefers abstinence; the former on account of the insidiousnesses of temptations, the latter on account of the straits of the times."

[17] In these cases, one should exercise moderation, or abstinence only for a limited time. Cf. Ambrose's warning against his sister's excessive fasting in *De virginitate* 1.95.

[18] See, for example, John Chrysostom, *De virginitate* 13.18, who insists on virginity being an advice, rather than command, which might save one from "a lot of toil and great struggles". Also, Barsanuphius and Joannes, *Epistulae* 368.10, which presents the counsels as advice about one's own interest (*sympheron*), indicating a prudential understanding of the counsels. Often, counsels such as virginity are presented in parallel with their corresponding precepts, such as upholding one's

Overall, the treatment of the counsels by the early Fathers of the Church provides little to indicate a theory of supererogation. While the early Fathers, for the first time, make an explicit reference to the two features that will become the core of supererogation—optionality and special/additional moral value, they do not engage into a detailed or systematic discussion of what sets the counsels morally apart from obligatory actions. In fact, most of the discussions of the counsels appear in the defense of celibacy by a number of authors, such as John Chrysostom, Tertullian, and Ambrose. While all authors are careful to note that celibacy is not for everyone, its presentation as a preferable or "better" state, especially when coupled with the strong recommendation against remarrying,[19] seems to indicate that such extraordinary yet not universally required choices can also be obligatory, for those who choose a particular kind of life, such as monks.[20] But, even in cases where one assumes extraordinary duties, there is an element of freedom, since the choice to become a monk is free and optional.

Christian theologians were keen to note that the Gospels expressed the Law of Liberty, which offers different ways of reaching the state of perfection and salvation, contrary to the Old Law that consisted of rigid rules and prohibitions.[21] The liberty that the Gospels allowed also seems to be the source of special value for extraordinary acts that are outside the proscriptions and prohibitions that everyone is subject to. In this sense, the evangelical counsels *stem out* of liberty, since there is no rule that prescribes them, and also *aim at* liberty, in the sense that they free one from the pull of earthly goods, such as money and sex, in favor of heavenly pursuits. The primary exercise of this liberty, in the case of Christianity, was through the taking up of vows and the self-imposition of the duties that certain forms of monastic life

---

matrimonial duties. According to the Catholic catechism (2.2.3.7.1), matrimony and virginity are different ways of repairing the moral damage caused by the original sin, which ruptured the original communion between man and woman. Virginity and the other counsels, therefore, seem to serve the same purpose as the precepts—restoring our relation to God and one another, although they do so in a particularly effective and praiseworthy fashion (cf. John Chrysostom, *De virginitate*, 10.2 for the superiority of virginity oven marrying).

[19] This is the case in Ambrose, *De viduis*, Augustine, *De bono viduitatis*.

[20] The idea that certain non-obligatory life choices also imply the assumption of duties might seem similar to cases where someone committing a patently supererogatory act, such as the hero risking his life by volunteering for a dangerous mission, while claiming that he merely did what he felt he had to do. But there seem to be some important differences. For one, being a monk *requires* that one practice celibacy *as a duty*, while being a hero does not require accepting life-risking actions as a duty. Also, it would be odd to think of being a hero as a life-choice, the way becoming a monk is. Actually, it seems that one *becomes* a hero by acting in certain ways (regardless of whether they think of them as something they had to do), while one is obliged to act in certain ways *after* becoming a monk.

[21] See Thomas Aquinas, *Summa Theologiae* IaIIae 108.1: "So to each one it is free to decide what he should do or avoid; and to each superior, to direct his subjects in such matters as regards what they must do or avoid. Wherefore also in this respect the Gospel is called the 'law of liberty' since the Old Law decided many points and left few to man to decide as he chose."

required.[22] Monasticism's progressive rise from the early Christian period into its heydays in the Middle Ages framed the discussion of the counsels, and the doctrine of supererogation in the Roman Catholic Church. The monastic life provides the background for Thomas Aquinas, who provides an extensive discussion of the counsels and supererogation more general, as part of his account of perfection, and the subsequent discussion of the religious state.

4 Aquinas' Opera Supererogationis

St. Thomas Aquinas was, in a number of ways, the first theologian to systematize the theology of the evangelical precepts, as well as the moral aspects of supererogatory action more broadly.[23] Aquinas uses the term "supererogation" throughout his works, primarily in his focused discussion of the difference between counsels and precepts in *Summa Theologiae* IaIIae 108.4, but also at various places in the *Summa* and other exegetical and polemical writings.[24] Much of what Aquinas says about the evangelical counsels, and supererogation in general, seems to reiterate the views of the early Fathers, who presented the counsels as optional life choices that have the practical benefit of making salvation more assured and expeditious by removing the temptations of earthly goods, but that also require a certain fitness of circumstance and strength of character, and are, therefore, not for everyone.[25]

Aquinas' remarks on wealth and the counsel of poverty are revealing of his general attitude toward supererogatory acts.[26] According to Aquinas' thought, wealth is not a bad thing by itself, in the sense that one does not need to denounce it in order to be saved. But rich people tend to cling too much to their money, to the detriment of

---

[22] A corollary of this idea is that the value of certain acts of supererogation lies in adopting a specific goal or role (which might last the whole of one's life), rather than the carrying out of the actions related to that goal (since carrying out the actions is necessary for having that role).

[23] On the role of Aquinas with respect to the theological aspects of the counsels, see Deák (2014).

[24] See for example, the polemical *Contra doctrinam retrahentium a religione* that discusses monasticism.

[25] On this canonical view, see *Summa Theologiae* IaIIae, 108,4: "we must understand the commandments of the New Law to have been given about matters that are *necessary* to gain the end of eternal bliss, to which end the New Law brings us forthwith; but that the counsels are about matters that render the gaining of this end *more assured and expeditious.*" For the practical argument in favor of adopting the religious life that consists in the observance of the counsels, Aquinas says: "Man is placed between the things of this world, and spiritual goods wherein eternal happiness consists, so that the more he cleaves to the one, the more he withdraws from the other, and conversely. [...] For man to gain the end aforesaid, he does not need to renounce the things of the world altogether, since he can, while using the things of this world, attain to eternal happiness, provided he does not place his end in them. But he will attain more speedily thereto by giving up the goods of this world entirely. Wherefore the evangelical counsels are given for this purpose." (*Summa Theologiae* IaIIae, 108,4) A presentation of the counsels along those lines is offered in the Catholic Encyclopedia, which presents the canonical views of the Roman Catholic Church (see Barnes 1913).

[26] On a detailed account of Aquinas' views on poverty and wealth, see Franks (2009).

the real spiritual goods. In fact, it is better to get rid of money altogether, and not be disturbed by it at all, than to try and attain eternal happiness while also attending to one's wealth.[27] In this sense, the supererogatory act of poverty assures a more secure and expedite course to salvation, since it gets rid of the potential temptation of wealth and the need to spend time concerned with money rather than spiritual goods. The role of counsels, therefore, is instrumental in the attainment of perfection, although perfection can also be attained by "merely" keeping the commandments.[28]

In general, Aquinas' treatment of the evangelical counsels is firmly rooted within the Patristic tradition.[29] But one can also notice, for the first time in such a clear way, two amplifications and developments in the notion of supererogation that is linked to the counsels. The first is an expansion of supererogation to include actions beyond the three counsels of chastity, poverty, and obedience.[30] Focusing mainly on their optional character, Aquinas presents as supererogatory actions that range from martyrdom to fasting to spreading the faith and refusing to accept alms.[31] The wide variety of actions that Aquinas presents as optional and supererogatory seems to be more in line with contemporary examples of supererogation that range from jumping on live grenades to tipping a waiter. It also signifies a broadening of supererogation outside the context of monasticism, where the evangelical counsels framed one's entire life, to "first-order" exceptional acts.

The second new feature is the amplification of the rewards associated with supererogatory acts. On top of making one' salvation more secure and expeditious, acts of supererogation can also accumulate moral merit that surpasses what is required in order to secure one's own salvation. The superabundant merit created by supererogatory acts is the source of an additional reward that befits those who demonstrated a great amount of determination and willpower in the abandonment of worldly goods, such as martyrs, virgins, and the preaching Doctors of the Church.[32] This reward was denoted by the Catholic Church through the "aureole", a luminous crown in depictions of saints, and accompanied by special honors reserved for

---

[27] See *Summa Theologiae* IIaIIae, 186,3: "Man is directed to future happiness by charity; and since voluntary poverty is an efficient exercise for the attaining of perfect charity, it follows that it is of great avail in acquiring the happiness of heaven."

[28] See *Summa Theologiae* IIaIIae, 184,3, where Aquinas presents both commandments and counsels as being instrumental to the attainment of perfection, which essentially consists in charity. For an instrumental interpretation of the counsels, see also Bonino (2002), 348.

[29] On the various aspects of this relationship, which has received a fair amount of disagreement among theologians and historians, see Deák (2014).

[30] In fact, Aquinas seems to leave open the list of potentially supererogatory acts. See *Summa Theologiae* Suppl. 96,11: "there are many works of supererogation in which one will rejoice in the life to come."

[31] The supererogatory nature of martyrdom and the spreading of the faith is discussed in *Summa Theologiae* Supp. 96, while the distinction between obligatory and optional/supererogatory almsgiving is mentioned in *Summa Theologiae* IIaIIae 32.5.

[32] The good that is abandoned in the case of preaching is freedom from the devil, in the sense that the Doctors of the Church choose to face the devil over and over again while preaching and tending to their flock, although they have managed to defeat the devil and temptation in their own lives.

martyrs and virgins.[33] Aquinas is careful to describe such rewards as non-essential, and to note that the supererogatory acts that merit the additional reward of the aureole belong to a "genus of the act that derives a certain praiseworthiness from its due circumstances".[34] But there exists an otherworldly benefit, and a corresponding praiseworthiness that accompanies non-obligatory acts of supererogation.

To this extraordinary reward associated with acts of supererogation, the Roman Catholic Church added a second controversial use for the good works of saints and martyrs. According to the Church, the superabundant merit created by supererogatory acts can be used not only toward the salvation of the person who accumulated it, but also for remitting the sins of others, at the discretion of the Pope and bishops. This was the basis of the institution of Indulgences in the Middle Ages. Ordinary believers could obtain some of the collected merit of the Church's saints in order to remit their sins, either by participating in the Crusades, or by contributing money to the Church.[35] Aquinas gives a brief defense of the theological validity of the Indulgences in *Summa Theologiae* Suppl. 25.1. The argument is based on four claims: (a) the merit of the saints' good work is more than is required for their own salvation; (b) this extra merit can be used for the salvation of others; (c) the extra merit is possessed by the "common body" of the Church, and (d) the Pope, as the head of the Church, has been given the power to distribute this extra merit to lessen the punishment of specific sinners (those that purchase the Indulgences).[36]

---

[33] Aquinas' discussion of the aureole and the acts that are awarded with it appears in *Summa Theologiae* Suppl. 96. Aquinas explicitly notes that the aureole is a *separate* reward that signifies a separate merit, *in addition* to that of perfection and the salvation that comes with it. "Accordingly, it must be said that an 'aureole' denotes something added to the 'aurea' a kind of joy, to wit, in the works one has done, in that they have the character of a signal victory: for this joy is distinct from the joy in being united to God, which is called the 'aurea'" (*Summa Theologiae* Suppl. 96.1).

[34] The reference to circumstance here is important. After all, it is only possible to become a martyr if there is a persecution going on, stay a virgin if you are not married off by your parents, etc. For a discussion of the role or circumstance in supererogatory actions, in ancient philosophy and early Christian writers, see Dentsoras (2014).

[35] Although some of the abuses of the institution of Indulgences have been recognized by the Roman Catholic Church, the practice itself was declared as having a solid scriptural and ecclesiastical foundation by the Council of Trent (Sess. XXV, 3–4 Dec. 1563). Its contemporary form is outlined by Pope Paul VI's bull *Indulgentarium Doctrina* of January 1st, 1967.

[36] "We must say on the contrary that indulgences hold good both in the Church's court and in the judgment of God, for the remission of the punishment which remains after contrition, absolution, and confession, whether this punishment be enjoined or not. The reason why they so avail is the oneness of the mystical body in which many have performed works of satisfaction exceeding the requirements of their debts; in which, too, many have patiently borne unjust tribulations whereby a multitude of punishments would have been paid, had they been incurred. So great is the quantity of such merits that it exceeds the entire debt of punishment due to those who are living at this moment. [...] Now one man can satisfy for another. And the saints in whom this super-abundance of satisfactions is found, did not perform their good works for this or that particular person, but they performed them for the whole Church in general. [...] Those things which are the common property of a number are distributed to the various individuals according to the judgment of him who rules them all. Hence, just as one man would obtain the remission of his punishment if another were to satisfy for him, so would he too if another's satisfactions be applied to him by one who has the power to do so." (*Summa Theologiae* Supp. 25.1).

Aquinas offers some scriptural support for two key aspects of the practice of Indulgences, which cover the last three claims above. The first is the common possession of moral merit, and the possibility of "satisfying for another" (i.e., serving the punishment for an action perpetrated by someone else).[37] The second is the authority of the Pope, which has been traditionally attributed to Jesus' giving the keys to heaven to Peter, in Matthew 16:19. On the other hand, Aquinas does not address the idea that became the focus of Luther's criticism during the Reformation, namely the thought that there is a threshold to salvation that can be exceeded (and the corresponding idea of superabundant merit). The idea of a moral threshold could make Aquinas' moral theory seem odd from a modern perspective. For example, Utilitarianism does not set a limit to the general obligation to increase common Utility. But the threshold idea seems to fit well with ancient eudaimonism, where the goal of human life and action, happiness, is only achieved once an appropriate amount of virtuous activity is achieved. Once that point has been reached (happiness, in the case of the ancients, salvation in the Christian case) Aquinas believes that one can assist and increase the happiness/salvation prospects of others, without any decrease of her own already achieved threshold. Supererogation is this acting in ways that go beyond the threshold (which is set by the moral precepts) and benefit others even when there is no such obligation.

We are now in a position to see at least a broad outline of an account of supererogation in Aquinas. Much of it is couched in and framed by theological notions related to salvation and divine authority, but we can identify some features that correspond to contemporary theories of supererogation. The main idea is that of a threshold associated with the essential reward of salvation, for which keeping the precepts is a necessary condition. On top of that, there is some extra moral credit that can be gained through acts of supererogation. These acts do not possess a sui generis value (after all, they also stem from a certain denial of worldly goods and aim at salvation), but are different with respect to their not being obligatory, since the withdrawal from worldly goods that is required for reaching the threshold of salvation can be reached by avoiding any trespass of the precepts. With these broad features in mind, we can assess the theological and philosophical merits of Aquinas' notion of supererogation and its broader effect on Roman Catholic moral theory.

5 Criticism of the Doctrine of Supererogation

Despite its canonical status, the doctrine of supererogation that we find in Aquinas was not without dispute, even within the Roman Catholic Church. The reasons for criticism varied: some were philosophical/theological—mirroring to an extent our contemporary debates about supererogation; some were scriptural; others political

[37] This possibility is explicitly discussed in *Summa Theologiae* Supp. 13.2. This possibility is a core tenet in Christianity, which makes it possible for Christ to suffer for our sins. For an interesting discussion of the concept from an analytic philosophical perspective, see Lewis (1997).

(especially those related to the role of the monastic orders and the Pope's granting of Indulgences). We can divide the challenges to the Catholic doctrine of supererogation into two broad groups. The first group, which we can label "internal," as it arose within the Roman Catholic Church, did not challenge the existence and scriptural authority of supererogatory actions, but rather focused mostly on their range, justification, and relationship to obligatory actions.[38] The second group of "external" attacks, was launched during the Reformation in the sixteenth century by Luther and his followers, and went further, in denying the very existence of acts that go beyond what is obligatory. Both kinds of challenges contributed to the steady decline of emphasis on supererogation, to the point that it became practically absent as a moral concept in the modern period. And while many of the reasons for this shift in emphasis are political and historical, having to do with the powerplay between the monastic orders, or the religious struggles of the Reformation and Counterreformation, one can find some interesting theological arguments that could be related to contemporary moral philosophical debates about supererogation.

The main internal disagreement regarding the doctrine of supererogation in the Roman Catholic Church concerns the relationship between the commandments and the counsels, as well as their corresponding goals. One apparent line of thought that seems to have gained some traction in the Roman Catholic Church essentially treats the counsels as an alternative moral code that only applies to the few that chose to adopt it, but is obligatory once one has chosen the life that is based on them (mainly, the monastic, or "religious" life). Unlike the obligatory precepts, which provide the necessary requirements for salvation, the counsels aim at perfection (cf. Jesus' comment "if thou wilt be perfect" in Matthew 19:21), and only apply to those who can and want to take them up as duties (cf. the suggestion "*he that is able to receive it, let him receive it*" in Matthew 19:12).[39] Precepts and counsels, in this case, form two parallel, and largely independent, moral schemes.[40]

Aquinas makes a distinction between salvation, which requires keeping the precepts, and perfection, which can be achieved through the counsels, in *Summa Theologiae* IIaIIae 186, where he discusses the "religious state" of monks. But he also indicates, elsewhere in the *Summa*, that the goals and justification of the monastic non-obligatory counsels are the same as those of the obligatory precepts. According to Aquinas, "Perfection is said to consist in a thing in two ways: in one way, primarily

[38] Konrad p. 25 places this conflict within the context of the dispute between the secular masters of the theological faculty of the European Universities (mainly Paris) and the mendicant orders of the Franciscans and the Dominicans. But there is a broader dispute here, I think. On this division, see also Pinckaers (1995), 255ff.

[39] On this, cf. Luke 14:33 where Jesus asks those who follow him to renounce all possessions in order to become his disciples ("whosoever he be of you that forsaketh not all that he hath, he cannot be my disciple").

[40] One feature of the evangelical counsels that points at the direction of a separate and autonomous moral scheme is the fact that they involve momentous lifelong commitments and not some momentary moral choice. In this sense, they differ from a lot of the paradigmatic cases of supererogation in contemporary literature, which focus on single moral actions, such as an act of generosity or self-sacrifice.

and essentially; in another, secondarily and accidentally." (*Summa Theologiae* IIaIIae 184.3) The primary aspect of perfection consists in charity, which requires the observation of the general obligation to love God and our fellow human beings (the two Great Commandments). Contrary to the law of the Old Testament, which gives rigid rules and prohibitions, the law of charity, as we saw in the previous sections, can be obeyed in a number of different ways, leaving us free to choose which of these best fits our circumstances and character. The open-endedness and fluidity of charity is the reason behind the proclamation that Christ in the Gospels ushers the new Law of Liberty, which allows for different paths to perfection. One path to perfection is through keeping the precepts, such as the Ten Commandments of the Old Testament without exception.[41] The other is through the counsels. The counsels, therefore, serve the same goal and derive their moral worth from the same source as the precepts, namely perfection, and their difference is not essential but accidental.[42]

The unified moral scheme that encompasses both commands and counsels is also presented in *Summa Theologiae* IaIIae 108, where the New Law is discussed. According to Aquinas, the New Law of the Gospels consists in the grace of the Holy Ghost, "which is shown forth by faith that works through love." The New Law subsumes the Old Law, in the sense that it maintains the commandments while showing why they should be kept (keeping the precepts is necessary for maintaining God's grace, since one could not claim to have faith in God and work through love, while stealing, cheating, or murdering). But there are also acts that do not necessarily contradict charity and the grace that comes with it, such as marrying or having property, although they might make it more difficult to keep the related precepts (which forbid usury, or adultery, for example). The counsels, in this case, are suggestions that have a heuristic function and help specify what the commands already imply, under special circumstances, and, in particular, once someone has assumed a life function (for example, has chosen the religious life as the one that reflects their true calling).[43]

According to Aquinas, precepts and counsels do not form two separate moral systems, but rather fit into a moral continuum, which is based on love and charity. This New Law of charity is open-ended. Acts of supererogation, therefore, are not an extra set of moral challenges that come on top of the observance of the commandments, but rather different ways of obeying the same commandments, which happen to be more

[41] It is important to keep in mind that one of the main points in the exegetical presentation of the New Law of Liberty that Jesus introduces in the New Testament is that the Old Law is not rejected, but simply subsumed under the Law of Liberty. On this relationship between the New and Old Law, and the effect it has on the notion of supererogation, see Hall (2002).

[42] Thomas Aquinas, *Summa Theologiae* IIaIIae 184.3: "Secondarily and instrumentally, however, perfection consists in the observance of the counsels, all of which, like the commandments, are directed to charity; yet not in the same way."

[43] On this, see Franks (2009), 124, Pinckaers (1995), 52. A parallel that illustrates this function can be drawn with medicine, where different ways of treating a disease are sometimes available, some of which might be more demanding but at the same time more effective, while some actions are strictly forbidden. Aquinas also draws a parallel between medicine and the counsels, in *Summa Theologiae* IIaIIae 184.3.

fruitful in producing the same kind of moral outcome, namely charity.[44] In many cases, the obligatory and the supererogatory might be parts of a spectrum regarding a particular material good. For example, fasting is obligatory on certain dates but supererogatory on others. The same is the case for almsgiving, which is to some extent obligatory, but can also be supererogatory beyond a certain amount.[45]

By viewing both precepts and counsels through the lens of the Law of Liberty, Aquinas does more than give a single moral scheme and justification for both kinds of actions. He is also able to defend the position that the counsels are not a separate set of rules that can be practiced only by those who make all-encompassing life commitments, such as joining a monastic order.[46] This is an important point within Aquinas' historical context of the dispute between the secular masters of theology and the mendicant orders, regarding the primacy of the commandments and counsels, and, correspondingly the two different types of life.[47] But, the move also comes at a cost, which became apparent in the following centuries, with the decline of the monastic orders and, more so, the advent of Protestantism.

One apparent weakness of Aquinas' position on the relation between precepts and counsels has to do with the fluid and open-ended nature of the law of charity. Rather than giving concrete commandments that are universally applicable, the New Law of charity, which includes both precepts and counsels, seems to leave a lot of its moral assessment indeterminate, sensitive to circumstance, or non-committal.[48] These are, of course, features that one could attribute to a number of moral theories, ancient eudaimonism being an obvious example. Clearly, Aquinas does not think that this is a weakness of his moral theory, especially since he understands morality as an inner state that has more to do with one's attachment to God (and the grace that comes from such a state), than compliance with rules. But a number of Aquinas' contemporaries, especially those teaching in the faculties of the Paris University, had

[44] *Summa Theologiae* IIaIIae 184.3: "the love of God and of our neighbor is not commanded according to a measure, so that what is in excess of the measure be a matter of counsel. [...] The reason of this is that 'the end of the commandment is charity,' according to the Apostle (1 Tim. 1:5); and the end is not subject to a measure."

[45] See *Summa Theologiae* IaIIae 108.4: "when anyone gives alms to a poor man, not being bound so to do, he follows the counsels in that particular case."

[46] *Summa Theologiae* IaIIae 108.4: "Now if a man observes these [the counsels] absolutely, this is in accordance with the counsels as they stand. But if a man observes any one of them in a particular case, this is taking that counsel in a restricted sense, namely, as applying to that particular case."

[47] On this, see Konrad (2005), ch. 1. Pinckaers (1995), 256.

[48] Of course, the New Law also contains some commandments that frame our relationship to God and to each other (fasting, attending mass, almsgiving, etc.). But it is not exhausted in those, unlike the Old Law (see Thomas Aquinas, *Summa Theologiae* IaIIae 108.1). A lot of the moral choices one has seem to be influenced by circumstance and vary from time to time. For example, loving God and others might be served by marrying in some cases (especially, for example, when there is a need to increase the population), or by staying celibate. Even the question of what is a better choice (generally, that is celibacy), might be indeterminate and sensitive to circumstance. This is not the case for the precepts, which are applicable at all times (with minor exceptions that are clearly stated, such as sickness absolving one from the obligation to fast).

a more legalistic view of morality that considered clear and definite prohibitions and commandments to provide a superior moral code.

A more concerning set of difficulties in Aquinas' presentation of supererogatory acts has to do with their presumed superior value and with their optional character. The standard argument for the moral superiority of the counsels over (merely) observing the commandments, is based on the counsels' difficulty and the fact that not all people are able to follow them.[49] But, if that is the case, then it seems that some people are able to receive a special benefit mainly because they have been gifted with a relative disdain for worldly goods and pleasures, or had the "good fortune" to be poor.[50] To this objection, one can respond in one of two ways: either claim that all people have the same "moral supplies" and that the way they follow (precepts or counsels) is only a matter of choice; or, accept that some people are better equipped for pursuing the counsels, and maintain that, for this reason, they (are made by God to) face greater challenges, commensurate to their abilities. While the first option of equal moral potential is not explicitly denied, much of the presentation of the counsels in the New Testament (especially the reference to "those who are able to receive it" in Matthew 19:12), suggests that not everyone is able to follow the counsels. This is, after all, part of the reason why they are not obligatory. The second idea also seems to be more in line with the flexible morality of the Law of Liberty, and the varying degrees of moral expectation (and the corresponding leniency) that Jesus shows in the Gospels. But, adopting such a flexible view does not seem to explain satisfactorily why the counsels are not obligatory *for those that are able to perform them*, even if they are not obligatory for all, and, thus, not obligatory in general.

As we saw in the previous section, the main idea behind the optional nature of the counsels is that they are not necessary for salvation. This, in turn, is based on a threshold idea that regards salvation as something that can be achieved once a certain amount of merit has been gained and the corresponding thought that observing all commandments without exception can get one to that threshold, leaving

[49] This is, presumably, a point accepted even by the critics of the monastic orders and the counsels more generally (see Konrad 2005, 34). But, if we were to accept difficulty and rarity as the main distinguishing feature of supererogation, then a lot of acts should no longer count as supererogatory. These include contemporary examples, such as tipping, and even some of the examples of supererogation Aquinas has in mind, such as giving more alms than is required, which might not be a great burden for a wealthy person. Even counsels such as celibacy may not be burdensome (for someone, for example, that is impotent by nature). On the other hand, there may be precepts that are very demanding and difficult to follow, under some circumstance (after all, the Fathers of the Church often advertised celibacy as an easy way out of the temptations of the flesh and the difficulty of resisting them).

[50] Cf. Jesus' claim that some people are eunuchs by nature, in Matthew 19:12, or the comment on the young man's wealth as a hindrance to perfection, in 19:22. In general, one should keep in mind that the inability to adopt the counsels does not only have to do with one's character, but also with circumstance. In some cases, for example, one cannot practice celibacy if their family has arranged for them to be married or to adopt poverty if they have to take over the family business and the lives of other people depended on that.

supererogatory acts as additional sources of merit that can take one beyond the salvation threshold. The idea of a salvation threshold is based on theological assumptions about the afterlife and about God's judgment. And it was these theological assumptions that seem to have been the central point of the Protestant rejection of opera supererogationis, starting with Luther's denunciation of the institution of the Indulgences.[51]

Luther's main criticism is based on the thought that there is no fixed threshold to salvation and, correspondingly, no way to accumulate extra merit (let alone being able to use it to save others). According to Luther, salvation cannot be assured by obedience to commandments, or even by the performance of supererogatory acts. Rather, salvation requires God's grace.[52] Otherwise, one is left with a position that Luther and other Protestants considered arrogant and blasphemous, namely the idea that God "owes" a life of eternal bliss to those who have not trespassed the commandments, and some extra merit to those who also performed supererogatory acts of self-denial.

The Protestant argument, like its Roman Catholic counterpart, is based on Scripture, which is the only source of knowledge regarding God's thoughts in general, and regarding salvation in particular.[53] Contrary to Aquinas, the Protestants interpreted Jesus' comment "he that is able to receive it, let him receive it" in Matthew 19:12, as giving a special command that applies only to some. Having higher expectations from those who have greater potential is not unfair or too demanding, if anything of that sort could be said about God.[54] Quite the contrary, one could interpret such high expectations as a sign of God's grace.[55] In fact, it would be blasphemous to claim that God has given each one of us anything other than the exact amount of obligations that each person is able to fulfill. We are all servants of God, and as such, each of our actions can never be more than the fulfillment of duty, and our reward can never surpass salvation.[56] In this case, where every action has a corresponding duty that it fulfills, supererogation becomes impossible, by definition.

---

[51] For an extensive account of the Protestant view on supererogation see Professor Mellema's entry in this volume.

[52] On this objection, see Luther's *Explanation of the Ninety Five Theses* (Thesis 58), and Article Fourteen of the Thirty Nine Articles of the Anglican Church. For a discussion of these views within the context of the theory of supererogation, see Heyd (1982), 26–29.

[53] Although the debate that took place during the Reformation and Counterreformation focused on soteriology, a version of this objection could be given, even if we were to accept that there is a salvation threshold, which the saints surpass, and that extra merit can be accumulated and used to help others. Why is there no obligation to help as many people as possible, by committing supererogatory acts? After all, God requires that we love and help our neighbor, and even forgive our enemies. On this, see Dentsoras (2014).

[54] Cf. the Parable of the Workers in the Vineyard in Matthew 20:1–16, where workers that start at different times of the day receive the same reward, although some have to do way more than the other.

[55] This could also be tied up to Luther's ideas about predestination. We should welcome the greatest amount of duty because it is a sign that we are among the chosen.

[56] See Luke 17:10: "when ye shall have done all those things which are commanded you, say, we are unprofitable servants: we have done that which was our duty to do."

6 Opera Supererogationis After the Reformation

As with other dogmatic disputes during the Reformation, the Protestant rejection of the Catholic Church's doctrine of supererogation was the result of different contributing factors, some of which were philosophical, while others were political and even personal. The connection of the doctrine of supererogation to the institution of the Indulgences, which was a focal point of the Reformation's early stages, played a role. So did Luther's and other Protestant leaders' dislike for the monastic life, a life that was primarily based on the counsels of celibacy, poverty, and obedience. Moreover, Luther's deep personal concern for the conditions of salvation and the extent to which it can be secured by one's own actions had a profound effect on his ideas on free will, moral merit, and the existence of actions that can exceed the threshold of salvation.[57] The combined effect of these factors was a marked move away from the vocabulary of supererogation and from the distinction between precepts and counsels in moral and legal philosophy, primarily among Protestant thinkers of the Renaissance and the Enlightenment, but also, to a smaller extent, even among Catholic thinkers. The reasons for this move are not always clear, and certainly involve more than philosophical argument, but I think some observations could give us a sense of the relevant developments.

Firstly, much of the Protestant criticism targets supererogatory actions indirectly, or only partially. For example, Luther's criticism of the Indulgences is primarily directed against the presumed authority of the Pope and the bishops, and their ability to affect one's fate in the afterlife. These issues are separate from the possibility of achieving superabundant merit. Similarly, the idea that even paradigmatic cases of supererogatory actions, such as martyrdom and voluntary hardships, might be *demanded* by God does not mean that these actions are not supererogatory in the way that reflects Aquinas' distinction between precepts and counsels, namely by making salvation more expeditious and assured, rather than being (merely) necessary requirements for it.[58] Even Luther's seemingly direct rejection of the evangelical counsels, which is tied to his criticism of the monastic life, is qualified (to allow for celibacy as an optional choice), and, in any case, would leave space for other saintly actions, such as martyrdom, to be supererogatory.[59] All this seems to have left ample theological and philosophical space for the defense of the existence of supererogatory acts.

As one would expect, the Roman Catholic Church reaffirmed the orthodoxy both of the doctrine of supererogation and of the Indulgences, based on scriptural and theological grounds. Moreover, Catholic theologians, such as Erasmus, addressed the philosophical grounds for the supererogation doctrine, focusing on the role of human

[57] On some of these aspects, see Konrad (2005), 132.

[58] On the Protestant argument, see Heyd (1982), 28. It should be noted that the Protestant argument does not use the language of necessity here, which was the main distinction between the precepts and the counsels in Aquinas, but rather that of duty. This is an important development that will be further discussed below (on it, see Konrad, 2005, 126, Hruschka, 1998).

[59] On this, see Konrad (2005), 130.

agency in securing salvation.[60] At the same time, the Catholic defense of the counsels and the corresponding doctrine of supererogation often seems to lack the rigor and determination that the Roman Catholic Church showed in defending other parts of its dogma.[61] This seems to be due to the fact that some of the misgivings of the leaders of the Protestant Reformation (who, after all, were themselves Catholics, to begin with) were shared also by some who chose to remain within the Catholic Church. The criticism of the excesses involved in the Indulgences is such an example. More importantly, the relative decline of the monastic orders and the development of the European universities and urban intellectuals (both clerics and laypeople) inevitably brought about a questioning of the superior moral value placed on the counsels of poverty and chastity that the monastic orders practiced. Sure, the evangelical counsels could claim some scriptural authority and they were supported by the practice of the Catholic Church, but this was not enough to demonstrate their moral superiority, at least to those outside monasticism. On the other hand, it would be hard to reject the belief that the actions of martyrs and saints were not only especially laudable, but also went beyond what one could expect and demand from everyone. This is a point that even the Protestants who denied the supererogation doctrine, in general, were willing to acknowledge. In sum, a second feature of the debate about supererogation during the Reformation is that there may have been points of convergence or, in any case, some common developments.

One of the most notable consequences of such a convergence is the move away from the language of charity and perfection that we find in Aquinas' moral philosophy, toward the idea that law and duty are the core concepts in ethics, and their observance is what conveys moral merit.[62] From the moral philosophical point of view, this seems to have been a key development of the Renaissance and the Enlightenment, which can be seen in Catholic scholarship's renewed emphasis on law and the codification of the Church's moral requirements. Catholic legal scholars viewed morality chiefly as compliance with the set of laws set out in the Bible, and presented the evangelical counsels as optional choices that go beyond what the law prescribes and that only become relevant for the few who choose to adopt the monastic vows. Once a duty has become the central moral concept, supererogatory actions need to

[60] See, for example, Erasmus' debate with Luther regarding free will (Erasmus and Luther 2012). The main point of contention was the ability of the human will to cooperate with divine grace and contribute to one's salvation. If it is possible for human beings to secure their salvation through their deeds, together with the divine grace, as Erasmus maintained, then it would make sense to think that there is a threshold of salvation, which might be achieved in multiple ways, and perhaps even surpassed (through supererogatory acts). This is not possible if human actions are incapable of securing our salvation, which only God can do, as Luther maintained.

[61] The defense of the Indulgences, for example, receives only a short mentioning in the decrees of the last day of the council of Trent. Supererogation and the distinction between precepts and counsels is not mentioned.

[62] On the Protestant side of this development, see Hruschka (1998). Pinckaers (1995), 287 ff. gives some relevant developments in Catholic theology. On the rise of the legalistic way of thinking about morality, within the Roman Catholic Church, see Pinckaers (1995), 271. Konrad (2005) also notes a related development, namely the shift of emphasis in Catholic moral philosophy from charity as the central moral concept to duty.

gain their moral value by satisfying a corresponding obligation. But such an obligation cannot be universally binding; after all, we cannot demand or expect that all people be martyrs or saints. Instead, the practices associated with the evangelical counsels should be considered as special duties that apply only to specific groups, such as the monastic orders, whose importance steadily decreased after the Middle Ages.

The emphasis on duty as the fundamental moral principle, which culminated with Kant, had a profound impact on modern moral philosophy. The effects of this development extended to the idea of supererogation, which fades away in moral philosophical discussions. An apparent reason for this is duty's ill fit for some aspects of supererogation, as Aquinas and his contemporaries understood it. Obedience to duty fails to account for the extraordinary merit that supererogatory actions deserve, and claiming that their merit comes from their satisfaction of a non universally binding duty seems inadequate and, perhaps, misleading.[63] In this respect, Aquinas' emphasis on charity and human perfection appears to have fared better in explaining the value of supererogatory acts as something connected to their optional character and the way in which they express one's love to God. But, by the time of the Enlightenment, Aquinas' moral philosophy has fallen from the forefront to the limited purview of Catholic scholars. We will need to wait until the second half of the twentieth century for fresh attempts to account for supererogatory actions, both within and outside the moral framework of duty and obligation.

References

Aquinas, St. T. (2006). *Summa theologiae* (Vol. 60). Cambridge: Cambridge University Press.

Barnes, A. (1913). Evangelical counsels. In: *The Catholic encyclopedia* (Vol. 4). New York: Robert Appleton Company.

Barsanuphius, John, St. (2006). *Letters*. Washington: Catholic University of America Press.

Betz, O. (1999). The Essenes. In W. Horbury, W. Davies, & J. Sturdy (Eds.), *The Cambridge history of Judaism* (pp. 444–470). Cambridge: Cambridge University Press.

Bonino, S. T. (OP). (2002). Charisms, forms, and states of life (IIa IIae qq. 171–189). In: S. J. Pope (Ed.), *The ethics of Aquinas* (pp. 340–354). Washington: Georgetown University Press.

Chrysostom, J. (1983). *On virginity; Against remarriage*. E. Mellen Press.

Deák, V. H. (2014). *Consilia sapientis amici: Saint Thomas Aquinas on the foundation of the evangelical counsels in Theological Anthropology*. Editrice Pontificia Università Gregoriana.

Dentsoras, D. (2014). The birth of supererogation. *Epoché: A Journal for the History of Philosophy* 18.2, 351–372.

Derrett, J. (1964). Law in the new testament: Fresh light on the parable of the good samaritan. *New Testament Studies, 11*(1), 22–37.

Erasmus and Luther. (2012). The battle over free will. In: C. H. Miller and trans (Eds.). Indianapolis: Hackett.

Forbes, G. (2000). *The God of old: The role of the Lukan parables in the purpose of Luke's gospel*. Sheffield: Sheffield Academic Press.

[63] Interestingly, this is an objection that Nietzsche seems to have noted. See Lopes (2020).

Franks, C. A. (2009). *He became poor: The poverty of Christ and Aquinas' economic teachings*. Grand Rapids: William B. Eerdmans Publishing Company.
Hall, P. M. (2002). The old law and the new law (Ia IIae, qq 98–108). In S. J. Pope (Ed.), *The ethics of Aquinas* (pp. 194–206). Washington: Georgetown University Press.
Heyd, D. (1982). *Supererogation: Its status in ethical theory*. Cambridge: Cambridge University Press.
Hruschka, J. (1998). Supererogation and meritorious duties. *Jahrbuch Für Recht Und Ethik / Annual Review of Law and Ethics, 6*, 93–108.
Joest, C. Once again: On the origin of christian monasticism: recent historical and exegetical insights and a new proposal with an ecumenical perspective. *The American Benedictine Review*, 61(2), 158–182.
Konrad, M. (2005). Precetti e consigli. Studi sull'etica di san Tommaso d'Aquino a confronto con Lutero e Kant (Dialogo di filosofia 7). Rome: Lateran University Press.
Lewis, D. K. (1997). Do we believe in penal substitution? *Philosophical Papers, 26*(3), 203–209.
Lopes, R. (2020). Nietzsche on the banishment of supererogation by Luther and its influence on modern ethical life and moral theorizing. In H. Heit & A. Sommer (Eds.), *Nietzsche und die reformation* (pp. 331–348). Berlin/New York: De Gruyter.
Luther, M. (2017). The ninety-five theses and other writings. In: W. R. Russell, trans (Eds.). New York: Penguin Classics.
Nel, M. J. (2015). The renouncement of possessions according to Matthew 19:16–30. *HTS Teologiese Studies/Theological Studies* 71(1), Art. 2928, 8. http://dx.doi.org/https://doi.org/10.4102/hts.v71i1.2928.
Pinckaers, S. (OP). (1995). *The sources of Christian ethics*. Washington: The Catholic University of America Press.
Plumptre, E. H. (1879). *The gospel according to St. Matthew with commentary*. London: Cassell.
Ambrose, St. (1996). *On virginity*. Toronto: Peregrina Pub. Co.
Augustine, St. (2010). *The excellence of widowhood. Treatises on various subjects* (The Fathers of the Church, Vol. 16). Washington: Catholic University of America Press.
Ambrose, St. (2016). *Concerning widows*. Aeterna Press.
Taylor, J. E., & Davies, P. R. (1998). The so-called therapeutae of de vita contemplativa: identity and character. *The Harvard Theological Review, 91*(1), 3–24.
Tertullian. (1951). *Treatises on marriage and remarriage: To his wife, an exhortation to chastity, monogamy*. Westminster: Newman Press.
Waterworth, J., ed. and trans. (1848). *The council of Trent: The canons and decrees of the sacred and oecumenical council of Trent*. London: Dolman.

Supererogation and Protestantism

Gregory Mellema

Abstract This chapter is divided into four parts. The first part describes the events leading up to the Protestant Reformation, including Thomas Aquinas' distinction between the commandments of God and the counsels of God. The second part describes the anti-supererogationist views of the Reformers, Luther, Calvin, and Melanchthon. The third part examines the views of a contemporary figure, Joseph Allen, whose Protestant commitment leads him to espouse anti-supererogationist views. The fourth part explains why contemporary Protestants are by and large open to acknowledging the possibility of supererogation.

Keywords Protestantism · Obligation · Praiseworthy · Supererogation · Commandments · Counsels

An act qualifies as an act of supererogation, as I understand the term, if and only if the performance of the act is morally praiseworthy, it does not fulfill a moral duty or obligation, and omitting the act is not morally blameworthy.

The Protestant Reformation took place during the Sixteenth Century, but fundamental to understanding the mindset of the Reformation is an understanding of the events leading up to it. For this reason, the first section of this essay sketches some developments in the Roman Catholic Church prior to and during the Sixteenth Century, beginning with Thomas Aquinas' noted distinction between the commandments of God and the counsels of God.

1 Background

Thomas Aquinas articulates the distinction in the following passage (a distinction that was explicitly drawn centuries earlier by St. Ambrose and St. Augustine).

G. Mellema (✉)
Calvin University, Grand Rapids, USA
e-mail: mell@calvin.edu

D. Heyd (ed.), *Handbook of Supererogation*,
https://doi.org/10.1007/978-981-99-3633-5_18

> The difference between a counsel and a commandment is that a commandment implies obligation, whereas a counsel is left to the option of the one to whom it is given. So in the New Law, which is the law of liberty, counsels are fittingly added to the commandments, but not in the Old Law, which is the Law of bondage (*Summa Theologia,* Ia IIae, Q. 108, Article 4).

The commandments of God, associated with the Old Law, the laws of bondage, are obligatory. By contrast, the counsels of God are optional. A person to whom they are directed has the freedom to decline the recommendations they embody. Accordingly, they are associated with the New Law, which Aquinas calls the law of liberty.

Aquinas goes on to say that the counsels of God benefit those to whom they are addressed. He describes them as "supremely useful and becoming" and believes it is in our best interests to respond positively to them. One who responds negatively to God's counsels is acting contrary to one's best interests. (Some may think this is a bit harsh. Do we really act against our best interests in not opting for the monastic life?).

The three areas in life to which the counsels of God pertain, according to Aquinas, are external wealth, carnal pleasures, and honors. We are encouraged to decline riches in favor of poverty, to forgo carnal pleasures in favor of chastity, and to renounce pride in favor of obedience. Those who have made vows of poverty, chastity, and obedience are of course required to renounce riches, carnal pleasures, and pride, but those who have not are under no obligation to renounce them completely. Nevertheless, to the extent that we follow these counsels, Aquinas believes, we will more quickly arrive at eternal happiness.

Clearly, the distinction between God's commandments and God's counsels is relevant to the possibility of supererogation. To qualify as supererogatory an act cannot be obligatory, and since God's counsels are not obligatory the possibility is created that following these counsels on a particular occasion can qualify as an act of supererogation. More will be said about this possibility in the following section.

There has been for centuries much opposition to the possibility of supererogation, and the story of the opposition which grew out of the Reformation begins with the sale of indulgences by certain individuals, such as the German Dominican Johann Tetzel (1465–1519), within the Holy Catholic Church. By purchasing an indulgence one could lessen the punishment in the afterlife of persons who have committed sins in this life, and the interest in purchasing them was considerable. Certain of the dioceses in the Church accumulated much wealth as a result of this practice. Over time these activities degenerated into corruption of the worst sort.

What made this practice possible was the idea that Jesus and various saints had built up a treasury of good works through their exemplary lives on Earth. Obviously, this treasury can be of no direct benefit to them, and thus the sale of indulgences was proposed to make it possible for others to benefit. Those whose lives are on balance less than exemplary are subject to penalties in the afterlife, such as a period of time in purgatory, By means of a purchase of indulgences, the Church contended, a measure of the treasury of merit built up by Christ and the saints could be applied to their account. As a result, their penalty could be lessened by an amount corresponding to the indulgence purchased.

The entire scheme is based on the assumption that disobeying God's commandments results in penalties, but the penalties affect only those whose lives produce more evil than good. A small number of people end their lives having generated more good than evil, and they make possible the treasury of merit which serves as the basis for reducing the penalties of those on whose behalf indulgences are purchased. In the eyes of the Church God demands no more than that balance is preserved in the entire scheme. The Church then administers financial transactions so as to ensure that the entire ledger is kept balanced.

Whether Aquinas endorsed the sale of indulgences is uncertain, but the Church unhesitatingly cited his authority in its defense of the institution. Distinguishing between God's commandments and God's counsels creates the possibility of affirming that people can act in such a way as to perform meritorious works that are not commanded by God. A person who follows God's commandments and in addition abides by God's counsels can possibly create a surplus of merit which, in turn, can be applied to the accounts of others on whose behalf indulgences are purchased. In other words, the possibility of contributing to a treasury of merit exists when one abides by God's counsels, but this is not the case when one follows God's commandments. In the next section, this point will be developed in greater detail.

To conclude this section, David Heyd points out that Aquinas does not believe that human law forbids all vices (only the more grievous ones). Religious law, on the other hand, forbids all vices since its concern is peoples' salvation. Nevertheless, supererogation is not ruled out by this view, the reason being that good acts, not evil acts, fall under its scope. Religious law for Aquinas rules out all evil acts, but it allows for non-obligatory good acts (Heyd, 1982, p. 26).

2 The Protestant Reformation

The two principal figures of the Protestant Reformation, Martin Luther, and John Calvin, embraced the idea that God's commandments are binding upon all persons, but they emphatically denied that there are counsels of God which have an optional status. By rejecting the idea that God has provided optional counsels to all persons, they have in their own way of thinking eliminated the possibility that human beings can create a treasury of merit.

I begin with the following passage from John Calvin's *Institutes of the Christian Religion*:

> These commandments -- "Do not take vengeance; love your enemies," which were once delivered to all Jews and then to all Christians in common – have been turned by the Schoolmen into "counsels," which we are either free to obey or not obey. What pestilential ignorance or malice is this! . . . The reason they assign for not receiving them as laws is that they seem too burdensome and heavy, especially for Christians who are under the law of grace. Do they dare thus to abolish God's eternal law that we are to love our neighbor? . . . Either let them blot out these things from the law or recognize that the Lord was Lawgiver, and let them not falsely misrepresent him as a mere giver of counsel (Calvin, 1960, p. 419).

Calvin states that it is pestilential ignorance to affirm the existence of God's counsels, laws which are optional to obey. He, of course, realizes that by affirming their existence the possibility of meritorious actions detached from obligation is created. When God's laws strike them as burdensome, people find it convenient to judge that they are optional to obey. But Calvin is steadfastly opposed to the idea that God's law can be divided into a portion that we are obliged to obey and another portion that we are allowed to disobey. Rather, God's law constitutes a unified whole and consists entirely of commandments that human beings are obliged to follow. (For Calvin and Luther we can never be sure that we have followed God's commandments, and hence we deserve salvation only by divine grace).

In Christ's Sermon on the Mount, we are told, "Love your enemies, pray for those who persecute you (Matthew 5:44)." The directive to love our enemies is mentioned by Calvin as an example of a commandment given by God which the Schoolmen have interpreted as a counsel. Regarding the directive to love our enemies as optional is certainly less burdensome, but Calvin cites the words of Christ as evidence which cannot be disputed that we are commanded to love our enemies. He then goes on to say the following: "Who will not here conclude with Chrysostom that the obligatory character of these utterances reveals them clearly to be exhortations by imperatives? (*Ibid.*, p. 420)" Calvin recognizes that some of God's commands are "indeed hard and difficult for our feebleness," but that does not give us a license to regard them as optional. They are still God's commands, and following them is exactly what we must do. In this context, David Heyd quotes Calvin as declaring that the doctrine of supererogation is a "patchwork of sacrilege and blasphemy" and that there can be no surplus of merit (Heyd, *Op. Cit.*, p.28).

Turning now to Martin Luther, the same antagonism directed to the so-called counsels of God is evident. In his *Treatise on Good Works,* he has this to say: ".... there are no good works except those which God has commanded. (Luther, 1943, p. 1:147)" A person who performs a good work is doing no more than fulfilling a commandment of God. To his way of thinking, the so-called counsels of God are actually commandments. Thomas Aquinas classifies renouncing wealth, carnal pleasures, and pride as counsels of God, but they are actually obligatory because, once again, Luther believes there are no good works except those which God has commanded.

It is worth pausing to note that Luther, in this short passage is articulating an anti-supererogationist position. Recall that an act qualifies as an act of supererogation only if the performance of the act is morally praiseworthy and does not fulfill a moral obligation. In saying that all good works fall under the rubric of what God has commanded, Luther is clearly rejecting the possibility of works of supererogation (given the assumption that whatever God has commanded is obligatory).

Ever since the appearance of Urmson's classic article 'Saints and Heroes,' which defends the possibility of supererogatory acts, the praiseworthy acts of saints have been commonly regarded as paradigm examples of supererogatory acts. In Luther, we find the opposite point of view forcefully articulated. Even the saints, he says, "have done nothing which is superabundant. Therefore, they have left nothing to be allocated through indulgences. (Luther, 1955, p. 213)" Since in his view, there are no good works except those which God has commanded, it follows that what the saints have done is nothing other than obligatory. Saints themselves, when praised for their saintliness, have been known to respond that they were just humbly fulfilling their obligations. Even when they perform acts of martyrdom, these acts are obligatory. As such, they cannot possibly constitute a surplus of merit, and accordingly, they cannot possibly contribute to a treasury of merit. As meritorious as a saint's actions might be, even to the extent of martyrdom, these actions can never exceed the status of obligatory in order to contribute to a treasury of merit.

Luther evidently declines to challenge the view that the works of Christ make possible the creation of a treasury of merit. Thus, there is no contradiction in holding that the works of saints fail to create a surplus of merit and maintaining that a treasury of merit exists, since without Christ's ultimate sacrifice no one could be saved. But the point to be emphasized at present is that no saint has successfully generated a surplus of merit.

Although the Reformers employ an anti-supererogationist line of argument to establish that indulgences lack legitimacy, one cannot conclude that undermining their legitimacy is the only reason why the Reformers are attracted to an anti-supererogationist position. At the bedrock of Reformational doctrine lies the idea that we are justified by faith alone, and this doctrine, in the eyes of the Reformers, did not comport well with the idea that people can perform meritorious acts that fall outside of what God has commanded. Being justified by faith alone entails that we cannot be justified by works, and to the Reformers that meant that performing acts that lie outside the scope of God's laws cannot possibly be meritorious. In other words, if one acknowledges the possibility of supererogatory acts, one is in danger of acknowledging the possibility of being justified by one's own works.

This, at any rate, was the position of the Reformers. In subsequent years various Protestant thinkers have developed an alternative stance. After all, the position that acts of supererogation are possible is arguably compatible with the position that God's creatures cannot be justified by works alone. An act which is meritorious and fulfills no moral obligation might still be insufficiently praiseworthy to justify one from a Divine perspective. It is not difficult to imagine a person who performs an act of supererogation from time to time but fails to be justified by virtue of performing them. Such a person could reasonably be described as justified by faith rather than works. Nevertheless, in spite of these considerations, the doctrine that people are justified by faith alone precluded the possibility of supererogatory acts in the eyes of the Reformers.

A third Reformer, Philip Melanchthon, less well known than Luther or Calvin, is also suspicious of the distinction between the commandments and the counsels of God. He makes reference to certain "Sophists" who have "taught that certain things are not necessarily demanded by God, but only recommended, so that if anyone cares to, he may obey, and they absolve the one who does not obey. (Melanchthon, 1969, p. 57)" Those proclaiming these teachings, Melanchthon declares, have "erred shamefully and godlessly," for the laws of God cannot be interpreted as recommendations. We must regard them as God's commandments.

Melanchthon then poses this question: Why haven't those who distinguish between God's commandments and God's counsels simply regarded God's laws as consisting entirely of optional counsels? By posing this question Melanchthon appears to be suggesting that the distinction between commandments and counsels is quite arbitrary. We can simply regard as commandments the laws of God we find convenient to obey, while those we find difficult can be regarded as optional counsels.

It might come as a surprise to discover that Melanchthon acknowledges one counsel among the injunctions of God: "As far as I know, in the Gospels, there is only one instance of a counsel, that of celibacy. (*Ibid.*, p. 59)" He believes that celibacy is recommended in Scripture as something favorable, but it is not a requirement binding upon us. To show that it is something favorable, he quotes the Apostle Paul as asserting that it is well for unmarried people to remain in that state. Those who have married have not violated a commandment of God by no longer remaining celibate, but those who are unmarried are recommended by Paul not to seek marriage (I. Corinthians 7:25–27).

Indisputably, Melanchthon recognizes a distinction between God's commandments and God's counsels, and here he departs from the teachings of Luther and Calvin. Nevertheless, he finds in Scripture only one clear reference to a counsel of God. All of God's laws must be recognized as obligatory with the exception of celibacy, and here his views diverge from the views of Aquinas who acknowledged multiple counsels. The only area in which God's injunctions can be legitimately regarded as optional, according to Melanchthon, is that of marriage. Once again, those who marry have not in doing so violated God's commandments.

The overall thrust of Melanchthon's doctrinal stance remains close to that of Luther and Calvin. It is only in a technical sense that Melanchthon acknowledges a distinction between God's commandments and God's counsels, since he recognizes only one counsel in all of Scripture. A person who remains celibate is thereby in a position to accrue merit in a manner that is not available to one who obeys God's commandments. But it is difficult to imagine that such a person is able to thereby build up a treasury of merit, and hence we can reasonably conclude that he is in agreement with the other Reformers in taking a hostile stance toward the Church's rationale for the sale of indulgences.

Two Protestant Reformation era confessional statements, the 1562 Belgic Confession and the 1563 Heidelberg Catechism, proclaim a doctrine known as total depravity. This doctrine teaches that human beings are so corrupt that they cannot perform any good acts and are inclined to all evil. According to the Heidelberg Catechism, God created human beings good, and our corrupt nature comes "from the Fall and disobedience of our first parents, Adam and Eve, in paradise. The Fall has so poisoned our nature that we are born sinners—corrupt from conception on (Answer to question 7)." The Belgic Confession describes our condition in these words: "So he made himself guilty and subject to physical and spiritual death, having become wicked, perverse, and corrupt in all his ways (Article 14)."

The doctrine of total depravity fits squarely with the anti-supererogationist tendencies of the Reformers. An act of supererogation is by definition morally praiseworthy to perform, and the doctrine of total depravity entails that human beings are incapable of performing any praiseworthy acts. We are wicked, perverse, and corrupt in all our ways. Nevertheless, both the Belgic Confession and the Heidelberg Catechism go on to teach that if we are born again, we are capable of rising out of this dire condition. Thus, some Protestants have argued that acts of supererogation are possible once individuals are born again. More will be said about this possibility in a later section.

This section concludes with a passage from the 1562 Book of Common Prayer, another Protestant Reformation era document that denounces the possibility of supererogatory acts.

> Voluntary works besides, over and above, God's commandments, which they call Works of Supererogation, cannot be taught without arrogancy and impiety: for by them men do declare that they do not render unto God as much as they are bound to do, but that they do more for his sake than of bounden duty is required: whereas Christ saith plainly, When ye have done all that is commanded to you, say, We are unprofitable servants. (Article XIV)

3 Contemporary Protestantism—A Skeptical View of Supererogation

Contemporary theists are not concerned with attacking the practice of selling indulgences, but a surprising number of theologians and philosophers of religion follow Luther and Calvin in maintaining that no one ever performs meritorious works that are not obligatory. It is significant that these contemporary opponents of the possibility of supererogation represent both Protestants and Roman Catholics. Karl Rahner is a paradigm example of a Roman Catholic who takes an anti-supererogationist position. In this section, I discuss the views of Joseph Allen, whom I take to be a paradigm example of a Protestant who is convinced that since we are subject to God's laws, there can be no room for supererogation.

Earlier reference was made to Urmson's article, 'Saints and Heroes.' There Urmson describes a soldier who saves his fellow soldiers by falling upon a live grenade, and he argues that heroic actions of this type cannot plausibly be categorized as obligatory. Urmson maintains that if the soldier had not fallen upon the grenade, it would not be reasonable to charge that the soldier failed to fulfill a duty. Does the soldier have an obligation to sacrifice his life for his fellow soldiers? Urmson answers in the negative.

In the following passage, Allen expresses disagreement:

> It is sometimes observed that we would not appropriately blame a person . . . for not being sacrificial in that way. . . . There is certainly an attraction in that way of thinking. From the standpoint of covenant love, however, there is something morally lacking, something that *ought* to be done, on the part of the soldiers who do not sacrifice their lives to save their comrades. . . . The idea that second-mile actions are not duties is a reflection of ordinary conventional morality, not of the requirements of covenant love. . . . . If by the strict requirements of covenant love a person ought to have gone the second-mile – ought in the case of the grenade to have attempted to fall upon it – then that person is to be blamed for not having done so (Allen, 1984, p. 127).

Allen distinguishes in this passage between ordinary conventional morality and the morality of covenant love. Optional second-mile actions are possible from the perspective of ordinary covenantal morality, but not the morality of covenant love. From the perspective of the latter, actions of this type are obligatory, from which it follows that declining to perform them amounts to the failure to carry out one's obligation. It is hard to see how acts of supererogation are possible within the morality of covenant love.

Thinking about second-mile acts from the perspective of ordinary covenantal morality is a natural response, Allen notes. While we would blame a soldier for omitting to fulfill "institutional duties," such as neglecting to perform guard duties, we would be reluctant to blame a soldier for declining to sacrifice his or her own life for the sake of fellow soldiers. It would seem harsh to do so. But from the perspective of the morality of covenant love, second-mile acts are obligatory and never optional. We must always go the second-mile in helping others because that is how we show love to them. Allen describes this type of morality as a "strict and demanding ethic."

Allen makes clear that in his view one is not given permission to admonish someone else who chooses not to perform a second-mile act.

> It is quite another matter whether anyone ought to *verbalize* that blame to a soldier who did not sacrifice himself. We have no call to be self-righteous, and what we say by way of blame in such a situation, even if properly thought out, may well be misunderstood, Nor do we mean by "blame" anything like self-rejection. Blame refers here to moral disapproval. . . . it does not entail rejection of the person (*Ibid*., p. 128).

Someone who declines to perform a second-mile act deserves moral blame for failing to discharge a moral obligation. But from this, it cannot be concluded that someone else is licensed to express moral blame concerning this failure. Allen points out that expressions of moral disapproval from others can be misunderstood. Sometimes people are eager to judge the faults in others, and Allen does not wish what he says to encourage such people.

The next question Allen considers is the question of how one can know when to be sacrificial. Sometimes it is not easy to know one's obligations and how to discharge them. Allen addresses this difficulty in the following passage.

> The distinction between commands and counsels is still useful, not in regard to the accumulation of merit, nor in distinguishing second-mile actions from basic institutional requirements, but in judgments about particular cases where it is not altogether clear what one's obligation is, and especially where the judgment is about someone else's action (*Ibid*., p. 130).

Thus, when one is not sure what he or she should do, according to Allen, the distinction between commands and counsels is still useful.

These words might appear puzzling to the reader, following a lengthy passage about the strict demands of the morality of covenant love. Why is Allen now conceding that the distinction between commands and counsels is still useful? Perhaps what Allen has in mind is something along the following lines. A woman approaches a stranger who appears hungry and asks if he would like something to eat. The stranger speaks a different language and shakes his head, signifying to the woman a negative response, but if he had known what the woman was asking he would have responded affirmatively. The woman meanwhile walks away, and clearly, she is not morally culpable for doing so.

Here I believe the point Allen is making can be construed as follows. The strict demands of covenant love are dependent upon knowing what these demands are in the context of a person's circumstances. The woman offering help to the stranger cannot be expected to know that the stranger was unaware of what she was asking, and here the distinction between commands and counsels is still useful. Suppose that she decided, just to make sure that the stranger could understand English, to verify this. Then her efforts could be judged as following a counsel rather than a command.

Thus, Allen allows that a person who is confused about what to do and happens to perform a second-mile act has fulfilled a counsel rather than a command. Nevertheless, Allen's overall position is clearly anti-supererogationist in nature since he maintains that performing second-mile acts is obligatory, even if it involves sacrificing one's life for the sake of others. No matter how commendable someone's actions, in Allen's view they become swallowed up by obligation. The common ground between what the Reformers argued and the position of Allen is that the boundaries of moral obligation are drastically enlarged beyond what is recognized by conventional ethics.

4 Contemporary Protestantism—The Acceptance of Supererogation

The Heidelberg Catechism and Belgic Confession both teach the doctrine of Total Depravity, as was observed in an earlier section. Nevertheless, they both go on to teach that those who are born again are capable of good works, and this is a teaching that is embraced by vast numbers of Protestants in the present age. The Heidelberg Catechism speaks about the "coming to life of the new self" and goes on to say, "It is wholehearted joy in God through Christ and a delight to do every kind of good as God wants us to do (Answer to question 90)." The Belgic Confession teaches that works "proceeding from the good root of faith are good and acceptable to God since they are sanctified by his grace (Article 24)."

The significance of the doctrinal statements lies in their making possible works of supererogation. If good works are possible by those who have been born again, the door is opened for those good works to be non-obligatory. The Heidelberg Catechism teaches that such works are done in gratitude to God, a teaching that strongly implies that they are not done because believers are commanded to do so. Protestants of the present age judge that there is a significant difference between acting to obey God's commands and acting in pure gratitude to God. Both Lutherans and Calvinists of the present age widely affirm that it is possible for believers to perform good works outside the scope of what God has commanded, and this view is shared across Protestantism.

Another important factor contributing to the widespread acceptance among Protestants of the possibility of performing works of supererogation is the recognition that the 'ought implies can' principle is true. This principle states that a person has a moral duty or obligation to perform a given act only if the person is able to perform it. Thus, a person is under no obligation to perform an act if the person is unable to perform it. If I am in the hospital (through no fault of my own) and unable to attend a family wedding, then I am under no obligation to attend it, and this despite the insistence of my furious mother-in-law that I have violated an obligation.

The relevance of this principle to the notion of supererogation can be illustrated by the following example. Walking down the street, I encounter two panhandlers. The only money I have with me is a ten-dollar bill. It would be morally praiseworthy for me to give the money to the first panhandler, and it would be morally praiseworthy for me to give the money to the second panhandler. Clearly, I am unable to give the money to both panhandlers, and by the ought implies can principle it follows that I have no moral obligation to give the money to both panhandlers.

The anti-supererogationist position embraced by Luther and Calvin maintains that every morally praiseworthy action is morally obligatory. According to this teaching, I have a moral obligation to do whatever is morally praiseworthy, and hence I have a moral obligation to give the money to the first panhandler and in addition I have a moral obligation to give the money to the second panhandler. In this manner, the version of the anti-supererogationist position held by the Reformers conflicts with the ought implies principle. Naturally, one could deny the ought implies can principle in favor of retaining solidarity with the Reformers, but in the present age the ought implies can principle has come to enjoy almost universal acceptance. Given the choice between denying the ought implies can principle and acknowledging that some morally praiseworthy actions are non-obligatory, I believe that few Protestants of the present age would find the first option attractive.

The ought implies can principle operates in many people's common sense thinking in ordinary circumstances. A foreman in a factory believes that each employee in his section has an obligation to carry out his or her work assignment. When one of his employees is severely injured on the job, he no longer believes that this employee has the obligation to carry out her work assignment. His belief has been altered due to a recognition of the ought implies can principle. Because the injured employee is no longer able to carry out her work assignment, she is no longer under an obligation to do so. The ought implies can principle has no particular connection with Protestant thought, but a growing recognition of its truth by the population at large has certainly led Protestants to find untenable the type of anti-supererogationist sentiment expressed by the Reformers. Simply put, the Reformers adopt a position that contradicts the ought implies can principle (as illustrated in the panhandler example), and hence contemporary Protestants find no reason to reject the possibility of supererogation.

Protestants affirm that God has laid down requirements that human beings are constrained to follow. Nevertheless, this constitutes no reason to deny the ought implies can principle. The principle's intuitive appeal is undeniable, and one can justifiably regard incompatible teachings with skepticism. One can view God's requirements in such a way that the principle is preserved. Naturally, somc Protcstants such as Allen will not be convinced by this line of reasoning, but Protestants, in general, can plausibly assume that the burden of proof rests with those whose views concerning God's requirements violate the ought implies can principle. One can appeal to Scripture, but for Protestants in general the verdict is inconclusive, and there is a strong non-Scriptural presumption in favor of affirming the ought implies can principle. In the end, Protestants can consider it reasonable to interpret God's requirements in such a way that they are not obligated to perform actions that are not within their power to perform.

Many Protestants of the present age affirm that God enables people to go above and beyond what obligation requires. God has given us the ability to go the extra mile and has given human beings the ability to demonstrate love in a manner that goes beyond adherence to obligation. We need not view these demonstrations of love as contributing to a treasury of merit, but they nevertheless take place outside the bounds of obligation.

A paradigm example of a Protestant who subscribes to this way of thinking, which views God's law quite differently than the Reformers, is theologian Reinhold Niebuhr (other contemporary Protestants who have argued in favor of supererogation are Michael Pritchard, Caroline Simon, and Gregory Trianosky). His point of departure is to argue that the laws of God are less fixed than have been commonly believed. He believes that an erroneous conception regarding God's law has taken place in both Catholic and Protestant thought. This is a view that he describes as based upon "a Stoic-Aristotelian rationalism which assumes fixed historical structures."

Niebuhr states his point of view in the following passage.

> All law, whether historical, positive, scriptural, or rational, is more tentative and less independent in its authority than orthodox Christianity, whether Catholic or Protestant, supposes. . . . The final dyke against relativism is to bc found not in thcsc allcgcd fixitics, but in thc law of love itself. This is the only final law, and every other law is an expression of the law of love (Niebuhr, 1986, p. 159).

We should not, according to Niebuhr, regard God's laws as constituting a fixed system of demands. In his opinion the law of love is foundational, and other laws, framed and interpreted according to one's historical circumstances, are expressions of this law.

The conflict between the views of Aquinas and those of the Reformers is largely diminished, according to Niebuhr, when the law of love is seen as central. The reason, he believes, is that "love means a perfect accord between duty and inclination in such a way that duty is not felt as duty (*Ibid.*, p. 144)." Love that is sacrificial by its very nature implies being prepared to sacrifice oneself for others, and this means that duty and inclination can be in perfect accord. A person for whom sacrificial love is operative need not engage in a cost/benefit analysis when making sacrifices.

> The final form of love is bereft of such calculation and meets the needs of the other without calculating comparative rights. Sacrificial love is therefore a form of love which transcends

> the limits of law. . . . it is a point which stands beyond all law, because the necessity of sacrificing one's life for another cannot be formulated as an obligation (*Ibid.*, p. 150).

Niebuhr believes that because sacrificial love goes beyond the realm of law it can make perfect accord between duty and inclination. Sacrificial love, accordingly, cannot be exhaustively described in the language of moral obligations, and attempting to do so is futile. If one asks whether a person is bound to show sacrificial love, one is asking something nonsensical.

Second-mile acts cannot for Niebuhr be captured within the confines of moral duty, and God's law does not require human beings to perform second-mile acts whenever they have the ability to do so. Giving up one's life for another person is not for Niebuhr a duty that God requires. Rather, it is an ultimate instance of sacrificial love; as such, it falls outside the scope of duty. Niebuhr is in agreement with the Reformers in regarding martyrdom as incapable of contributing to a treasury of merit. But by the same token, he regards martyrdom as an act of sacrificial love that cannot be judged as satisfying or fulfilling God's commandments.

5 Conclusion

Fundamental to Protestant thought is the idea that good works can be an expression of thanksgiving to God. They are not a means to earning salvation, nor a means of building up a treasury of merit, nor primarily a means of escaping God's wrath. Protestants believe that God, through his abundant grace, has made salvation possible to those whose faith is genuine. The good works of those for whom salvation is possible can then function as a means of expressing gratitude. This is not to say that all good works are intended by Protestants to express thanksgiving to God, but Protestants affirm that at least some good works some of the time are expressions of gratitude for what God has done.

These good works have a strong potential to qualify as acts of supererogation. Sometimes the good works performed by people fulfill moral obligations, but at other times the good works performed by people are not obligatory and are not morally blameworthy to omit. This is true in particular of situations in which people are acting in gratitude to God. It would in fact be unusual for the good works performed out of thanksgiving to God to fulfill moral obligations. Calvin, Luther, Melanchthon, and Allen would presumably hold that good works performed out of thanksgiving to God are invariably obligatory, but I believe the vast majority of contemporary Protestants would disagree and would hold the door open to the possibility that sometimes these good works qualify as supererogatory.

It would not be surprising to discover that some contemporary Protestants regard acts of supererogation to be rare, especially by people who are self-centered or victims of narcissism. But a wholesale rejection of supererogation strikes me as something which few contemporary Protestants would endorse. Such a strict and demanding approach to ethics characterized the spirit of the Reformation, but it does not seem to capture the spirit of contemporary Protestantism.

David Hume once posed this question: "What hopes can we ever have of engaging mankind to a practice which we confess full of austerity and rigor (Hume, 1957, p. 100)?" Virtue, in Hume's opinion, "... talks not of useless austerities and rigors, suffering and self-denial (*Ibid.*, p. 99)." Suffering and self-denial can certainly be present in one's attempt to show gratitude to God, but the vast majority of contemporary Protestants would agree with Hume that these are not demanded of us in practicing virtue. When one suffers or denies oneself in showing gratitude to God, one is not typically doing so out of a sense of duty or obligation.

References

Allen, J. (1984). *Love and conflict.* Abingdon Press.

Aquinas, T. (1894). *Summa theologia.* P. Marietti.

Calvin, J. (1960). Institutes of the christian religion. J. T. McNeill (Eds.) trans, *Ford lewis battles.* Philadelphia: The Westminster Press

Heyd, D. (1982). *Supererogation.* Cambridge University Press.

Hume, D. (1957). *An inquiry concerning the principles of morals.* Bobbs-Merrill.

Luther, M. (1943). Treatise on good works. In: *Works of martin luther.* Philadelphia: Muhlenberg Press.

Luther, M. (1955). Explanation of the ninety-five theses. In: H.T. Grimm (Eds.), *Works of martin luther* (Vol. 31, Thesis 58). Philadelphia: Muhlenberg Press.

Melanchthon, P. (1969). In: W. Pauck (Eds.), *Loci communes theologici.* Philadelphia: The Westminster Press.

Niebuhr, R. (1986). Love and law in protestantism and Catholicism. In: R. M. Brown (Ed.), *The essential reinhold niebuhr: Selected essays and addresses.* New Haven and London: Yale University Press.

Rahner, K. (1965). The theology of the religious life. In: G. Hughe et al. (Ed.), *Religious orders in the modern world.* Gerald Westminster, Maryland: The Newman Press.

Urmson, J. O. (1969) Saints and heroes. In: J. Feinberg (Ed.), *Moral concepts.* London: Oxford University Press

Does Judaism Recognize the Supererogatory?

Samuel Lebens

Abstract This chapter puts forward a *prima facie* argument for a Jewish form of anti-supererogation before finding that no such argument can do justice to the Jewish tradition. Instead, the question becomes: what form of supererogation can Jewish law recognize? Qualified forms of supererogation would allow the Jewish philosopher to preserve certain theological and philosophical desiderata, but an unqualified form of supererogation sits more easily with a central approach to the nature of Divine revelation. Accordingly, the shape of a Jewish supererogation has deep consequences for Jewish philosophy at large.

Keywords Supererogation · Halakha · Judaism · Maimonides · Nachmanides

1 Introduction

Readers of this book will know already that "supererogation" is a technical term referring to that class of actions, if there be such a class, that go beyond the agent's ethical responsibilities, duties, or obligations. One standard way to make sense of our question, therefore, is to ask whether—from a Jewish perspective—the following list is exhaustive of all possible actions[1]:

1. Actions that are good to do and bad not to do.
2. Actions that are neither good to do nor bad not to do.
3. Actions that are bad to do and good not to do.

If Judaism doesn't recognize the supererogatory, then Judaism thinks that our three-part division of actions is exhaustive. If, however, Judaism *does* recognize the

[1] In posing the question this way, I'm following (Urmson, 1958), (Chisholm, 1963) and (Heyd, 2019).

S. Lebens (✉)
University of Haifa, Haifa, Israel
e-mail: slebens@univ.haifa.ac.il

D. Heyd (ed.), *Handbook of Supererogation*,
https://doi.org/10.1007/978-981-99-3633-5_19

supererogatory, then it recognizes a fourth category of action, beyond the three we've already listed, namely:

4. Actions that are good to do but not bad not to do.

This way of framing the question helps us to understand what we mean by "recognizing" or "failing to recognize the supererogatory." Fine. But we should, of course, ask, what is Judaism? This is a much harder question to answer. The adage, that for any two Jews there are at least three views, is borne out by experience and applies even to the question as to what Judaism is. But for our purposes, we don't need a particularly sharp definition. We can relate to Judaism, in the context of this paper, as a multivalent intellectual tradition set around a canon of texts: the Hebrew Bible, the works of the classical Rabbis—namely, the classical Midrashim, Mishna, Tosefta, and the Babylonian and Jerusalem Talmuds—in addition to the medieval, modern, and contemporary commentaries to these texts, whose authors viewed Jewish law (*halakha*) as authoritative, even as they disagreed as to its content, and the responsa literature (in which Rabbis respond to questions posed to them in correspondence). Ultimately, the question of this paper is whether or not that literature makes room for the notion of supererogation.

Of course, as a multivalent tradition, with multiple voices, Judaism doesn't always have just one clear answer to a question but often makes room for a variety of different approaches. Perhaps there are strands of the tradition that recognize the supererogatory and strands that don't. But sometimes you do find consensus, for example—nobody in this literature sanctions idol worship (in any circumstance) or the consumption, by Jews, of pork (except for in extreme cases, in which life and death hang in the balance). Does the recognition of the supererogatory fall within such a consensus? Many people, familiar with the Jewish tradition, would instinctively say, "yes." After all, the halakha recognizes four categories of actions:

1. Actions that are obligatory;
2. Actions that are not obligatory, forbidden, encouraged, or discouraged;
3. Actions that are forbidden; and,
4. Actions that are neither obligatory nor forbidden, but which are nevertheless encouraged.

The fourth category includes (but isn't limited to) those actions which the Rabbis describe as *lifnim mishurat hadin* (literally: *beyond*, or *before, the line of the law*). The existence of this fourth category is reason to think that Judaism does recognize the supererogatory. And yet, despite the fact that it's ultimately doomed to fail, a *prima facie* case can be made for a Jewish anti-supererogationism.

2 Jewish Anti-Supererogationism

Supererogation is, if it exists at all, a phenomenon that relates to the ethical. Its existence entails that, alongside the ethically mandated, the ethically forbidden, and the ethically neutral, there's a category of actions that are praiseworthy *without* being ethically mandatory.

It should go without saying that what is ethically mandated might not always be *legally* mandated. That all depends upon the system of law in question. Accordingly, the fact that Jewish law divides actions into four categories—the mandatory, the forbidden, the neutral, and the encouraged—doesn't entail that Jewish *ethics* makes a similar four-way division. Perhaps Jewish ethics is more demanding than Jewish law. Jewish law doesn't command and probably doesn't even specifically encourage, a person to jump on a live hand grenade to save the life of others. Perhaps this would be a good candidate for a supererogatory deed. But then again, the fact that Jewish law doesn't pronounce on this issue doesn't mean that Jewish ethics is similarly silent.

Perhaps Jewish ethics makes obligatory anything that's worthy of ethical praise, even if Jewish law is less exacting. The Babylonian Talmud (tractate Baba Metzia 30b) cites Rav Yochanan, who said, "Jerusalem was destroyed because [its inhabitants] judged [each other in accordance with] Torah law within it." The Talmud then asks, incredulously, "should they have followed the law of the Magians instead?!" Ultimately, the Talmud interprets Rav Yochanan to have meant the following: Jerusalem was destroyed "because they based their judgements solely upon Torah law and did not act *lifnim mishurat hadin* [Literally: *they did not* go *beyond*, or *before, the line of the law*. More colloquially, it means: *they did not go beyond the letter of the law*.]".

At least on a first reading, the temptation would be to see Rav Yochanan's statement, and its Talmudic gloss, as distinguishing between Jewish law and Jewish ethics. The problem in Jerusalem wasn't that the Jews were judging their cases of civil conflict according to the wrong law books. Jewish law was the appropriate legal system for that. The problem was that they were too litigious to begin with and were unwilling to compromise. They demanded what was rightfully theirs by dint of Jewish law but neglected to treat one another kindly, ethically, going beyond the letter of the law.

But, if that's the right reading, and if acting *lifnim mishurat hadin* literally means acting in accordance with an ethic that's higher, or more demanding, than Jewish law, then we have no reason to think that Judaism does or doesn't recognize the supererogatory. Supererogation, if it exists at all, isn't about going beyond the simple demands of the *law*; it's about going beyond the simple demands of *ethics*. In encouraging us to act *lifnim mishurat hadin*, and therefore, to go beyond the letter of Jewish *law*, Jewish texts are not taking a stance on whether we should, and whether it's even possible, to go beyond the minimal demands of *ethics*. The category of *lifnim mishurat hadin* is, on this reading, a red herring. It is a notion that relates to Jewish law, rather than to ethical obligation, and thus tells us nothing about Jewish attitudes to supererogation.

Admittedly, the category of *lifnim mishurat hadin* is just one candidate for supererogation in Jewish literature. In the Catholic tradition, and in much of the philosophical literature, post-Urmson (i.e., after Urmson, 1958), the supererogatry tends to be associated with saintly and heroic action, rather than going just a little further than one is strictly obligated to go.[2] But still, to the extent that halakhic Judaism is so centrally dedicated to the articulation of Jewish law, we're likely to find the same problem emerge with other Talmudic or Rabbinic candidates for supererogation. They're likely to tell us to go further than the law requires, whilst remaining silent as to whether one should or could go further than ethics demands.

But then, when we reflect upon the notion that Jewish law, according to Jewish sources, is based upon Divine revelation; when we reflect upon the idea that Jewish law was designed to govern a "priestly kingdom and a holy nation" (Exodus 19:6) so as to give rise to a people that would be a "light unto the nations" (Isaiah 42:6), it becomes extremely counterintuitive to imagine that there could be much of a gap between the demands of Jewish law and the demands of ethics. The idea that Jewish law is the product of Divine revelation might lead us to think that it would differ from secular law, which is the product of human decision-making.

Depending upon the aims and objectives of a secular legal system, we can imagine a lacuna of varying size between the demands of the law and the demands of ethics. But in a system where the laws have been given by God, for the above-stated purposes, you might think that any such gap would evaporate. Accordingly, an extreme position emerges, in the literature, according to which Jewish law itself is the key determinant as to what is ethical and what isn't. Indeed, this was the position of Rabbi Avraham Yeshaya Karelitz (1878–1953, otherwise known as the *Chazon Ish*). He declared that "it is the halakhah which determines the forbidden and permitted in ethics."[3]

If that's true, then how could there possibly be actions that are demanded by ethics, but not called for by the halakha? On the other hand, what do we make of Rav Yochanan's statement, and its Talmudic gloss, which *do* imply a great gulf between the demands of Jewish law and the demands of Jewish ethics? Such a gulf that it led to the destruction of Jerusalem.

One way to proceed would be to recognize that the halakha itself enjoins us to go *lifnim mishurat hadin.* We are, on this suggestion, *commanded* by Jewish law, to go further than its own explicit demands. This suggestion is most famously associated with Rabbi Moses Nachmanides (1194–1270), though it could be argued that this was the point that Rav Yochanan was trying to make about the destruction of Jerusalem. In the book of Leviticus (19:2), God tells the Children of Israel that, "you shall be holy, for I, the Lord your God, am holy." In his commentary to this verse, Nachmanides writes:

---

[2] One important exception to this trend has been the work of David Heyd (cf. e.g., Heyd, 1982, §7.4).

[3] (Karelitz, 1996, p. 19).

> The meaning thereof is as follows: The Torah has admonished us against immorality and forbidden foods, but permitted sexual intercourse between man and his wife, and the eating of [certain] meat and wine. If so, a man of desire could consider this to be a permission to be passionately addicted to sexual intercourse with his wife or many wives, and be among winebibbers, among gluttonous eaters of flesh, and speak freely all profanities, since this prohibition has not been [expressly] mentioned in the Torah, and thus he will become a sordid person within the permissible realm of the Torah![4] Therefore, after having listed the matters which He prohibited altogether, Scripture followed them up by a general command that we practice moderation even in matters which are permitted... And such is the way of the Torah, that after it lists certain specific prohibitions, it includes them all in a general precept. Thus after warning with detailed laws regarding all business dealings between people, such as not to steal or rob or to wrong one another, and other similar prohibitions, He said in general [Deuteronomy 6:18], "And thou shalt do that which is right and good," thus including under a positive commandment the duty of doing that which is right and of agreeing to a compromise [when not to do so would be inequitable]; as well as all requirements to act "beyond" the line of the law [i.e., *lifnim mishurat hadin*].[5]

Nachmanides expands upon this point in his commentary to Deuteronomy 6:18, and its injunction that the Israelites do what is "right and good in the eyes of the Lord." There he explains that it isn't possible for a finitely long law book to set out explicit instructions to cover every eventuality. For that reason, once a sufficiently large set of laws have been delineated, we find more general injunctions to be holy, and to do the right and the good.

Is Nachmanides calling for supererogation here, and—somewhat paradoxically—encoding it as a legal *obligation*? Perhaps he's merely encoding an obligation to act *equitably*. Equity, as Aristotle defines the notion (*Nicomachean Ethics*, 1137a-1138a), is what's called for when the abstract demands of justice, applied to a specific situation, will lead to an injustice, because the explicit law code fails to be sensitive to the very specific details of the case at hand. In other words, equity is "a correction of law where it is defective owing to its universality." Accordingly, to act equitably isn't really to act beyond the demands of justice. It is rather to enact, in a given situation, "a correction of law where it is defective owing to its universality." Aristotle's notion of equity, therefore, distinguishes between the formal requirements of the law—which we could call *legal justice*—and what might be called *true* justice. It follows therefore, that equity isn't to go *beyond* the demands of justice. It is merely the "correction of legal justice" in the name of true justice.

Nevertheless, David Heyd argues that equity can "in principle be viewed as one type of supererogation," even if only a "degenerate" case of it: "[I]ts corrective character makes it supererogatory only in the *qualified* sense: ideally it should have been part of justice, as Aristotle himself implies" (Heyd, 1982, p. 48). So, if Judaism commands us to be equitable in our dealings with others—as Nachmanides implies—there is thereby room to say that it *commands* the supererogatory (even if only a degenerate case of it). Moreover, Nachmanides sees the injunction to be holy, and to

[4] Rabbi Lichtenstein and Shmuel Shilo translate this phrase—*naval birshut hatorah*—as "a scoundrel with Torah license" (Lichtenstein, 2004, p. 39; Shilo, 1978, p. 370).

[5] As translated by Charles Chavel (with one very minor amendment of my own), see: https://www.sefaria.org.il/Ramban_on_Leviticus.19.2.2?lang=en.

do the right and the good, as having two elements: (1) to seek compromise, and (2) to go beyond the letter of the law. The first element seems to point in the direction of equity—the second element seems to push further—in the direction of more standard forms of supererogation (e.g., doing good deeds that we're not obliged to do) and suberogation (e.g., refraining from doing certain permitted deeds, or at least from indulging in them, when to do so would be dishonorable).[6]

Interestingly, in his commentary to Maimonides' *Sefer Hamitzvot*, in which Maimonides lays out what he takes to be the 613 commandments of the Torah, Nachmanides doesn't go so far as to list either equity or supererogation, in the form of acting *lifnim mishurat hadin*, among them (neither does Maimonides). Nevertheless, it's clear from what he wrote in his commentary to the Torah, that Nachmanides viewed these things, if not as fully-fledged *commandments*, then at least, to quote Rabbi Lichtenstein (2004, p. 41), "as a normative duty, incumbent upon—and expected of every Jew as part of his [or her] basic obligation." But it was Rabbi Isaac of Corbeille (d. 1280) who extended Nachmanides' logic to its ultimate conclusion when he did go so far as to include, in his own count of the 613 commandments, the duty to act *lifnim mishurat hadin.*[7]

Rabbi Isaac's inclusion of *lifnim mishurat hadin* within the body of Jewish law allows us to recover the theologically appropriate link between halakha and ethics. Sometimes, the explicitly codified laws of the Torah fall short of what's ethically demanded, but at those times, the Torah's more general injunction to act ethically kicks in. What Rav Yochanan, and his Talmudic gloss meant, wasn't that there's a gulf between the demands of halakha and the more exacting demands of ethics. There can be no such gulf. Rather, what they meant was that the explicit and specific laws of the Torah, when unencumbered by the additional Torah law of acting *lifnim mishurat hadin*, is what gives rise to such a gulf. The Jews of Jerusalem were particular regarding 612 of the 613 commandments, but they didn't act *lifnim mishurat hadin.*

Where does this leave us regarding the question of supererogation? In one fell swoop, we seem to have made the halakhic system relevant to the study of ethics again, by insisting upon its encompassing every demand of ethics, but we also seem to have undone any connection between the notion of *lifnim mishurat hadin* and the notion of supererogation. After all, if there is a *commandment* to act *lifnim mishurat hadin*, then it turns out that we have an obligation so to act. And if we have an obligation so to act, it can hardly be considered supererogatory. Indeed, if the category of *lifnim mishurat hadin* is truly to be viewed as *obligatory*, then we need to revise the four-fold division of actions into the following new five-fold schema:

1. Actions that are explicitly mandated in the codes of Jewish law;
2. Actions that are neither halakhically obligatory nor forbidden;
3. Actions that are explicitly prohibited in the codes of Jewish law;
4. Actions that are obligatory, even though they are not explicitly mandated in the codes of Jewish law, but fall under the more general exhortations of the law, to be holy and good; and

---

[6] For more on the notion of suberogation, see (Driver, 1992).

[7] *Sefer Mitzvot Katan*, §49.

5. Actions that are forbidden, even though they are not explicitly prohibited in the codes of Jewish law, but whose prohibition falls under the more general exhortations of the law, to be holy and good.

The fourth and fifth lines of this list, on this understanding of the concept, are what maps onto the category of *lifnim mishurat hadin*. You might think that line 4 refers to the supererogatory, and that line 5 refers to the suberogatory. But that would be a mistake. Both lines delineate legally binding obligations. So, either we say, perhaps with Nachmanides, that Jewish ethics really is broader than Jewish law; that it *is* possible to be a scoundrel with the full sanction of Jewish law, but that with its notion of going *lifnim mishurat hadin*, Jewish *ethics* demands much more from us. But if that's the case, then the category of *lifnim mishurat hadin* is merely telling us to go further than a legal system that doesn't fully accord with the demands of ethics. It isn't telling us to go further than the demands of ethics themselves. It therefore tells us nothing about supererogation; not to mention it's giving up on the moral perfection of Jewish law. Or, we say with Rabbi Isaac of Corbeille that there is no room, within the parameters of Jewish law, for unethical behavior, and that acting *lifnim mishurat hadin* is itself a halakhic obligation. But, if that's the case, then there's no room for the supererogatory. If an action is good and right, then it's obligatory. So understood, and contrary to popular opinion, the category of *lifnim mishurat hadin* is either silent about, or amounts to an outright denial of, the supererogatory.

Having made this Jewish case against the supererogatory, we can immediately see a number of problems. In the history of religious philosophy, anti-supererogation tends to be associated with a Lutheran or Calvinist ethic according to which, as summarized by Heyd (2019), "The demands of God are so extensive that human beings have not the slightest chance of ever satisfying them, let alone going beyond them." If every possible good is actually obligatory for us to perform, would we have time to sleep, or earn a living? This Protestant resignation in the face of God's extensive demands runs out of kilter with the Jewish sources. Indeed, the Talmud, on four separate occasions, invokes as a principle of Jewish law that, "The Torah was not given to ministering angels."[8] This principle is used to limit how demanding our interpretation of the law can be, since the law was given to human beings; not to angels. The law should be possible to keep. God doesn't make unrealistic demands. Rabbi Eliezer Berkowits recognized this to be one of the meta-principles at the heart of Jewish law. He called it *the wisdom of the feasible* (Berkovits, 1983, p. 8). Note that this potentially goes beyond the claim that *ought implies can*. It tells us that a halakhic ought implies a "can-with-relative-ease."[9]

---

[8] Babylonian Talmud, tractates Berachot 25b, Yoma 30a, Kiddushin 54a, and Me 'ilah 14b.

[9] Daniel Statman rightly notes, in correspondence, that (a) "observing halakha is often not that easy," and that (b), "the examples offered by philosophers for [the claim that] *ought implies can* are usually not examples in which the agent literally *can't* do X." Accordingly, I may have made too much of the distinction between the doctrine that ought implies can, and the halakhic notion of feasibility. On the other hand, different halakhic authorities place different weights upon the notion of feasibility. Observing the halakha as understood by one halakhic authority can often be much easier than observing the halakha as understood by another halakhic authority; depending upon how

As codified in the most authoritative code of Jewish law, the *Shulchan Aruch*, of Rabbi Yosef Karo (1488–1575), Jews are commanded, where feasible, to give 10% of their income to charity. It is then codified as "laudable" to give up to 20%. Rabbi Moshe Isserlis (1530–1572), in his gloss on these laws, warns us not to give more than 20%, lest we squander our wealth and become dependent on charity ourselves (Shulchan Aruch, Yoreh Deah 249:1). But if anything good is obligatory, then surely it would be obligatory for us all to give every penny we earn that isn't required for our own subsistence. If something were laudable, wouldn't it automatically be obligatory? Likewise, we're commanded to stay up until midnight on the first evening of Passover, and then we're told that those who stay up longer are to be praised.[10] But if it's praiseworthy, shouldn't it be obligatory? Could that really be what Rabbi Isaac meant, when he included *lifnim mishurat hadin* in the 613 commandments? What happened to the Torah not being given to ministering angels?

When Rabbi Isaac includes *lifnim mishurat hadin* in the 613 commandments, he explicitly cites Rav Yochanan's statement about the destruction of Jerusalem. In that context, as in a number of other contexts in the Talmud,[11] going *lifnim mishurat hadin* seems to mean nothing more than being willing to compromise upon that to which one is strictly entitled in cases of civil law. In fact, as Shmuel Shilo demonstrates, when the Talmud invokes the principle, in cases of civil law, to act *lifnim mishurat hadin* tends to means nothing more than to waive a special exemption given to you, and not to the general population, by Jewish law (Shilo, 1978, p. 364).[12] Perhaps that's all that Rabbi Isaac ever meant. He didn't mean to encode, as one of the 613 commandments, an obligation to do anything and everything that could be considered to be laudable. Perhaps he merely meant to encode a willingness to compromise upon one's legal rights, and to waive special exemptions, in the name of equity in civil law. True, equity may be a (degenerate) form of supererogation, but it's a far cry from transforming every laudable deed into an obligation. If that's the case, then the inclusion of acting *lifnim mishurat hadin* within the body of the law shouldn't be viewed as the denial of the supererogatory.[13]

---

they understand the notion of feasibility, and what weight they place upon it. On the more liberal ends of the spectrum, therefore, you might find "feasibility" given a wider scope than the "can" of *ought implies can*. But I concede to Statman that this isn't always, and certainly not obviously, the case.

[10] See, e.g., *Mishne Torah, Hilkhot Hametz U'Matza* 7:1.

[11] Babylonian Talmud: tractates Ketubot 97a, Baba Metzia 24b, Ibid., 30b, and Baba Kama 99b.

[12] Having said that, in the Bablyonian Talmud, Ketubot 97a, *lifnim mishurat hadin* is used to waive not a special exemption, but a regular right. Based on his own analysis of all of the primary Talmudic sources, Newman (1998, p. 29) is inclined to define *lifnim mishurat hadin* in terms of a waiver: "Whether we are concerned with an elder who has a right to refrain from unloading animals (but does so anyway), or a man who has a right to keep the property that has been sold to him (but returns it to the seller), the term *lifnim mishurat hadin* designates a willingness to waive voluntarily some benefit or right to which one is entitled by law. In each case, it is implied that the party who waives the right in question does so out of a concern for the other party, who would be harmed or disadvantaged if the right were exercised."

[13] In correspondence, David Heyd makes the following point: Calling for people to waive their legal rights is sometimes equivalent to a call for equity, but it might sometimes be a call for something

Indeed, the centrality of *feasibility* to the edifice of Jewish law will make it very hard to sustain a robust anti-supererogationism. But if we really can't make the case for a Jewish anti-supererogationism, does that mean that we have to accept that the demands of Jewish law really are less than ideal; since the laudable isn't always obligatory? The answer to this question will depend upon the shape that Jewish supererogation might take. And you won't be surprised to hear that this has been subject to debate.

3 Qualified Supererogation

The one pleasing consequence of Jewish anti-supererogation is that it resolutely refuses to believe that the Torah leaves out any ethical imperative; and, like any form of anti-supererogation, it refuses to believe that there could be good deeds that we're not obligated to perform. In other words, it zealously guards the connection between axiology (value) and deontology (duty). If there's something good that you could do, then you *should* do it. One way to respect the halakhic wisdom of feasibility,[14] without breaking the link between axiology and deontology, and without conceding that the Torah demands less than ethical perfection, is to adopt a qualified form of supererogation.

Qualified forms of supererogation insist that value generates duty, but that duties are not uniformly distributed. What we call the supererogatory, on this qualified view, are the duties that only certain people have, or the duties from which many people are exempt. Supererogatory actions are still, on this account, *duties*—but they are *qualified* duties because they only apply to certain people.[15]

---

more obviously supererogatory. "If the letter of the law says that you have in a particular case to do X but it goes against the *spirit* of the law to do X (since the law could not foresee the particular unethical implications of this case), then it is a case of equity" but, he continues, "if you compromise with a thief who stole your property because he is poor and you leave him with half of the property, this is not equity but an act of supererogation." I concede the point. To gauge its relevance would require us to better understand the situations in which Rabbi Isaac would demand that a person waive his legal rights, and when he would make no such demand. Moreover, even if Rabbi Isaac always demands the supererogatory waiving of rights in cases of civil law, it's still a very limited inclusion of the supererogatory into the body of Jewish law. Remember: to include it as a demand of Jewish law is to rob it of its supererogatory status and render it an obligation. So, even if Heyd is right that Rabbi Isaac is demanding more than mere equity here, he has still left plenty of room for the truly supererogatory beyond the body of Jewish law.

[14] For more on this notion, and its roots in Jewish literature, see (Berkovits, 1983, p. 8) and the Bablyonian Talmud, tractates Berachot 25b, Yoma 30a, Kiddushin 54a, and Me 'ilah 14b.

[15] The three-way distinction between anti-supererogation, qualified-supererogation, and unqualified-supererogation is owed to Heyd (1982). As examples of different forms of qualified supererogation, Heyd (2019) cites: Richards (1971), who speaks of "supererogatory duties"; John Rawls, who relates to supererogatory actions as duties falling under an exemption, when the agent waves the exemption (Rawls, 1971, p. 117); Raz (1975), who similar to Rawls, thinks of supererogatory actions occurring when an agent disregards a second-order reason that counts as a permission not to act, in favor of a first-order reason so to act (1975); and Shlomo Cohen's idea of "forced

One might think that Maimonides adopts a qualified supererogationism in his treatment of *lifnim mishurat hadin.* In his codex, the *Mishne Torah* (*Hilkhot De'ot* 1:5), Maimonides presents two character types—the pious and the wise. He writes:

> One who controls himself with increased restrictions, and distances himself even from the mean tendency [i.e., the golden mean], a little one way or a little the other way, is called devout [i.e., pious]. How may one do it? One who will distance himself from the extreme point of arrogance and become exceedingly humble is called devout for, this is the tendency of piety. If one distances himself to the mean point only and practices meekness he is called a wise man for this is the tendency of wisdom. A like pathway exists in all the rest of the tendencies. The ancient devotees diverged their tendencies from the middle-path facing the two extremes; some tendency they bent opposite to the last extreme and some tendency they bent opposite the first extreme. This is more than being within the limits of the law [i.e., this is acting *lifnim mishurat hadin*]. As for us, we are charged to walk in these middle-paths, which are the good and straight paths, even as it is said: "And thou shalt walk in His ways" (Deut. 28 9).[16]

In other words, the Torah enjoins us to calibrate our character traits to the golden mean—don't be too serious, but don't be too jocular either. This is the wise path of the sage. It is, of course, one of the main teachings of Aristotle's *Nicomachean Ethics*. The pious, by contrast, act *lifnim mishurat hadin* and go beyond the mean, towards extreme piety. The reason that this can, and has, been confused with qualified supererogation is that it looks as if there's a special duty, the duty to go *lifnim mishurat hadin*, and that this special duty falls only upon an ethical elite: the pious; but not the rest of us.

One reason for thinking that this can't be the right interpretation is that it would be uncharacteristic of Maimonides to place sagacity below any other characteristic, even piety. Indeed, for Maimonides, sagacity and true virtue come hand in hand with one another. Another reason for revising our understanding of this passage is how Maimonides uses the phrase *lifnim mishurat hadin* in other places in his legal code.

Basing himself on the Talmud (*Baba Metzia* 24b), Maimonides rules that there are certain circumstances in which a person isn't obliged to return an item of lost property. But, in such a case, "He who wants to take the good and the straight road and act *lifnim mishurat hadin* should return the lost article" (*Mishne Torah, Laws of Theft and Lost Property,* 11:7). This isn't put forward as an obligation so much as a choice that's open to anybody to make, without compulsion. Similarly, basing himself on the Talmud's use of "*lifnim mishurat hadin,*" Maimonides rules that a respected elder is exempt from various obligations that may be socially embarrassing for a person of his station, but rules that if this respected elder wants to "follow the good and straight road," he should act *lifnim mishurat hadin* and waive his exemption (Ibid., 11:17; Ibid., *Laws of Murder and the Preservation of Life,* 13:4). As Shilo (1978, p. 372) notes:

---

supererogation" (2015), according to which some actions are such that one isn't blameworthy for failing to do them, but one has still done something wrong.

[16] As translated by Simon Glazer, https://www.sefaria.org/Mishneh_Torah%2C_Human_Dispositions.1.5?lang=en&with=all&lang2=en.

> We can conclude from Maimonides' phraseology that not only is the ordinary individual free from having to act *lifnim mishurat hadin* but even those persons—including sages and scholars—who are exempted from returning a lost article under certain circumstances, and from helping with loading and unloading, are not duty-bound to act *lifnim mishurat hadin.*[17]

As Shilo notes, every instance in which the Talmud invokes the principle of *lifnim mishurat hadin*, it is discussing the actions of sages (and, we could add, in some instances, God Himself)[18] rather than the actions of the everyman. Bringing all of these considerations together, I want to suggest my own understanding of Maimonides' view.

---

[17] In the very next page, Shilo confuses matters. He argues that, as far as Maimonides is concerned, and in conflict with certain episodes quoted in the Talmud, we do not demand even from the *pious* that he act *lifnim mishurat hadin*, when doing so would incur a monetary loss. But, if what we've said until now is correct, then, strictly speaking, Maimonides never demands that the pious (or anyone else) act *lifnim mishurat hadin*. Instead, acting in that manner is sometimes offered as an option, but never a duty.

[18] For an example of the concept applied to God, see Babylonian Talmud, Tractate Brachot 7a, where—somewhat bizarrely—God is depicted as praying ("to whom?" you may ask) that He will go *lifnim mishurat hadin* in His interactions with the Jewish people. In many Rabbinic texts, God's attributes of justice and mercy are presented as being in *prima facie* tension with one another. God is here depicted as praying for his Mercy to overcome his justice.

David Heyd suggests, in correspondence, that God doesn't pray for his actions to accord with justice, since his nature is to be just, and thus he necessarily acts in accordance with justice. By contrast, Heyd suggests, God prays to act mercifully, in this Talmudic story, because mercy is thought of, by the Rabbinic authors of the text, as falling within the realm of Divine discretion. The idea is, therefore, that God's mercy—unlike his justice—is supererogatory.

If the tension between God's justice and his mercy is more than *prima facie*, and turns out to be *ultima facie*, then—when God acts mercifully, there's a sense in which he isn't acting justly. This would undermine Heyd's suggestion that God's mercy is discretionary while his justice is necessary. On the other hand, there are good reasons for thinking that the tension, however vividly portrayed by the Rabbis, between God's mercy and his justice can only *really* be *prima facie*.

Indeed, the Rabbis (in Tractate Avoda Zara 4b) seem to make a distinction between the times in which God is learning Torah, and the times in which he sits in judgment. When he is learning Torah, they suggest, his mercy isn't manifest, and he isn't inclined to act *lifnim mishurat hadin*. When he sits in *judgment*, by contrast, his mercy *is* manifest because he is inclined to act *lifnim mishurat hadin*. One way to understand what's being said, by this conceit, is that we should distinguish, as we did when discussing Aristotle's treatment of equity, between *legal justice* and *true justice*. When God is studying law as an abstract text, there isn't room for mercy. At that point, he relates to justice in terms of *legal justice*, which creates the *prima facie* conflict between God's mercy and his justice. But, when he sits in *judgment*, no longer pontificating on the law as an abstract concept, but engaging with the concrete realia of people's lives, the relevant species of justice is *true justice* which *is* compatible with God's manifestation of mercy.

And yet, despite there being no *ultima facie* conflict between God's justice and his mercy, it certainly seems that, in the Rabbinic imagination, and in consonance with Heyd's reading of God's prayer, God's mercy is discretionary, in a way that his justice isn't. Indeed, Rabbi Meir understands Exodus 33:19 to be saying that God exercises a discretionary prerogative to show grace and mercy to whomever He will, even if the recipients are not worthy of it.

Going further that this distinction between God's non-discretionary justice, and his supererogatory mercy, one might think, along with Adams (2002), that there is some sort of conceptual confusion in thinking that God has any obligations whatsoever. If that's the case, one might think that all of God's deeds go beyond His duties—since he has no duties—and thus, all of his actions are supererogatory!

The only circumstances in which Maimonides sanctions the cultivation of extreme character traits are (a) against haughtiness, and in favor of extreme humility, (b) against anger, and in favor of extreme calm, and (c) when the agent in question is seeking to fix a prior imbalance. For example, even though the ideal is to seek moderation, a person who is struggling with carnal desire, might be best advised to adopt, if only temporarily, a life of extreme asceticism, in order to fix the imbalance, and steer the agent back to the golden mean.[19] It is in this context that we should understand the Talmud's stories of pious sages waiving their legal exemptions, mucking in with the common-folk (so to speak), by acting *lifnim mishurat hadin.*

The exemptions granted to sages, not to trouble themselves with people's lost property, and not to involve themselves with loading other people's donkeys, are well motivated. They are designed to underline the important social status of a sage, to protect his precious time, and to inspire respect for the learned. But these exemptions can lead the sage himself away from extreme humility, towards haughtiness and pride. In such cases, it may be wise for the sage (if only from time to time) to wave their exemptions, and to act *lifnim mishurat hadin,* so as to keep any incipient haughtiness in check. This is a good thing to do, but very much, *ex post facto.* Moreover, the tendency to act *lifnim mishurat hadin* would be no part of a Maimonidean description of the ideal virtues, because a person only needs to act that way if he's worried that the privilidges of his station will go to his head. A more virtuous person wouldn't need to act *lifnim mishurat hadin* because their humility would be more secure than that.

On my reading of Maimonides, to act *lifnim mishurat hadin* is never a duty, not even for an elite. Moreover, there's a sense in which recourse to it, though sometimes appropriate, is always far from ideal.

Even though Maimonides, on my reading, doesn't treat the category of *lifnim mishurat hadin* as a form of qualified supererogation, there certainly are thinkers in the Jewish tradition who did. For example, Rabbi Azaria Piccio (1579–1647), also drawing from the fact that all of the human Talmudic examples of people acting *lifnim mishurat hadin* concern pious sages, concludes that the halakha itself makes two sets of demands regarding ethical behavior: "One is for the ordinary and the other for the exceptional individuals."[20]

One of the central Talmudic cases in discussions of *lifnim mishurat hadin* doesn't even invoke the phrase itself but certainly looks to be an example of the same phenomenon. I quote:

> Rabba bar bar Ḥanan [hired] some porters who broke his barrel of wine [which he had hired them to transport]. He took their cloaks [as payment for the lost wine]. They came and told Rav.
>
> [Rav] said [to Rabba bar bar Ḥanan], "Give them their cloaks [back]."
>
> [Rabba bar bar Ḥanan] said to him, "Is this the *halakha*?"
>
> [Rav] said to him, "Yes, [as it is written,] "That you may walk in the way of good men" (Proverbs 2:20)."

---

[19] *Mishne Torah, Hilkhot De'ot* 2:2–3.

[20] *Sefer Binah Le'ittim,* §12.

[Rabba bar bar Ḥanan] gave them [back] their cloaks.

[The porters] said to [Rav], "We are poor people, and we toiled all day and we are hungry, and we have nothing."

[Rav] said to [Rabba bar bar Ḥanan], "Go and give them their wages."

[Rabba bar bar Ḥanan] said to him, "Is this the *halakha*?"

[Rav] said to him: "Yes, [as it is written,] "And keep the paths of the righteous" (Proverbs 2:20)."

(Tractate Baba Metzia 83a)

Interestingly, a very similar story is related in the Jerusalem Talmud, but there, the employer is not a distinguished sage.[21] That should, at least, raise the prospect that there's a difference of opinion between the Jerusalem and (the more authoritative) Babylonian Talmud on this issue. Perhaps the reason that the Babylonian Talmud concentrates on a case in which the employer is, himself, a pious sage is that acting *lifnim mishurat hadin* can only be called obligatory for them. Perhaps Rav wouldn't demand such action from regular people, but only from the elite.

Maimonides doesn't cite this case in his code but in a reading that agrees with Rabbi Piccio, Rabbi Shlomo Ha-Cohen of Vilna (1828–1905) cites this incident at the end of a responsum, in order to share a "wonderful thing that I heard from Rabbi Shlomo Ze'ev, our Rabbi and teacher, of blessed memory."[22] On Rabbi Shlomo Ze'ev's reading of the story, Rav is trying to send a subtle message to Rabba bar bar Hanan. Rav quotes a verse from Proverbs that actually refers to two types of path: "the way of the good" and "the paths of the righteous." These two types of path, according to Rabbi Shlomo Ze'ev, symbolize the two levels of ethical injunction to be found in the Torah. One level binds the regular person, keeping them on the way of the good. The other level binds the righteous. Of course, it would be good for the ordinary people to emulate the righteous and walk in their paths too, but only the righteous are so obligated.

Even if Maimonides doesn't generally think of *lifnim mishurat hadin* as a category of qualified duties that fall only upon an ethical elite, it does seem that Maimonides believed in the existence of duties like that. In fact, there's one instance in which Maimonides tells us that a person "*must* take particular care" to act *lifnim mishurat hadin*. That is to say, he recognizes one sort of case in which to act *lifnim mishurat hadin* isn't a choice, but an obligation. When delineating the contours of a particular sin, called *hilul Hashem,* sometimes translated as blasphemy, and the related commandment of *kiddush Hashem,* the sanctification of God's name, he writes:

There are other things included in blasphemy, although they are not of themselves either among the mandatory or prohibitive commandments, as for example, when a great man, famed for his learning and piety, will do something that the public will suspect him on account thereof, even though such deeds be not transgressions, yet he has committed blasphemy, as for example: if he makes a purchase and does not pay for it at once, although he has the money and the vendors are claiming it and he delays them; or if he indulges in frivolity, or doth eat and drink with and among the ignorant, or if his speech with his fellow men be

[21] Jerusalem Talmud, Baba Metzia 6:6.

[22] *Azei Broshim, Hilkhot Nedarim* §53.

> not polite, or if he does not receive them pleasantly, but acts as one looking for strife and shows anger. In such and like matters, all measured by the standard of the greatness of such scholar, he must take particular care, and act exceedingly better than the law requires [*lifnim mishurat hadin*].

Admittedly, it's not immediately clear why refraining from behaving rudely and from being mean should count as cases of supererogation. But Maimonides continues:

> Conversely, if the scholar restrains himself, speaks politely to his fellow men, and when among them acts like one of them, and receives them pleasantly, takes abuse from them but never gives abuse to them in return, respects them, even those who do not respect him, acts in business honestly, does not remain in company with the unlearned, nor visits their assemblies, and is rarely seen otherwise than to be engaged in the study of the Torah, wrapt in the prayer-garment and adorned with phylacteries, and performs his duties exceedingly more than the law requires [*lifnim mishurat hadin*], provided he does not go to extremes and does not act so ridiculously, so that all praise him and love him and crave to imitate his actions, behold he doth sanctify the Lord [thereby fulfilling a commandment of the Torah], and concerning him, the verse speaks, saying: "And He said unto me, thou art my servant, Israel, in whom I am glorified" (Isaiah 49.3).[23]

And thus, it seems that Maimonides uses the phrase *lifnim mishurat hadin* in two different ways. Generally, he uses it to refer to a non-obligatory category of actions in which a person waives a legal exemption, or deviates from the golden mean, as a precautionary procedure to counter various spiritual and psychological dangers. That form of action, *lifnim mishurat hadin,* is neither an ideal virtue nor obligatory, even for some subsection of humanity. But the second sense of the phrase, as it appears here, in his discussion of blasphemy, comes much closer to the view of Rabbis Piccio, and Shlomo Ha-Cohen. As a person becomes more well-known for his sagacity and piety, he becomes a walking ambassador for the Torah and its way of life. To the degree that this is true, that person comes to be obligated by more and more duties.

And thus, Maimonides, like Rabbis Piccio and Shlomo Ha-Cohen, allows for the existence of supererogatory actions. It's not as if these actions slip through the net of the halakhic system. Rather, these actions exist as qualified halakhic duties that only come into force when a person has come a long way in their journey towards ethical perfection. In the meantime, the masses can continue to discharge their lesser obligations, as they walk along the path of the good, all the time aspiring to elevate themselves to the paths of the righteous, upon which the righteous are obligated to walk. In this way, a Jewish qualified-supererogationism allows for us to respect the wisdom of the feasible, which doesn't place unrealistic obligations upon the masses, whilst insisting that every single good deed is actually a Torah-mandated obligation, even if these obligations don't fall on every person equally.

Even if Maimonides thinks of *lifnim mishurat hadin,* on rare occasions, in terms of a qualified duty, falling only on certain ethically advanced individuals, he does sometimes uses the injunction of *imatatio Dei* to exhort us, all of us, and not just the ethical elite, to go beyond the minimal requirements of Jewish law.[24] And thus,

---

[23] Mishne Torah, Hilkhot Yesodei Hatorah, 5:11, as translated by Simon Glazer https://www.sefaria.org/Mishneh_Torah%2C_Foundations_of_the_Torah.5.11?lang=en.

[24] See, for example, *Mishne Torah, Laws of Slaves* 9:8.

his use of *imatatio Dei* might constitute the introduction of a non-qualified form of supererogation into the legal world of Maimonides. What would a Jewish, unqualified supererogationism look like?

4 Unqualified Supererogation

We've already mentioned, in passing, the phenomenon of Jewish laws that conflict with our ethical sensibilities. This is not a new phenomenon. The Rabbis themselves were, seemingly, perturbed by certain laws. For example, Deuteronomy 21 grants a victorious Jewish army certain sexual rights over female captives. The soldier must first of all take the woman he desires back to his home, leave her to grieve her losses, shave her hair and pare her nails, and more. Once the process is complete, the soldier has a right to make this woman his wife, via a consummation of the relationship, with or without her consent. If he no longer desires this, instead of becoming his wife, she becomes a free and equal citizen of Israel.

The Rabbis were shocked by the depravity of these laws. They had pioneered a legal system in which conjugal rape was forbidden.[25] By contrast, it wasn't made illegal in Finland until 1994. Lichtenstein didn't make it illegal until 2001. No wonder the Rabbis—given their views on this issue—were dismayed by the provisions of Deuteronomy 21. This was their defense: these laws are a concession to the evil inclination of men.[26] In times of war, soldiers are prone to act in horrendous ways. God knew that if he commanded an ancient people not to rape in times of war, they would not listen. But a larger number of potential rapists *would* listen if they were told that they could have their evil way, but in a restrained and delayed fashion. The picture that emerges still isn't pretty, but the Rabbis comforted themselves with the notion that this was merely a compromise with the evil inclination of mankind.

But, if that's the case, if Jewish law sometimes compromises with the cultural situation of the community to whom it was revealed, then we shouldn't too quickly endorse an equivalence between the halakha and the ethical. To respect the Rabbinic notion that the Torah sometimes addresses itself to the evil inclination of man, we might want to temper the *Chazon Ish*'s insistence that the halakha alone determines what is ethical and what isn't. A more moderate attitude recognizes that the demands of ethics, at any given time, might differ in various ways from the demands of halakha (perhaps in ways that echo the differences that John Rawls recognized between ideal and non-ideal political theory). The halakha might be aimed towards inculcating virtues and bringing society ever closer to the ethical ideal without being identical to a code of ethics. I would associate this position with Rabbi Aharon Lichtenstein

[25] So rules Maimonides (Mishne Torah, Laws of Forbidden Relations 21:12), based upon numerous Talmudic injunctions, for example, the Babylonian Talmud, Tractate Eiruvin 100b. The Israeli Supreme Court appealed to such passages in Jewish law in order to affirm that conjugal rape is illegal in the State of Israel (Cr. A. 91/80, Cohen v. State of Israel).

[26] Babylonian Talmud, Tractate Kiddushin 21b.

(1933–2015), who maintained that the religious and the ethical are "inextricably interwoven," without going so far as to claim, with the *Chason Ish*, that the ethical and the halakhic are indistinguishable.[27]

We can recognize, with Rabbi Lichtenstein, that the halakha cannot be entirely divorced from the ethical, since the very *raison dêtre* of the Torah and its law is to shape a people and a culture and to bring them closer to God. But we can recognize that there may be a gap between the legal requirements of the halakha at any given time, and the requirements of ideal ethics.

Indeed, and as I argue at length elsewhere (Lebens, 2020, pt. II), an abundance of Jewish sources gives rise to the notion that the revelation of the Torah is a work in progress. One way to express this thought is to talk of two different Torahs.[28] There is the heavenly Torah, which—according to the Midrashic tradition, and hinted at in the book of Proverbs (Chap. 8)—existed before the creation of the universe. This Torah is perfect. The Torah that was revealed to us on Mount Sinai, by contrast, is the best approximation of that heavenly Torah that we are ready and able to receive. As Torah literature continues to proliferate, and the body of Jewish law expands and evolves, guided by the hand of providence, the thought is that the earthly Torah inches closer to its heavenly paradigm.

In the book of Deuteronomy (5:19), the theophany at Sinai is described with these words: "The Lord spoke those words to your whole congregation at the mountain, with a mighty voice out of the fire and the dense clouds, and no more." The words which we have rendered "and no more" are, in the original Hebrew, ambiguous. They could be read as attaching to the words that God spoke. The meaning would then be that God spoke those words and no more. That is to say: he said nothing else. Alternatively, they could be read as attaching to the mighty voice. Their meaning would then be different. They would mean: "it did not cease," which is to say, the voice heard at Sinai is still echoing today. Siding with the second reading, Rabbi Isaiah Horowitz (1555–1630, known as *the Shla*) argues that every generation of rabbinic law was, in a sense, uttered at Sinai. It was somehow latent in the voice we heard there, as a potential; a potential that is only actualized when the time is right, by Rabbis who hear the echo of Sinai, and have a great deal of understanding concerning the specifics of their time, in terms of "earthly affairs," and the status of the "souls of the generation."[29]

With this sort of picture in mind, and with the Rabbinic notion that the Written Torah made certain compromises with the non-ideal cultural situation into which it was revealed, we make room for the notion that Jewish law itself is (a) a work in process and (b) aimed at bringing an imperfect society progressively closer to the ideal. In this vein, we can begin to understand the argument of Rabbi Nahum Rabinovitch (1928–2020), for example, according to which the Torah was never resigned to the institution of slavery, despite appearances to the contrary, but was

27 (Lichtenstein, 2004, p. 38).

28 Although a less radical way to express a similar thought is simply to distinguish between the Torah, which is unchanging, and its interpretation, which is subject to evolution.

29 *Shnei Luchot Habrit, Toldot Adam Beit Chachma (Telitaah).*

specifically designed so as to wean us off of it (Rabinovitch, 2003). Rabbi Eliezer Berkovits makes similar claims about the Torah's attitude to women's rights, a process which he thought was far from over (Berkovits, 1990).

With this theory of a cumulative relation, and Torah as a process, we move towards a much less qualified form of supererogation. This form of supererogation says the following:

- At least for a Jew, to do one's duty, at any given time in history, is to do the minimal baseline required by Jewish law. To think that God commands us to do less than our duty is unthinkable. But there will often be more that can be done, and the Torah encourages us to do it. It encourages us, all of us, without qualification, to act *lifnim mishurat hadin.*

This takes us back, it seems, to the position of Nachmanides. On the one hand, in his commentary to the Pentateuch, he talks of a commandment to be holy, and a commandment to do the right and the good, and he interprets these injunctions in terms of acting *lifnim mishurat hadin* (which demands equity, in addition to more standard forms of supererogation). But, at the same time, and unlike Rabbi Isaac of Corbeille, he didn't include any of these injunctions in his authoritative list of the 613 commandments of the Torah. Rabbi Lichtenstein suggests that, in order to make sense of this sort of position, we could distinguish between Halakha with an upper-case H, and halakha with a lower-case h (Lichtenstein, 2004, p. 51).

Expanding upon this suggestion, we could say the following. Lower-case-h halakha refers to the explicit injunctions of Jewish law as they are codified at any given time. It is those injunctions that calibrate matters of formal duty. But the Halakha with an upper-case H includes all of the exhortations of the Torah to go further than the halakha with a lower-case h. The Halakha with an upper-case H doesn't add any new *duties* to the law books, but it does calibrate our ethical aspirations, and the ideals to which we should try to conform, even if the wisdom of feasibility entails that these cannot be considered to be fully-fledged obligations. Or, if you prefer, you could say that Nachmanides' commandment to act *lifnim mishurat hadin* is a commandment of the ideal and heavenly Torah, even if it isn't one of the 613 commandments that are halakhically obligatory in the hear-and-now of the earthly Torah.

This unqualified Jewish supererogationism denies that every value has to come along with an associated duty. Instead, we could distinguish between formal duties and quasi-duties. The former are those actions that we are really obliged to perform—and all of those actions, without remainder, are prescribed by Jewish law. The quasi-duties, by contrast, constitute something like a set of aspirations, or regulative ideals.[30] Perhaps as we progress, or as society progresses, some of the quasi-duties become fully fledged formal duties, either for individuals, or for the community (as happened with monogamy in Jewish law, for example)[31]—but there will always be

[30] I'm grateful to David Heyd for pointing out how similar this suggestion is to Henri Bergson's idea of a two-tiered morality, of which the higher tier was called "the morality of aspiration" (Bergson, 1956).

[31] The captive woman of Deuteronomy 21 is probably another good example of this phenomenon, since it's hard to imagine any mainstream or responsible halakhic authority argue that this Biblical

space (at least until the messianic age) to do more than our formal duties call upon us to do. There will always remain (at least until the messianic age) the domain of the supererogatory; which Nachmanides calls *lifnim mishurat hadin*, and which Maimonides might include under the category of *imitatio Dei*.

In fact, you might think that something is sometimes lost in transforming an aspiration into an obligation. Rabbi Lichtenstein notes that if husband and wife consult the halakhic sources in search of a clear prescription for how their relationship should unfold, and what dynamic they should stive to achieve in their relationship: "How intense, how superficial, how cordial?" They will be disappointed. The "Halakha does not tell you" (Lichtenstein, 2007). The situation is the same for those looking for halakhic guidance as to what sort of parent to be:

> Do you intend the relationship to be formal or chummy? The [Babylonian Talmud] ([tractate] Kiddushin 32a) teaches that a father who foregoes the honor due him *may* do so; does it say anywhere whether a parent *should* do so? There are differences between cultures and families. When we are at home, my children can poke fun at my wife and at me. It is part of the scene, and we take it in stride and with joy. One would never have spoken in that way in my parents' home, and it would never even have occurred to anyone to speak that way in [Rabbi Joseph Soleveitchik's] family.[32] It is not that the degree or quality of the love is different, but the manifestation is different.
>
> (Ibid.)

Some things simply shouldn't be legislated, either by legal or ethical norms. To legislate them would be to formalize something that should be left intuitive, and to render lifeless and impersonal something that should be personal and alive. To think that every value can be translated into a maxim and then codified as a duty is to do violence to the world of value. That is the cost of anti-supererogation and even qualified supererogation—it wants to turn every value into the sort of duty that could be codified into a law book (or an ethics manual). But some things shouldn't be reduced to the formulaic prose of a law book (or an ethics manual); so says the unqualified supererogationist. It's not just, as Nachmanides notes, that no law book could be long enough to legislate for every situation. It's not just, given the cultural, technological, and environmental changes that would inevitably accompany the unfolding of history, it was desirable for the Torah only to give general guidance on various issues. It's also that some values are ruined by being transformed into obligations or duties.[33]

---

right is still in force, since today a military can and should be expected to enforce better discipline (see, for example, *Penini Halakha, Sefer Ha'am ve'Ha'aretz* 4:18—I'm grateful to Rabbi Jeffrey Saks for this reference).

[32] Rabbi Soloveitchik (1903–1993) was the most prominent Rabbinic proponent of Modern Orthodoxy in America. He was the teacher and father-in-law of Rabbi Lichtenstein.

[33] Thanks to Rabbi Jeffery Saks for discussion of this issue. In correspondence, David Heyd rightly points out how important it is to reflect on why some values shouldn't be transformed into duties. After all, "such transformation could raise the level of happiness in the world." Heyd's own view is that "the value underlying the distinction between supererogation and obligation is in the *personal freedom* of each individual to express his or her altruistic intentions." I agree that this is *a* value that underlies the distinction between supererogation and obligation. I imagine that there are others too.

More can be said about the category of *lifnim mishirat hadin*. For example, there is an ongoing debate as to whether its injunctions are halakhically actionable; that is to say, whether Jewish courts are authorized to enforce the injunctions of *lifnim mishurat hadin*.[34] It would also be interesting to investigate the Talmudic distinction between a *chova* (literally: an obligation) and a *mitzva* (literally: a commandment). Since, on some occasions, the Talmud reports that a certain good deed isn't a *chova* but is still a *mitzva*.[35] What is the difference? Is the later supererogatory? Or do both categories actually refer to an obligation, and the distinction is only to do with whether or not the obligation is actionable? But as an unqualified supererogationist might say, one could always do more!

What I hope to have achieved is this. I have argued that very few Jewish sources can be interpreted as being against supererogation, given the Torah's wisdom of feasibility. Thus, having established that Judaism must recognize some form of supererogation or other, the question becomes, what form?

On the one hand, a qualified supererogation championed by Rabbis Piccio, Shlomo Ha-Cohen, and perhaps Maimonides, allows us to hold onto the intuitive connection between axiology and deontology, and makes room for the claim that Jewish law, even as it currently stands, includes every ethical duty, even if it doesn't extend those duties to all people. On the other hand, an unqualified supererogation, as championed by Nachmanides, avoids reducing intuitive and living values into rigid lifeless formulas, and makes room for a different theory of the revelation; a theory that gives us more room for making sense of the ethically troubling facets of the Jewish legal tradition. Then again, this theory of revelation might be disquieting to those, like the *Chazon Ish*, who think that the Torah as we have it today, in our hands, is the final word on what is ethical and what isn't. And thus, the form of supererogation that Judaism adopts has all kinds of downstream consequences and opens up questions about the very nature of the Torah itself.[36]

References

Adams, R. (2002). *Finite and infinite goods: A framework for ethics*. Oxford University Press.
Bergson, H. (1956). *Two sources of morality and religion*. Doubleday.
Berkovits, E. (1983). *Not in heaven: The nature and function of halakha*. Ktav Publishing House.
Berkovits, E. (1990). *Jewish women in time and torah*. KTAV Publishing House.

I have the sense that Rabbi Lichtenstein was worried about a species of authenticity and sensitivity to the specific contours of the ethical moment that are sullied and/or obscured when certain values are transformed into duties. Perhaps this authenticity and sensitivity are merely a corollary of the personal freedom of which Heyd speaks, but perhaps there's something else that's also at stake. As the great Rabbinic sages say, when they're not yet certain that they have the answer to their own question—*tzarich iyyun* (literally: this requires further thought!).

[34] For more on that debate see (Shilo, 1978).

[35] See, for example, Babylonian Talmud, Tractate Ketubot 49b.

[36] Sincere thanks to David Heyd and Daniel Statman, whose comments on a previous draft saved me from a good deal of embarrassment. Any remaining infelicities are my responsibility alone.

Chisholm, R. (1963). Supererogation and offence: A conceptual scheme for ethics. *Ratio, 5*, 1–14.
Cohen, S. (2015). Forced supererogation. *European Journal of Philosophy, 23*(4), 1006–1024.
Driver, J. (1992). The suberogatory. *Australasian Journal of Philosophy, 70*, 286–295.
Heyd, D. (1982). *Supererogation: Its status in ethical theory*. Cambridge University Press.
Heyd, D. (2019). Supererogation. In: E. N. Zalta, (Ed.), *Stanford encyclopedia of philosophy.*
Karelitz, A. Y. (1996). *Sefer Hazon Ish: Emunah u-Vitahon*. Bnei Brak: Sifriyati.
Lebens, S. (2020). *The principles of judaism*. Oxford University Press.
Lichtenstein, A. (2004). Does Jewish tradition recognize an ethic independent of Halakhah? *Leaves of faith: The world of jewish living* (pp. 33–56). KTAV Publishing House.
Lichtenstein, A. (2007). *On raising children*. https://etzion.org.il/en/philosophy/great-thinkers/harav-aharon-lichtenstein/on-raising-children
Newman, L. E. (1998). *Past imperatives: Studies in the history and theory of Jewish ethics*. State University of New York Press.
Rabinovitch, N. (2003). The way of torah. *The Edah Journal, 3*(1), 1–34.
Rawls, J. (1971). *A theory of justice*. Harvard University Press.
Raz, J. (1975). Permissions and supererogation. *American Philosophical Quarterly, 12*, 161–168.
Richards, D. (1971). *A theory of reasons for action*. Clarendon Press.
Shilo, S. (1978). On one aspect of law and morals in Jewish law: Lifnim Mishurat Hadin. *Israel Law Review, 13*(3), 359–390.
Urmson, J. (1958). Saints and heroes. In: A. Melden (Ed.), *Essays in moral philosophy.* Seattle: University of Washington Press.

Islamic Narratives of Duty, Supererogation, and *Ithar*

Mustafa Abu Sway

Abstract This paper discusses the categories of action in the Islamic worldview: obligatory (*fard*), recommended (*mandub*), permissible (*mubah*), detested (*makruh*), and prohibited (*haram*). It will also refer to their subdivisions, where applicable, such as individual obligation (*fard 'ayn*) and community obligations (*fard kifayah*) in which individuals can be exempted if the obligation is fulfilled by other members of the community.

The notion of supererogation will be explained against this blueprint for action. The Islamic notion of *ithar* will be interpreted.

Keywords Islam · Supererogation · Charity · Al-Ghazali · *Ithar*

Islam divides action into five major categories: obligatory (*fard*), recommended (*mandub*), permissible (*mubah*), detested (*makruh*), and prohibited (*haram*).[1] The first category of obligation is divided into two subcategories: *fard 'ayn*, which is an obligation for every mature and capable individual Muslim, and *fard kifayah*, which is an obligation for the Muslim community, where if the obligation is fulfilled by enough people, one or more, it relieves the whole community from the burden and moral responsibility. Otherwise, the community at large is sinning. The literal meaning of *kifayah* is sufficiency.

The recommended (*mandub*) is a rewarded action, where the person who does not do it is not punishable. This concept is close to the supererogatory. The detested (*makruh*) action is not punishable, yet the person who consciously avoids doing it is rewarded. The latter might be comparable to the supererogatory category as having

[1] Abd al-Wahhab Khallaf, *'Ilm Usul al-Fiqh* (EDA Neshriyat: Istanbul, 1968) 8th ed., pp. 105–115.

Integral Chair for the Study of Imam Al-Ghazali's Work at the Holy Al-Aqsa Mosque and Al-Quds University.

M. A. Sway (✉)
Holy Al-Aqsa Mosque and Al-Quds University, Jerusalem, Occupied Palestinian Territory, Palestine
e-mail: abusway@staff.alquds.edu

D. Heyd (ed.), *Handbook of Supererogation*,
https://doi.org/10.1007/978-981-99-3633-5_20

a negative value, except that the one who abstains from performing a detested act is rewarded (not just *not* punished). If divorce is detested but one has to get a divorce for a legitimate reason, then there is a negative value to it. If the divorce takes place with the intention to harm one's spouse, the divorce itself is legitimate, but harming the spouse is forbidden, and might entail divine punishment. As for the prohibited (*haram*), it could become temporarily permissible if it is intended to save life.

Imam Al-Ghazali (d. 505 AH/1111 CE) argues in the Book of Knowledge, the first chapter of his magnum opus *Ihya' 'Ulum Ad-Din* (The Revival of the Islamic Sciences), that medicine and math are *fard kifayah* (community obligations). The same applies to what he described as the "primary industries of agriculture, making fabric, politics, cupping and tailoring."[2]

How to assess the needs of the community and what is a duty in this regard? *Maqasid al-Shari'ah*, the aims of Islamic law, include the preservation of life, religion, progeny, property, and sanity/reason. A typical discussion of *Maqasid* focuses on the individual. Dr. Jasser Auda, who is a contemporary expert on *Maqasid*, quotes Ibn al-Qayyim (d. 748 AH/1347 CE):

> Shari'ah is based on wisdom and achieving people's welfare in this life and the afterlife. Shari'ah is all about justice, mercy, wisdom, and the good. Thus, any ruling that replaces justice and mercy with its opposite, common good with mischief, or wisdom with nonsense, is a ruling that does not belong to the Shari'ah, even if it is claimed to be so according to some interpretation.[3]

The technical juridical narrative does not usually address people's welfare, which is broad enough to include every sphere of human needs, including political, socio-economic, social, health etc. Jasser Auda himself argues in an interview[4] that the *Maqasid* should be applied to the community. As an example, the aim that reflects the obligatory protection of human life on the level of the individual includes the duty of taking care of public health.

I would argue, as I did already in the concluding remarks of my paper "Towards an Islamic Jurisprudence of the Environment"[5] that I presented at An-Najah University in Nablus about 25 years ago, that protection of the environment should be included in *Maqasid al-Shari'ah*. Humanity's survival and welfare are organically tied to protecting the physical environment without ignoring emotional and spiritual dimensions. It is a personal and communal duty to prevent pollution to any component of the environment. There are those who are very dedicated and would do heroic acts to protect the environment. The latter could be construed as supererogatory. Individuals have a carbon footprint and contribute to cumulative climate change and eco-disasters. Those who are decision-makers and pass negative laws protecting large industries or cause large-scale damage to the rainforests are more culpable.

---

[2] Imam Al-Ghazali, *Ihya' 'Ulum Ad-Din* (Dar Al-Ma'rifah: Beirut, 1982) vol. 1, p. 16.

[3] Jasser Auda, *Maqasid Al-Shari'ah as Philosophy of Law* (The International Institute of Islamic Thought: Herndon, 2008) pp. xxi–xxii.

[4] https://www.youtube.com/watch?v=Bvbp4OMbdqo. Accessed on Oct. 2, 2022.

[5] http://environment-ecology.com/religion-and-ecology/487-towards-an-islamic-jurisprudence-of-the-environment-.html.

The opposite is true: To maintain a sustainable environment individuals can make a positive impact. Those who design and initiate large-scale programs and pass laws to protect the environment are commended for their action. They often challenge large polluting industries and their lobbies. The heroes here are those who mobilize others to join them. Afroz Shah, an Indian lawyer and environmentalist, led the effort to clean Versova Beach in Mumbai, India, removing, with hundreds of volunteers, 11 million pounds of plastic. Following Imam Al-Ghazali's logic that students seeking knowledge are engaged in jihad, this is definitely eco-jihad and it is supererogatory.

Imam Al-Ghazali made it very clear in his above-mentioned book that he aimed at reviving the original meaning of five major concepts including *fiqh*, *'ilm* (/knowledge/ science), *tawhid* (monotheism), *tadhkir* (recollection), and *hikmah* (wisdom), noting that, in classical use, *fiqh* means "jurisprudence." The revived meaning in the *Ihya'* means "understanding" as in understanding the true meaning of this life, and the path to the Hereafter. An example from the *Ihya'* is the Book of Fasting. A typical scholar of Islamic jurisprudence (i.e., *faqih*) would define fasting as the obligation to abstain from food, drink, and intimate relations between spouses between dawn and sunset. Imam Al-Ghazali would say that this is the level common to all those who are fasting. A superior level, fasting of those who are special, would entail abstinence of the external organs and limbs so that they wouldn't engage in prohibited actions. These include the lowering one' gaze, the tongue refraining from all 20 ills that Al-Ghazali lists such as backbiting, the ears not listening to these ills of the tongue, the hands refraining from harming innocent human beings, the legs not walking in the direction of that which is prohibited, and "all of you" which I think Imam Al-Ghazali uses as a metaphor out of politeness to mean not engaging in prohibited sexual relations. Does it mean that the concept of duty entails a broader spectrum for the God-conscientious?

Al-Ghazali became wealthy and very famous at a young age, being the scholar par excellence in the Muslim world, with an appointment at the Nizamiyah college of Baghdad, with excellent relationship with power. He distributed his wealth, left his job, and shunned fame in his quest for spiritual fulfillment on the path of the hereafter. As a Sufi, he aimed at *fana'*, self-annihilation, as the antithesis of egoism. He promoted this way of life. This telos would render traditional discourse of jurisprudence reductionist, for the classical fiqh deals with external actions only, while the revived broader understanding includes internal actions of the heart.

Three of the four major Islamic schools of jurisprudence, namely, the Maliki, Shafi'i, and Hanbali have one category for obligatory action (i.e., *fard*) and they do not differentiate between obligation and duty (i.e., *wajib*) as the Hanafi school does. The difference between both, according to the Hanafis, is that the *fard* is deduced from *direct* indubitable texts of the Qur'an and/or the Prophetic traditions, and *wajib* is derived from *indirect* references, a probabilistic proof.[6] Imam Al-Ghazali used

[6] Maryam Muhammad Salih Adh-Dhafiri, *Mustalahat al-Madhahib al-Fiqhiyyah wa Asrar al-Fiqh al-Marmuz*, (Dar Ibn Hazm: Beirut, 2002) p. 37.

only *wajib* for both, without using *fard*, in *Al-Mustasfa fi Usul al-Fiqh* when he dealt with the above five categories of action.[7]

The supererogatory acts belong to a category that surpasses duty. Surpassing duty in material giving, as an example, could be performed by a person who either has more than enough, or not having enough but gives preference to others over himself/herself. The latter is *ithar*.

Intention (*niyyah*), which is a necessary condition for all actions, whether obligatory or not, should always be for the sake of God. There is no room for ulterior motives. When one gives charity, if it is egoistic, so that people would say the giver is generous, the action is not accepted, and the duty is not fulfilled. The Prophet said: "The first one among people to be judged on the Day of Judgement is a man that God brought him, and He reminded him of His favors to him and he recognized them. He asked him: What did you do with them? He said: I fought for your sake until I was martyred! He said: You lied! You wanted that it would be said: So, and so is courageous, and it has been said. He gave orders and he was dragged on his face until he was thrown into hellfire. And a man who acquired knowledge and studied the Qur'an, and God brought him, and He reminded him of His favors to him and he recognized them. He asked him: What did you do with them? He said: I acquired knowledge, recited the Qur'an, and taught it for your sake! He said: You lied! You wanted that it would be said: So, and so is a scholar, and so and so is a reciter, and it has been said. He gave orders and he was dragged on his face until he was thrown into hellfire. And a man that He gave him different kinds of money, and God brought him, and He reminded him of His favors to him and he recognized them. He asked him: What did you do with them? He said: I didn't leave a venue that you love that it would be spent in it, except that I spent in it for your sake. He said: You lied! You wanted that it would be said: So, and so is generous, and it has been said. He gave orders and he was dragged on his face until he was thrown into hellfire".[8]

The Prophetic tradition about intention (*niyyah*) is the first one to be mentioned in the compendia of *Sahih al-Bukhari*. The Prophet (peace be upon him) said: "Verily, deeds are only [judged] according to intentions and every person will have only what they intended. Whoever emigrated to Allah and His Messenger, his emigration is for Allah and His Messenger. Whoever emigrated to get something in the world or to marry a woman, his emigration is for that to which he emigrated."

In his *Critique of Practical Reason*, Kant speaks about "self-love" tainting motivation:

> Whenever we bring any flattering thought of merit into our actions, the incentive is already mixed with self-love and thus has some assistance from the side of sensibility.[9]

In other words, egoism has a compelling presence. Kant wants action to be guided by the pure dictates of duty (i.e., without "self-love"). And he warns that those actions

[7] Al-Ghazali, *Al-Mustasfa fi Usul al-Fiqh*, Muhammad Suleiman al-Ashqar, ed. (Mu'assasat al-Risalah: Beirut, 2015) p. 70.

[8] Muslim, *Sahih*, # 1905.

[9] https://archive.org/stream/in.ernet.dli.2015.223575/2015.223575.Critique-Of_djvu.txt. P. 159.

that we extol as displaying greatness and hence appear to be *supererogatory* are the most susceptible to display "the self-complacent imagination of merit." The Muslim Sufis advocate a total *fana'* (self-annihilation) where there is no room for egoism or self-love. If one gives to the needy, for example, then it is God who is using him/her as an agent to do that. One should not allow the ego to feed on the action. A pure theological perspective on *tawhid* (i.e., monotheism), according to Al-Ghazali, is to see everything in the universe as coming from God. In addition, the most essential condition for the person who is required to fulfill a religiously sanctioned obligation is that he/she must become of age, sane, and has the capacity to perform the required action. Therefore, a child, or a cognitively impaired adult, or a person in a coma is not subject to the same demands of Islamic law, except if the child or cognitively impaired person has financial wealth, typically through inheritance, then the legal guardian should pay regular almsgiving, 2.5% on money, including gold and silver, that has reached a certain amount and has been in one's possession for a lunar year, or tithe on certain crops such as wheat. Here, the financial duties are still applicable, yet they are performed as a duty by the legal guardian who is prohibited from supererogatory acts on behalf of the child or the cognitively impaired adult.

The "normative" narrative about reaching legal age is centered in general around reaching puberty. In the Maliki school, according to Khalil's compilation, it is the age of 18. The Hanafi school sets the age of puberty at 18 for boys and 17 for girls. The Shafi'i and Hanbali schools used the hadith of 'Abdullah Ibn 'Umar to set it at 15 years of age.[10]

There are numerous contexts where supererogatory acts, including *Ithar*, could be performed, including potentially self-sacrificial altruistic acts. This is not restricted to defending one's country in bellicose circumstances:

"He who is killed defending his property is a martyr, and he who is killed defending his religion is a martyr, and he who is killed defending his own life is a martyr, and he who is killed defending his family is a martyr."[11] The criterion distinguishing between obligatory and supererogatory acts is whether the act serves the public good and the benefactor is society at large or whether it is done for the sake of the personal welfare of an individual. In the former case, the act constitutes a pure duty to God whereas in the latter it is a matter of individual choice and discretion, i.e., does not impose a duty.[12]

When it became apparent that one of the two sons of Adam intended to kill his brother, there was a passive reaction:

> Relate to them in truth [O Prophet] the story of Adam's two sons—how each offered a sacrifice: [Abel's] offering was accepted while [Cain's] was not. So Cain threatened, "I will kill you!" His brother replied, "Allah only accepts ⌜the offering⌝ of the sincerely devout.

---

[10] Wahbah al-Zuhayli, *Al-Fiqh al-Islami wa Adillatuh*, (Dar al-Fikr: Damascus, 2008) vol. 6, pp. 173–176.

[11] At-Tirmidhi, *Sunan*, # 1421.

[12] Abd Al-Wahhab Khallaf, *'Ilm Usul Al-Fiqh* (E.D.A. Nesriyat: Istanbul, 1968) 8th edition, pp. 210–215.

> If you raise your hand to kill me, I will not raise mine to kill you, because I fear Allah—the Lord of all worlds. I want to let you bear your sin against me along with your other sins, then you will be one of those destined to the Fire. And that is the reward of the wrongdoers. Yet [Cain] convinced himself to kill his brother, so he killed him—becoming a loser. (Qur'an, 5:27–30).

Based on this story in the Qur'an, Jawdat Saeed, the late Syrian Muslim scholar, developed an Islamic theology of non-violence, published in 1964 in Arabic as *Madhhab ibn Adam al-Awwal* (*The Doctrine of the First Son of Adam*), also translated as *The Way of Adam's Upright Son.*[13]

Globally, the essential problem is that state actors want exclusive legitimacy for the use of brutal force, committing massacres, genocides, and ethnic cleansing against the natives, the aborigines, and the indigenous. State terrorism, imperial, classical or neocolonial, or home-based, is the womb that gives birth to the reaction from the victims that no one wants. Picasso's Guernica captures the moment of death and destruction delivered by state terrorism, this time Nazi, uncompromisingly screams on behalf of all victims, all times, all places!

One could extend the duties to the protection of communal property, the protection of synagogues, churches and monasteries, and mosques,[14] and to the protection of human life (vaccination for all, prohibition of WMD, etc.)

The concept of martyrdom itself is broad enough to include the following:

The Prophet (peace be upon him) said: "While a man was walking on a road, he found a thorny tree branch on the road and removed it. God deemed it a thankful act to him, and He forgave him. Then [the Prophet] said: The martyrs are five: The one who dies of a plague, the one who dies of abdominal diseases, the one who drowns, the one who dies under collapsed [structures], and the martyr for the sake of God…".

While one might think that only the last category constitutes a supererogatory act, it is possible for a health worker to voluntarily go beyond the call of duty in dealing with a deadly communicable disease. The same applies to volunteering to go inside a collapsed structure because of earthquakes, and still tremors cause further collapsing. As for drowning, there are many stories of volunteers who drowned while rescuing others.

As to whom a duty applies, the following verse makes one's capabilities the criterion for being obligated or not:

> Allah does not require of any soul more than what it *is able [to do/give]* … (Qur'an, 2:286)[15]

While this verse makes it imperative for those who can perform their duties because it is within their reach, it might go beyond the "legal" minimal obligations. There are minimum religious duties that one must fulfill. At the same time

---

[13] https://www.jawdatsaid.net/en/images/d/d2/The_Way_of_Adams_Upright_Son.pdf.

[14] "…Had Allah not repelled ⌜the aggression of⌝ some people by means of others, destruction would have surely claimed monasteries, churches, synagogues, and mosques in which Allah's Name is often mentioned…" Qur'an, 22:40.

[15] I use Dr. Mustafa Khattab's translation *The Clear Quran*, except where there are differences, the translation/interpretation is mine. His translation of this part of the verse is "Allah does not require of any soul more than what it can afford…".

there are human needs, beginning with oneself, one's society, and then lending a hand to humanity at large. These needs could be reflected in the following verse that comes after reminding the tribe of Quraysh of God's favors, asking them to worship Him, the Lord of the Sacred House in Mecca:

> Who has fed them against hunger and made them secure against fear? (Qur'an, 106:4)

Food and security, but also shelter, clothing, education, and healthcare, are essentials, while keeping the door open for the *needy*. It is the duty of the state to make sure that basic human needs are met. If and when the state fails, NGO's might step in to patch the work of governments, and the contribution of individuals will be either *ithar* or supererogatory.

To love for others that which you love for yourself indicates a sense of equality. Supererogatory actions put the other ahead of oneself:

> As for those who had settled in the city [of Medina] and [embraced] the [Islamic] faith before ⌜the arrival of⌝ the [Muslim] emigrants [from Mecca], they love whoever immigrates to them, never having a desire in their hearts for whatever ⌜of the gains⌝ is given to the emigrants. And they *give preference (yu'thirun)* [to the emigrants] over themselves even though they may be in need. And whoever is saved from the selfishness of their own souls, it is they who are ⌜truly⌝ successful. (Qur'an, 59:9)

The Qur'an in the above verse describes a remarkable widespread phenomenon of supererogation, of selflessness, and altruism (i.e., *ithar*) and praises the people who provide for others while being in a state of need. The Qur'an declares supererogation a rewarding act, a source of merit, and a commended higher ideal. Every good act, including *ithar* and supererogatory acts, along with the right intention, warrants divine reward. Those who act more than required will be rewarded more, elevating their status in the hereafter. In addition, good deeds could be used to offset bad ones:

> Surely good deeds wipe out evil deeds… (Qur'an, 11:114)

The good deeds might benefit others as in the story of Prophet Moses (peace be upon him) with a man, possibly Al-Khidr, who had performed three actions that Prophet Moses didn't approve. In one of them, this man repaid inhospitality with a supererogatory good act of rebuilding a wall where two orphans live, with the raison d'etre being the goodness of their deceased father:

> So, they moved on until they came to the people of a town. They asked them for food, but the people refused to give them hospitality. There they found a wall ready to collapse, so the man set it right. Moses protested, "If you wanted, you could have demanded a fee for this." (Qur'an, 18:77)
>
> And as for the wall, it belonged to two orphan boys in the city, and under the wall was a treasure that belonged to them, and their father had been a righteous man. So your Lord willed that these children should come of age and retrieve their treasure, as a mercy from your Lord. I did not do it ⌜all⌝ on my own. This is the explanation of what you could not bear patiently. (Qur'an, 18:82)

The specific context for the above-quoted verse is that a companion of the Prophet (Peace be upon him) gave preference to his guest to have the very little food he had for his children. Al-Bukhari narrated that Abu Hurayrah (May Allah be pleased with him) said: A man came to the Prophet (Peace be upon him), and he sent [someone] to ask the women in his household [if they had any food to offer the guest]? They said: We have nothing but water! Then the Prophet said [asking his companions]: Who can accommodate or host this guest? A man from among the Ansar (i.e., the original residents of Medina who rendered help to the migrant Muslims, whose name is Abu Talha) said: I do! He took him to his [house and said to his] wife: Honor the guest of the Messenger of God (Peace be upon him)! She said: I only have the food of my children! He said: Prepare the food [for the guest], light the lantern, and put the children to sleep when they want dinner. She prepared the food, lit the lantern, and put the children to sleep. Then [when the guest arrived] she pretended as if she was fixing the lantern and turned it off, and they made him believe they were eating [with him], but they went to bed hungry. In the morning, [the host] went to the Messenger of God (Peace be upon him), who said: God laughed or was astonished from what both of you [along with his wife] did last night. Then God revealed: ("…and they *give preference* over themselves even though they may be in need. And whoever is saved from the selfishness of their own souls, it is they who are ⌜truly⌝ successful."[16]

This was not an isolated case. The indigenous people of Medina literally offered sharing everything with the migrant Muslims from Mecca who left their livelihood and belongings behind them. Today, there are refugees in different parts of the world, Muslims and non-Muslims alike, who are forced to leave war zones, collapsing economies in former colonies because of mismanagement or corruption, or fleeing natural disasters because of climate change. Even within traditionally "healthy" economies, the hardships were exacerbated by the recent COVID-19 pandemic. Thus, supererogatory action is in dire need, to friend and foe.

Other Prophetic traditions provide other details to this story such as the "guest" expressing to the Prophet his state of exhaustion from hunger. Though it cannot be Abu Talha Zayd bin Sahl (d. 34 AH) who was wealthy being the largest owner of palm trees, and donated his best property, the orchard which contains the water well of Bayruha' (Bi'ru-Ha'). He said that it was a response to the following verse:

> You will never achieve righteousness until you donate some of what you cherish. And whatever you give is certainly well known to Allah. (Qur'an, 3:92).[17]

When the Mosque of the Prophet in Medina was expanded, the land of this orchard became part of it in the northern section, near King Fahd Gate.

To "give preference to others" is *ithar*, an act of altruism or supererogation which could be defined as going beyond the call of duty, though the latter could evade simplistic definitions. The following tradition could be used to explain the different reactions that go beyond the call of duty within the same context:

Umar ibn al-Khattab reported: The Messenger of Allah, peace and blessings be upon him, ordered us to give charity and at the time I had some wealth. I said to myself, "Today I will outdo Abu Bakr, if ever there were a day to outdo him." I went

---

[16] Al-Bukhari, *Sahih* # 3587.

[17] Al-Bukhari, *Sahih* # 2758.

with half of my wealth to the Prophet, and he said, "What have you left for your family?" I said, "The same amount." Then, Abu Bakr came with everything he had. The Prophet said, "O Abu Bakr, what have you left for your family?" Abu Bakr said, "Allah and his messenger." I said, "By Allah, I will never do better than Abu Bakr." (*Sunan al-Tirmidhī* # 3675).

Abu Bakr and Umar ibn al-Khattab's actions were supererogatory when they put the needs of the nascent Muslim community ahead of those of their own families, with Abu Bakr not leaving them anything except his trust in God and, therefore, His messenger. The fact that both gave different amounts to charity is a clear proof that supererogation cannot be quantified, except that both went beyond *zakat*, the obligatory almsgiving. The latter becomes an obligation once a lunar year has passed with the minimum wealth of 81 gr. of gold or its equivalent in silver not decreasing at all during the same year. Of course, today it includes fiat money, stock shares, etc. As for certain crops, such as wheat, the *zakat* is tithe (10% of the harvest).

Supererogatory acts should also be contextualized. Any supererogatory act could be measured against the same act being done by others of different means and capabilities within the same time-space context, or against those who lived in other times and socio-political conditions. The Prophet (peace be upon him) said: A dirham (a silver coin) outweighs one hundred thousand dirhams. They said: O Messenger of Allah! How? He said: A man has two dirhams took one of them and he gave it as charity, and a man who has abundant money took one hundred thousand from part of his money and gave it as charity.[18]

The same ethos will always be present as compared to what the first generation of Muslims did. The Prophet said (Peace be upon him): "Do not insult my companions, for if one of you spends like [Mount] Uhud[19] in gold, he would not match the two hands full of one of them, or even half of that."[20]

Of course, not all supererogatory actions are monetary. The highest form of sacrifice is to offer oneself, or to a lesser degree to offer a part of oneself, as in donating one's organ in an act of altruism. To add a personal touch that could help dispel stereotyped images, I have donated blood on three continents: Africa, America, and Asia. I never knew the recipients. Did I receive anything in return? In Africa, I was given a can of condensed sweetened milk to drink before leaving the hospital where I made the blood donation.

Supererogatory actions could be easier to fulfill once the relationship with the material world is understood properly. Everything in the universe belongs to God and we really do not "own" anything. We are rather tested with that which passes through our hands, which is a very temporal relation.

[18] An-Nasa'i, *Sunan* # 2528.

[19] A mountain in the vicinity of Medina. It is 7 km long, with an elevation of 1077 m.

[20] Bukhari, *Sahih*, # 3673.

1 Pre-Islamic Supererogation Heritage

The Arab cultural heritage highlights the generosity of Hatem at-Ta'iy who died just before advent of Islam. Hatem didn't find anything to feed his guest during famine, and he had no animals to slaughter except for his only horse. His story is passed down, generation after generation to Arab children. The Prophet (peace be upon him) met with Hatem's daughter Saffanah and his son 'Uday. The Arabs before Islam had their moral standards. The Prophet said: "Verily, I have been sent [as a prophet] to augment good behavior."

2 Timing of Duty

There are duties that have specific times, such as the five daily prayers, fasting during the lunar month of Ramadan, or performing hajj (pilgrimage) to Mecca in the month of Dhul Hijjah (about 2 months and 10 days after Ramadan). This obligation is required once in the lifetime of a Muslim if one is able to perform the hajj, but sometimes people might die without being able to perform this pillar of Islam. Ibn Hazm (d. 456 AH/1064 CE), the famous Andalusian scholar, could not perform hajj. He was a vociferous critic of other Muslim scholars who must have influenced Al-Mu'tadid bin 'Abbad, the Emir of Seville, who ordered the confiscation of Ibn Hazm's property, burning his books, preventing him from issuing religious rulings, limiting him to the village of his ancestors, and prohibiting people from visiting him until he died! There are reports of at least three prominent contemporary Muslim scholars who performed *hajj* on behalf of Ibn Hazm, a pure act of supererogation.

The Qur'an recognizes the limitations of human capacity. Therefore, one is required to act according to his or her ability, be it intellectual, physical, financial, or any condition that could limit human agency:

> Allah does not require of any soul more than what it can afford. All good will be for its own benefit, and all evil will be to its own loss. ⌜The believers pray, "Our Lord! Do not punish us if we forget or make a mistake. Our Lord! Do not place a burden on us like the one you placed on those before us. Our Lord! Do not burden us with what we cannot bear..." (Qur'an, 2:286)

A concrete example of a permission not to fulfill a religious duty is when life is at stake or simply when it could exact great effort such as in the case of exempting travelers and patients from fasting:

> Ramadân is the month in which the Quran was revealed as a guide for humanity with clear proofs of guidance and the standard ⌜to distinguish between right and wrong⌝. So, whoever is present this month, let them fast. But whoever is ill or on a journey, then [let them fast] an equal number of days [after Ramadân]. Allah intends ease for you, not hardship, so that you may complete the prescribed period and proclaim the greatness of Allah for guiding you, and perhaps you will be grateful. (Qur'an, 2:184).

It should be noted that supererogation is not extended to spirituality and worship. An example is that one should not sacrifice his or her chance to perform pilgrimage to Mecca and give it to someone else. There are limited numbers of pilgrims from each country. The Saudi government is right in restricting the numbers. Out of almost two billion Muslims, only 2.5 million pilgrims were permitted before COVID-19 in 2019, and close to 900,000 in 2022.

Righteousness does not lie in turning your faces towards the east or the west. Rather, the righteous are those who believe in Allah, the Last Day, the angels, the Books, and the prophets; who give charity out of their cherished wealth to relatives, orphans, the poor, ⌜needy⌝ travelers, beggars, and for freeing captives; who establish prayer, pay alms-tax, and keep the pledges they make; and who are patient in times of suffering, adversity, and in ⌜the heat of⌝ battle. It is they who are true ⌜in faith⌝, and it is they who are mindful ⌜of Allah⌝. (The Quran, 2:17).

3 Atharah(t) Versus Ithar

It is remarkable that *atharah(t)*, the opposite of supererogation, has the same root. It means selfishness. The Prophet (peace be upon him) said:

> You will see, after me, [the rise] of selfishness and other issues that you will denounce. They said: O Messenger of Allah, what do you command those who among us who reach that? He said: Pay your dues and ask Allah that which is yours.[21]

This hadith speaks volumes about the spread of egoism among rulers, who prioritize themselves over the interest of the people rather than the other way around. One can think of members of parliaments around the world who vote to increase their own wages while the average citizen lives in appalling conditions.

Our relationship with money is problematic. On the one hand, we are stewards in this life, and money should serve a higher goal, including the welfare of others. On the other hand, keeping our money to ourselves is one of the deepest expressions of human beings' egoistic nature.

Ithar requires prioritizing the other over oneself, while one is conscious of one's own needs.

4 Forgiveness as Supererogatory

God is both the Just (Al-'Adl), the Forgiver (Al-Ghafour), and the Oft-Forgiving (Al-Ghaffar). There is room for both justice and forgiveness in the Islamic worldview, where if one insists on justice, there is no blame. Yet, one is encouraged to be merciful and forgive those who caused him wrong, accidentally, or intentionally.

[21] Muslim, *Sahih*, #1843.

The reward of an evil deed is its equivalent. But whoever pardons and seeks reconciliation, t their reward is with Allah. He certainly does not like the wrongdoers. (Qur'an, 42:40).

And hasten towards forgiveness from your Lord and a Paradise as vast as the heavens and the earth prepared for those mindful [of Allah]. [They are] those who donate in prosperity and adversity, control their anger, and pardon others. And Allah loves the good-doers. (Qur'an, 133–134).

The above two verses present three charitable traits, all supererogatory. When the harmed person controls anger, the offender takes notice. The second step is pardoning the offender. The third act, which is higher than the other two, needs some explanation. The "good-doers" in Arabic is "Al-Muhsinin," which is accusative. The original root is h-s-n, the etymology of which is about beauty. The third act is doing something beautiful. Around the year 1995, when I was teaching at the International Islamic University in Kuala Lumpur, Malaysia, an Iraqi professor of law's car was rear-ended by a double-stroke motorcycle that has pedals like a bicycle, usually the mode of transportation for the poor. This colleague took the first two steps. As for the third, he gave money to the driver to fix his motorcycle which sustained some damage, knowing very well that legally it should be the other way around.

The hadith of the Prophet (peace be upon him) confirms the excellent status and the divine reward for those who forgive, he said:

> Never a charity decreases money, and Allah increases His servant for [his] forgiveness [of others] in dignity, and no one humbles himself for the sake of Allah except that Allah will raise him [in status in this world and the hereafter].[22]

In another hadith with the same message, the Prophet (peace be upon him) added his oath to emphasize the validity and importance of these three acts.

To conclude, the division of human action in the Islamic worldview into five categories, obligatory (*fard*), recommended (*mandub*), permissible (*mubah*), detested (*makruh*), and prohibited (*haram*), doesn't cover all possible commendable proactive actions. While there is always room for the minimalist who will not go an iota beyond *fard*, the higher moral grounds are for those who go beyond the line of duty, religious acts or otherwise. They engage in altruistic supererogatory acts from a position of wealth, or *ithar* where one prioritizes the *other* knowing fully that they themselves are in need of help.

Our world suffers primarily from policies that reflect self-interest, geopolitical considerations, and greed, leading to impoverished and marginalized societies. It is a duty of the highest order to end all forms of xenophobia, discrimination, apartheid, despotism, and colonialism, beginning at home. An ethos of convivencia and supererogation beyond borders, not a "clash of civilizations" should prevail.

---

[22] Muslim, *Sahih*, #2588.

Supererogation in Buddhism

Soraj Hongladarom

Abstract Supererogation in Buddhist philosophy is a rather neglected topic. Among the questions to be investigated are: "Is there supererogation in Buddhism?" "Can one explicate the examples of apparently supererogatory acts performed by bodhisattvas and other enlightened beings in terms of supererogation?" "Is there room in Buddhist ethics for acts which are neither obligatory but still meritorious?" The answer that I aim to defend here is that there is a place for supererogation in Buddhism, as exemplified, among others, by the acts of Buddhist saints and practitioners who perform action that exceeds normal expectation. Supererogation depends on the existence of minimal norms, and I argue that in the Sigalovāda Sutta the Buddha did lay out such minimal norms which are expected of lay people who are not practitioners. This discussion will shed light not only on Buddhist ethical theory itself but also serve as a contribution to the ongoing discussion of the nature and possible justification of supererogation itself.

Keywords Buddhism · Supererogation · Bodhisattvas · Saints

According to Heyd (1982, p. 115), an act is considered supererogatory if it fulfills all of the following four conditions: (1) it is neither obligatory nor forbidden; (2) its omission is not wrong and is thus not censured; (3) it is morally good; and (4) it is performed voluntarily. In other words, an act is supererogatory if it is done "beyond the call of duty." According to J. O. Urmson, who started the modern philosophical discussion of the nature of supererogation, the traditional classification of actions into three deontic types, namely, the obligatory, the permissible, and the forbidden, is incomplete, and one should add the fourth category of actions that are neither obligatory nor forbidden, but whose performance is nonetheless meritorious (Urmson, 1958). His example consists of a soldier who, without regard to his own life, throws himself on an exploding grenade in order to save the life of his fellow

S. Hongladarom (✉)
Chulalongkorn University, Bangkok, Thailand
e-mail: Soraj.H@chula.ac.th

D. Heyd (ed.), *Handbook of Supererogation*,
https://doi.org/10.1007/978-981-99-3633-5_21

soldiers while killing himself. Such act is beyond the call of duty, and, presumably, is not obligatory, yet it is highly commendable and virtuous.

Urmson's paper gave rise to a large number of subsequent works aiming at explicating the notion of supererogation, either defending or attacking the notion. However, almost all of these works operate within the context of the ethical theories of the West. There is almost no work that substantively discusses supererogation in the context of non-western philosophy or ethical theory. Those that touch upon supererogation in Buddhism do so only tangentially. For example, Keith Neigenfind Jr. (2020) asks whether the concepts of nonviolence and pacificism in Christian and Buddhist ethics are obligatory or supererogatory. In his Ph.D. dissertation, however, he explicitly argues that the existence of monasticism is evidence for supererogation in Buddhism, since monks are supposed to follow stricter rules than laypeople, and there is no obligation for laypeople to become monks (Neigenfind, 2021). Furthermore, Harris (2018) argues that the story of *bodhisattvas* (beings who have made a vow to become a Buddha in order to help sentient beings) descending to hell in order to save the beings there shows that the *bodhisattvas* are performing supererogatory acts. Furthermore, in discussing the difficulty of the Buddhist injunction for universal compassion, Seth Segall argues that Buddhism does not have the conception of supererogation, and that the injunction holds the practitioner to an impossible standard (Segall, 2020, p. 159). These studies, however, touch upon the topic of supererogation only in passing, and none discusses the concept in full, taking supererogation itself as its object of analysis.

Thus, the present chapter aims at filling this gap. The main questions are: "Is there a place for supererogation in Buddhism?" "Can one explicate the examples of apparently supererogatory acts performed by *bodhisattvas* and other enlightened beings in terms of supererogation?" "Is there room in Buddhist ethics for acts which are neither obligatory but still meritorious?" The answer that I aim to defend here is that there is a place for supererogation in Buddhism, as exemplified, among others, by the acts of Buddhist saints and practitioners who perform actions that exceed normal expectation. Supererogation depends on the existence of minimal norms, and I argue that the Buddha did lay out such minimal norms which are expected of laypeople who are not practitioners in the *Sigalovāda Sutta*. I hope that this discussion will shed light not only on Buddhist ethical theory itself but also serve as a contribution to the ongoing discussion of the nature and possible justification of supererogation itself.

The brief review of the scant literature on Buddhism and supererogation shows a wide-ranging view on the issue. On the one hand, Neigenfind argues that supererogation is there in Buddhism, evidenced by the ordaining of monks who take their oath as a wholly voluntary act; on the other hand, Segall argues that the teaching of Mahayana Buddhism for universal compassion and for "putting ourselves last" (Segall, 2020, p. 159) demonstrates that the distinction between what is obligatory and what is supererogatory is blurry at best because the teaching seems to require that followers perform actions that are very difficult and is incompatible with the requirements of daily life (Segall, 2020, p. 160). We can look at these diametrically opposing viewpoints as a framework on which we will try to build a clearer picture of whether the concept of supererogation can be found in the Buddhist teaching.

Let us look first at the context in which the talk of supererogation began. In the standard Christian context, a supererogatory act is an act that is not required but would be good to perform nonetheless. According to the Catholic doctrines, what a person ought to do is divided into "precepts" and "counsels," where the former are required but the latter are only recommended (Heyd, 2019). This twofold injunction originated from the New Testament. When Jesus was asked what one should do in order to gain eternal life, he answered that the minimal condition was to follow the Commandments, and added that if one were to become perfect, one should perform optional acts such as helping the poor (Heyd, 2019). This set the stage for the discussion of the supererogation in the West. Actions that are above and beyond what is required are considered supererogatory.

However, when we try to apply the concept to Buddhism, there is a problem. The problem arises when one realizes that in Buddhist ethics there is no counterpart to moral requirement as in the Divine Command Theory in western ethics. Since there is no God in Buddhism, there is no Commander. What is operational as the arbiter of normative judgments is the Law of Karma, which says roughly that one is always responsible for one's actions. The law is a species of the natural law of causation, adopted in particular for intentional action. Someone who does a beneficial act will get beneficial treatment in return, and the same goes for harmful action, which will bring about harmful results to the perpetrator in return. There are criticisms of the Law of Karma, notably the law is allegedly refuted by pointing out that there are many people who prosper and lead a highly successful life even though they have done egregious acts in the past, and that there are many who suffer tremendously even though they have done good deeds. The charge is usually countered by pointing out that one lifetime is often insufficient for the Law of Karma to take effect, and in these cases, one must expand one's scope to multiple lifetimes to see how the Law takes effect. In any case, what is clear in how the Law of Karma operates is that the Law is impersonal. There is no personal God in Buddhist thought who metes out punishments or rewards. The Law works only on the cosmic level: A harmful action will create effects of the same kind, and sooner or later the perpetrator will suffer those effects in return. The Law is thus the normative justification of actions: You should not do harmful deeds because you yourself will suffer harmful results in return.

However, this leads to a problem when it comes to the question whether supererogation is present in Buddhist thought. Supererogation is founded upon the distinction between what is required and what is optional. But when normativity is based, as in Buddhism, on the Law of Karma, it is difficult to see where the line between the two can be drawn. In Christianity the issue is straightforward: Jesus tells his followers to follow the Commandments if they are to have any chance of entering the Kingdom of God; this much is the *requirement*. However, he also adds that if they want to be perfect, they should perform supererogatory acts such as volunteering to help those in need. This seems to be possible only because there is God who issues the Commandments and who in effect requires His followers to follow them. But in Buddhism, the Law of Karma functions only impersonally. It is a blind law that does not see who is doing what, but functions only as part of the working order of the Buddhist cosmos. In that case, there is absolutely speaking no one who requires or commands a Buddhist to act in this or that way. In this strict sense, there is then no separation between what is obligatory and what is permissible (including what is supererogatory). It is in this strict sense that there is no space for supererogation in Buddhism.

The question of the source of moral duty is not limited to the existence of God, however. According to Kantian ethics, such a source can be found in the realization of individuals, based on their autonomy and rationality, that they must follow moral rules that are universalizable. This deontic logic underlies that moral requirement that functions as the basis for (deontological) ethics. Kant's Categorical Imperative is couched in the form of an *imperative* statement. Although there is no one who issues the command, the nature of the Categorical Imperative is based on the recognition of one's autonomous condition, which serves as the source of commands. Kantian ethics is based on moral requirement and moral duty no less than Christian ethics is based on Divine Command. The question of whether this basis covers all acts in such a way that no supererogatory act is possible in Kant's system is, however, another complex issue best dealt with in more detail elsewhere. My point is to show that, in Kantian ethics, one can also find the source of the moral command. If moral command or moral duty is the basis for the intelligibility of supererogation, then the absence of a clear source of moral duty or moral command (as in Buddhism) would mean that one must find the basis for supererogation elsewhere. In the following, I will show that Buddhist ethics does not in itself possess a theoretical underpinning for moral command but that does not have to preclude Buddhist ethics from having a concept of supererogation.

As for utilitarianism, Heyd has the following to say: "Classical utilitarianism may also be interpreted as denying any space for supererogation. If promoting the overall good in the world is the fundamental principle of action, there can be no (non-utilitarian) exemption from the duty to do so" (Heyd, 2019). This presupposes that in utilitarianism one always has the duty to maximize utility; thus a supererogatory act might not be considered morally good in case it does not promote the overall happiness or utility. Nonetheless, in utilitarianism, we can also find the source of normativity, namely, the utilities or the happiness itself. One *must* act in such a way that happiness is maximized. But in Buddhism since such a source cannot, strictly speaking, be found, Buddhist ethics seems to be distinctly different from the mainstream ethical theories in this regard.

All this, nonetheless, does not provide a complete picture. Saying that in the strict sense, there is no place for supererogation in Buddhism does not mean that there are no saints or heroes. In fact there have been countless saints and heroes in Buddhism, namely, those who devote their entire lives to spiritual purification and have accomplished very difficult and arduous deeds for that purpose. At least Neigenfind is on the right track when he argues that the existence of monks shows that supererogation is somehow present because becoming a monk is not an obligation.

In Thailand, for example, one can become a monk for a period of time and can disrobe and return to lay life as soon as one wishes. The existence of Buddhist saints and heroes also shows that there are those who, out of their own voluntary decision, devote their lives to the very stringent regimen of rules which are required for those who want to accomplish great deeds. Their deeds are thus clearly supererogatory. The story by Harris (2018) mentioned earlier of a *bodhisattva* who willingly enters the realm of hell to help sentient beings who suffer there is a clear example of someone whose act is exemplary and supererogatory. The *bodhisattva* does not have to enter hell to help others; this decision arises purely out of his or her genuine desire to help sentient beings out of their suffering the best they can. In this case, both the actual existence of monks (which number in hundreds of thousands in Thailand alone) and the exemplary nature of the *bodhisattva* entering the realm of hell show that there indeed are acts in the Buddhist tradition, which are supererogatory. Moreover, the existence of numerous Buddhist volunteer organizations throughout the entire world also shows that on the mundane, day to day level supererogatory acts are common. Members of these organizations are not required to work there, yet a very large number of them volunteer their time, effort and money to working through these organizations for the benefit of fellow sentient beings (some such volunteer organizations work for animal care).

So where do we stand? Perhaps an answer can be found when one realizes that the primary motivation for Buddhists to continue to work hard to purify themselves or to volunteer to help others is that they want to make progress in their own spiritual path. Helping others is an act of compassion (*karuṇā*), and compassion is a key ingredient in one's spiritual journey toward the Supreme Goal, or *nirvāṇa* (total cessation of suffering and therefore total freedom from the cycle of births and rebirths or *saṃsāra*). That is the ideal goal. However, those who engage in volunteer work often do not think of themselves as entering nirvana soon. On the contrary, they would say that they feel good to be a part in the community effort to help fellow human beings or non-human animals in need. They gain merit (*puṇya*), which enables them to progress along the spiritual journey even though they might not be thinking about it when they do the work. The good feelings that they get when they have helped other beings and when they have accomplished something so that their lives are meaningful constitute their merit. An important thing is that the Buddhist is not required to work to gain merit. Since there is no Commander as previously mentioned, there is no authority figure which requires Buddhists to follow certain sets of rules. Certainly, there are rules in Buddhism, most notably of which are the Five *Sīlas*, which consist of guidelines for behavior such as that one should refrain from killing, stealing, and so on. But the Buddhist *sīlas* are not Commandments. No God is there to punish those who transgress. There is only the impersonal Law of Karma.

So, the issue boils down to how we should understand the concept of supererogation itself in Buddhism. Does the fact that there is no distinction between what is required and what is obligatory show that there is no place for supererogation in Buddhism, no matter what the existence of monks, saints, and volunteer organizations might indicate otherwise? I would like to argue that there is in fact a kind of moral requirement in Buddhism, even though the Law of Karma is an impersonal one and as such does not impose any obligatory requirements on individuals. In other words, there does not seem to be any minimum requirement that is expected of a Buddhist. All there is the Law of Karma, which, being impersonal, does not set any minimum standard for persons in the way Jesus did for his followers. Furthermore, Kantian ethics also clearly sets at least a minimum standard of morality: One is expected to follow the moral rules that arise from within oneself as a rational being, and the rules do not appear to require extreme sacrifices, being the kind of rules that are universal and applicable to everyone. One might argue, on the contrary, that Kant's moral rules exhaust all the possibilities of one's moral activities, such that supererogatory action is not possible under his scheme, but that is a topic for another paper. The same goes for utilitarianism as mentioned earlier. It seems that there being a minimum requirement is a necessary condition for the existence of supererogatory action, but if the Law of Karma by itself does not set any minimum standard, saying only that such and such effects will follow such and such action, then how can a Buddhist claim to be able to perform supererogatory action? What about the deeds of the *bodhisattvas* mentioned earlier? If there is no minimum, then logically speaking any meritorious action, no matter how trivial, would be supererogatory. That would deprive supererogation of any meaning.

One way to handle this problem is to see that in Buddhism one acts according to one's overall goal in life. A *bodhisattva*, being *bodhisattva*, naturally aims at performing actions that would enable her to progress along the path toward becoming a fully enlightened Buddha, and this naturally results in her performing acts that ordinary persons would regard as supererogatory. An ordinary person, in contrast, does not set her goal to be either an enlightened Buddha or a *bodhisattva*, but is content to merely live her life as a normal person, taking care of her family, getting a job, trying to be happy from time to time, and so on. For the ordinary person, then, the minimum requirement is established in this way—the normal, day-to-day living of ordinary person who is not intent on being or becoming a saint or a hero. And we seem to share this minimum standard in our society. One lives as a law abiding person, does not cause harm to others, does not commit crimes, works for one's own living in the normal way, and so on. Thus, even though the Law of Karma itself does not set any minimum moral standard, this community-based standard established through the normal living conditions of ordinary persons seems to suffice. Thus, any action that goes over and beyond this minimum standard and expectation can be regarded as supererogatory in Buddhism.

In this sense, then, Segall's criticism of Buddhist compassion as being overly demanding misses the point. The point is not that compassion is over-demanding—the Buddha does not expect each and every one of us to attain *nirvāṇa* or Buddhahood in this lifetime. The point is that there are many levels of compassion that one can show, ranging from the compassion of the *bodhisattva* who voluntarily enters the fires of hell in order to preach to the beings there, to the more mundane kind of helping others in need when our resources permit. Neigenfind's point that simply being ordained as a monk is itself a supererogatory act can thus be explained. The minimum standard set by the Buddhist community is that one lives one's life in a normal way as a layperson. Being a monk, however, requires effort, sacrifice and a total change in one's way of life. A monk in the Theravada tradition is not allowed to eat anything after noon time; he (it is always a he at the moment, though the movement for Buddhist nuns is growing) is not permitted any sexual contact with women; he is expected to devote full time to study and practice the Buddha teachings, and so on. This in itself can be considered a supererogatory choice. Though simply becoming a monk is not as dramatic as the deeds of the *bodhisattva* that one finds in the Scripture (such as entering hell as mentioned, or letting oneself be eaten by a starving mother tigress which has no milk to feed her cubs, see Junko, 2010), it is supererogatory nonetheless. Moreover, the actions of laypeople who are commonly regarded as supererogatory, such as entering a burning house in order to rescue a child who is trapped in there, can be regarded also as supererogatory in the Buddhist ethical system because this goes well beyond what is normally expected of an ordinary person.

Viewed in this way, then, it is possible for there to be many levels of supererogatory action, depending on the context in which it is performed. Among the laypeople, dashing into a burning house is a paragon of good action, clearly something that goes beyond the call of duty, and giving up lay life to become a monk is similarly so. And among the monks, all of whom practice asceticism laid down in the monastic rules as a matter of course, their supererogatory action must go one step further. For the layperson simply becoming a monk is itself supererogatory, but for a monk to perform a supererogatory act, an act that is praised and admired by fellow monks, he must do something that is not expected of a normal monk. His act must go beyond "the call of duty" for an average monk. There is a practice, both in the Theravada and in the Tibetan traditions, for a monk to enter a long retreat of three years, three months and three days, sealing himself up in a cave, only to be given food by his supporters through a small hole. This is a highly meritorious act and is in no way expected of every monk to perform. Nonetheless, some monks do choose to perform this difficult act in order to further purify themselves. The point is that in Buddhism an act would be regarded as supererogatory depending on the context in which it is performed. But this means that supererogation is more a relative or contextual concept than perhaps hitherto recognized.

It would be instructive to compare and contrast the conception of supererogation I am arguing for here with that proposed by Heyd. In "Can Virtue Ethics Account for Supererogation?" Heyd says:

> …I want to suggest that supererogation and duty are mutually dependent. They are correlative concepts. Supererogation cannot be conceptually articulated without reference to duty and duty cannot be normatively justified without reference to what lies beyond duty. In that respect, supererogation is essentially a deontic concept or a phenomenon which is naturally accounted for in deontological theory (Heyd, 2015, p. 31).

And he further argues that, since there is no counterpart to the concept of duty in virtue ethics, the deontic theory is to be preferred to virtue ethics theory when it comes to supererogation. I think the issue here hinges on how we understand the concept of duty or obligation. In the account of supererogation in Buddhism that I have offered so far there is no direct discussion of duty or obligation either. In fact, many scholars have noted certain similarities between Buddhist ethics and virtue ethics. In the case of Buddhism, since there is no God and no conception of duty-based ethics as in deontological theory, it seems that the Buddhist conception of supererogation I am proposing here might be the object of Heyd's criticism too.

However, I believe that the two accounts here can be reconciled. Duty perhaps does not have to be construed only in terms of either Divine Command Theory the Kantian deontological theory, or utilitarianism. The reason is that one can have a sense of duty or obligation when one has set for oneself a goal and is committed to accomplishing it. I play the piano in my free time, and this year I have set a goal of being able to play a prelude and fugue by Bach fluently before the end of the year. Having set the goal, it seems that I have some sense of an obligation, at least to myself. Having an obligation in this way seems to leave no room for the Divine Command or the Kantian moral duty, or the utilitarian maxim, since it happens at a mundane, day-to-day level. The point is that if I want to play the Bach pieces well, then I need to undertake a regimen of practice, setting up time for practice each day, and so on. In other words, I now have a duty to practice the pieces in order to achieve the goal. This is consistent with the Buddhist practice in that in practicing the teaching one is not coerced to do so by anyone. It is one's desire to become free from suffering and the understanding that by practicing and cultivating oneself one will eventually achieve the goal. So, perhaps there is a way to decouple supererogation from the requirement of the deontic moral theory, or the other standard moral theories in the West.

This connection between the conception of goal-related duty here and supererogation can also be found at the more social level. I have mentioned earlier the expectation of a community that constitutes the ordinary norm of day-to-day living, such as abiding by the law, living one's life ordinarily without causing trouble, and so on. At this level, this expectation can be regarded as constituting the norms that govern the daily living of individuals in their society or community. Toward the end of the chapter, I will cite one of the teachings of the Buddha where he discusses how the lay members of society should behave toward others. For example, the Buddha teaches that the husband has certain duties to his wife and vice versa the wife to her husband, such as being faithful to each other. Furthermore, sons and daughters are expected to help their parents and the parents are expected to give their children their due inheritance (*Sigalovāda Sutta*, 2013). These social norms together constitute the duties and obligations that an average person is expected to follow. These norms thus constitute the line that separates actions performed out of duty and actions that go beyond the call of duty. It is as if society has set up a goal for itself—living together harmoniously and prospering together, or something of that kind. Being faithful to one's spouse and helping and honoring one's parents are all part of the social norms that are expected of each member of society. However, as social and moral norms can change as society changes, practices that used to be followed and respected may be abandoned and relegated to history as time passes. In any case, there are still norms, and it is these norms that function as the dividing line between the normally ethical and the supererogatory action.

The connection between the conception of duty proposed here and Buddhist ethics is that the Law of Karma is a *natural* law. It describes the working of nature without any regard to whether there are human beings or not. As the Law is the key element in the Buddhist conception of normativity, the latter is natural too in the sense that it is related to how nature works. In short, one's action is judged to be good and ethical just in case it follows the course of nature and is bad otherwise. This is an ancient wisdom found not only in Buddhism but also in other ancient philosophical and religious traditions. When normative judgments are based on natural law, the former take essentially the form of an if then statement: If you want certain goals to be realized, then take such and such a course of action, and to the extent that you take the goals to be valuable, the course of action that leads to their fulfillment will be good, but otherwise will be bad. In Buddhist language, this is tantamount to saying: If you want to achieve certain goals, then take such and such a course of action, and since you are a Buddhist practitioner intent on making progress in your spiritual journey, then any action that promotes your progress will be judged to be good, and any action that does the opposite will be judged to be bad. This, moreover, is of course an account of a *normally* good action and one can extrapolate from that to form an account of supererogatory action by saying that an action that goes beyond this normal expectation is thereby supererogatory.

Therefore, we can articulate a formula for supererogatory action: If you want to achieve a high goal such that ordinary persons do not think of achieving, then take such and such a course of action, and since your action is meritorious, your exceeding the normal expectation will be (very) good. In Buddhist terms, this type of supererogatory action will speed up the individual even further along her spiritual path. However, there are conditions attached. The quality of mind of the individual performing the action must be pure, in that the motive must be truly altruistic (This, however, does not apply to my more mundane example of practicing the piano; in that case the goal is not to become enlightened, but only to play the piece better.). There is an old saying ascribed to Buddhist Masters: As long as you want to reach the goal, then you will not reach it because your mind is distracted by the *desire* to reach the goal, and desire is an enemy of reaching the goal. You have to pay attention to the present, not thinking of wanting to reach the goal. Instead, the journey itself is the only thing that matters. Pay attention only to the present moment, and sooner or later you will realize that you have arrived at the goal without knowing it.

In discussing supererogation in Buddhism there is a question whether it is the action or the choice of adopting a goal that is supererogatory. In the first case, one exceeds what is required or what is normally expected, such as when somebody rushes into a burning house to save a child. In the second case, it is the intention or the choice that one makes and the subsequent obligation to oneself in trying to realize the intention that is supererogatory, such as when one makes a commitment to oneself of progressing along the spiritual path, which is normally not expected for laypeople. The answer is that, since both types do incur positive karmic results, both types can be supererogatory. That action can be supererogatory is clear. As for the choice of a kind of life, such as the life of a monk and the subsequence course of action required of the monastic life, it can be supererogatory too. However, in the second case, only the choice alone may not be enough; it seems rather that it is the course of action that must follow from the choice that qualifies it to be supererogatory.

So far we have discussed the Buddha teachings in terms of what a practitioner should do in order to progress along the spiritual path leading eventually to Enlightenment. In this case, supererogatory acts are those that go even beyond what is normally expected of a practitioner, or in the case of living together harmoniously in society, what is normally expected of normal members of the community. It is indeed the case that most of the Buddha teachings that are found in the Buddhist Scripture are intended for monks. However, in one of his teachings, the *Sigalovāda Sutta* (2013), the Buddha presents a teaching directly intended for ordinary people. In this *Sutta* (a teaching), the Buddha chanced upon a man named Sigalo, who was wandering around paying homage to all the six cardinal directions, including the direction of the zenith (up) and the nadir (down). The Buddha asked him why he did that, and Sigalo answered that he was taught by his late father to do so. The Buddha then gave Sigalo a teaching that one should instead pay homage to the six directions in another way, namely by practicing kindness and respect toward those who were around. Thus, instead of merely paying respect or prostrating to the six directions, one should actually practice acts of kindness and generosity toward others. For example, one should avoid taking life, stealing, committing sexual misconduct, and lying. These are four types of action that every householder should avoid. More specifically, the Buddha also said that each of the six directions represented the six kinds of people that surrounded oneself; that is, the North represented one's friends; the West one's spouses; the East one's parents; the South one's teachers; the Zenith represented the ascetics; and the Nadir one's workers. The householder should, for example, treat one's friends (North) with kind words and generosity, and the friends should also reciprocate with support and loyalty. He should treat his own parents (East) by supporting them and fulfilling their wishes, and the parents should support the householder (their own child) with teaching skills and leaving inheritance. He should also treat his workers and subordinates (Nadir) with just wages, appropriate time off, health care, and perks, and the subordinates should reciprocate with avoiding stealing, working well, and diligence. There are more details concerning how one should treat those around oneself, and the point of the Buddha is that one takes on numerous roles in life. A person may be both a parent and a child, both a subordinate and a manager, and they should treat others according to the nature of their particular relations. What is important is that persons live together in a society, assuming various roles toward one another. Acting toward one another according to the appropriate roles that define a relation is a good way of maintaining social relations in general.

This teaching is one of the few places where the Buddha directly addresses how laypeople should treat one another. One could regard this teaching as a basis for an account of supererogation in Buddhism. The account I presented earlier about maintaining cordial relations among members of society can find textual support here, as the teaching on the Six Directions constitutes the Buddha teaching on how members of society should behave toward one another. The North direction, the direction of one's friends, is the most general of all the six directions, because it addresses one's friends, and one can interpret "friends" here to include everybody, including strangers. The Buddha said that one of the things that a friend should do is to use kind words to one's friends. This reflects the need for members of society to be courteous toward one another, not harming the feelings of others. This is the basis for a well-functioning society since kind words indicate decency and respect toward one another. It must also be reciprocal, because, unlike the other directions, being a friend is a mutual relationship: If I am someone's friend, she is my friend too.

However, supererogation does not only arise when one becomes a monk and leaves the life of the householder. Even in the context of the householder, supererogatory acts are still possible. An employer is normally expected to provide "just wages" to their employees, but we can imagine a supererogatory employer who goes out of the way to provide more than minimum or more than "just" wages for their workers. The employer does not change her householder life by performing this supererogatory act. In the same vein, a worker may also perform a supererogatory act when she does her job well beyond what is required of her as part of her job description, without expecting increased pay from her employer. She may find her job particularly rewarding and beneficial, but we would say of such a case that the worker here is acting supererogatorily. In this case, there is then a line between what is normally expected (such as what is expected of a worker in her relation toward her employer) and what exceeds that expectation. The expectation here is an empirical condition and varies according to time and place; this does not imply that the Buddha teaching is morally relativistic because the emphasis on kindness and generosity is still there and there are objective measures for those qualities supported by the Law of Karma. Thus, supererogation can take place at various levels. It can take place entirely on the level of a householder life, or when a householder wants to exceed what is expected of the lay life and chooses monasticism, or when one performs extraordinary acts that few monks do. In each case, one must find where the level of expectation is that is appropriate to each level and context.

In conclusion, I have argued that supererogation has a place in Buddhist ethics. Although the concept itself is not found in the Scripture itself, we can still find it expressed through stories of monks and laypersons who perform extraordinary deeds that exceed what is normally expected of either monks or laypersons. Accounts of supererogation are not limited only to the deontic ethical theory as Heyd suggests. Supererogation in Buddhism occurs when someone performs a deed that exceeds the expectation that is normally the case for the roles taken by that person. A supererogatory act for a monk is not in the same class as a supererogatory act for a layperson. Here one can see that Buddha's own decision to renounce his princely life and to become a wandering ascetic in search on the real meaning of life is also an act of supererogation because he was absolutely not required to become an ascetic. On the contrary, his father actively forbade him to leave his palace and to renounce his princely life, having been told by a Brahmin that his son would either become a world-conquering emperor or a world-conquering spiritual leader and obviously he wanted his son to become the former. Nonetheless, the Buddha, having seen the pointlessness of life after his sojourn outside of his palace with his trusted attendant to see the sick, the dying, and the dead people, decided to renounce his status as a prince and to become a beggar. His life can, thus, be a very good example of supererogation.

References

Harris, S. E. (2018). A nirvana that is burning in Hell: Pain and flour-ishing in mahayana buddhist moral thought. *Sophia, 57*(2), 337–347.

Heyd, D. (1982). *Supererogation.* Cambridge University Press.

Heyd, D. (2015). Can virtue ethics account for supererogation?. *Royal Institute of Philosophy Supplements, 77*, 25–47.

Heyd, D. (2019). Supererogation. In E. N. Zalta (Ed.), *Stanford Encyclopedia of Philosophy.* https://plato.stanford.edu/entries/supererogation/

Junko, M. (2010). The Vyāghri-Jātaka known to Sri Lankan Buddhists and its relation to the Northern Buddhist versions. *Journal of Indian and Buddhist Studies, 58*(3), 1164–1172. https://www.jstage.jst.go.jp/article/ibk/58/3/58_KJ00006159730/_pdf

Neigenfind Jr, L. K. (2020). Is Nonviolence and Pacifism in Christian and Buddhist Ethics Obligatory or Supererogatory?. *Buddhist-Christian Studies, 40*(1), 387–401.

Neigenfind Jr, L. K. (2021). Pacifism and nonviolence in Christianity and Buddhism applied to a Thai context [Unpublished doctoral dissertation]. Assumption University.

Segall, S. Z. (2020). *Buddhism and human flourishing: A modern western perspective.* Palgrave Macmillan.

Sigalovāda Sutta: The discourse to Sigala (DN 31). (2013). Translated from the Pali by Narada Thera. *Access to Insight (BCBS Edition),* 30 November 2013, http://www.accesstoinsight.org/tipitaka/dn/dn.31.0.nara.html.

Urmson, J. (1958). Saints and heroes. In A. Melden (Ed.), *Essays in Moral Philosophy.* Seattle: University of Washington Press.